Microglia in the Regenerating and Degenerating Central Nervous System

Springer
New York
Berlin
Heidelberg
Barcelona
Hong Kong
London
Milan
Paris
Singapore
Tokyo

Wolfgang J. Streit
Editor

Microglia in the Regenerating and Degenerating Central Nervous System

With 44 Illustrations

 Springer

Wolfgang J. Streit
Department of Neuroscience
University of Florida College of Medicine
100 Newell Drive
Gainesville, FL 32611
USA
streit@ufbi.ufl.edu

Library of Congress Cataloging-in-Publication Data
Microglia in the regenerating and degenerating central nervous system / [edited by]
Wolfgang J. Streit
     p.   cm.
   Includes bibliographical references and index.
   ISBN 0-387-95301-9 (hbk.: alk. paper)
     1. Nervous system—Degeneration.   2. Nervous system—Regeneration.   3. Microglia.
 I. Streit, W.J. (Wolfgang J.)

RC365.M53 2001
616.8'047—dc21
                                                                                     2001032813

Printed on acid-free paper.

© 2002 Springer-Verlag New York, Inc.
All rights reserved. This work may not be translated or copied in whole or in part without the written permission of the publisher (Springer-Verlag New York, Inc., 175 Fifth Avenue, New York, NY 10010, USA), except for brief excerpts in connection with reviews or scholarly analysis. Use in connection with any form of information storage and retrieval, electronic adaptation, computer software, or by similar or dissimilar methodology now known or hereafter developed is forbidden. The use of general descriptive names, trade names, trademarks, etc., in this publication, even if the former are not especially identified, is not to be taken as a sign that such names as understood by the Trade Marks and Merchandise Marks Act, may accordingly be used freely by anyone.

Production managed by Jenny Wolkowicki; manufacturing supervised by Jeffrey Taub.
Typeset by Impressions Book and Journal Services, Inc., Madison, WI.
Printed and bound by Edwards Brothers, Inc., Ann Arbor, MI.
Printed in the United States of America.

9 8 7 6 5 4 3 2 1

ISBN 0-387-95301-9   SPIN 10838683

Springer-Verlag New York Berlin Heidelberg
A member of BertelsmannSpringer Science +Business Media GmbH

Preface

This book is focused on the functions of microglial cells during neuronal regeneration and degeneration. Regeneration and degeneration in the Central Nervous System (CNS) are influenced profoundly by the activities of microglia, and neuron-microglia interactions are increasingly becoming the subject of intense research activity, Elucidating the molecular mechanisms of neuron-microglia interactions is of vital interest for understanding the consequences of acute traumatic/ischemic injury, as well as the pathogenesis of chronic neurodegenerative processes, in particular Alzheimer's disease (AD). This book is meant to provide an up-to-date summary of microglial cell function and their relevance to these acute and chronic pathological states.

So what's new with microglia? We have learned a lot about normal physiological functions of microglia from studying their activities in CNS pathology. The general concept that emerges is that microglia have a role in determining neuronal survival and death and that their primary role in the normal CNS is to support and protect neuron viability. Microglia are our neurons' friends. They have been called "sentinels" and "sensors of pathology," terms that underscore this physiological function. However, what would happen if microglia ceased to function properly and became pathological? This book introduces the idea of microglial dysfunction as a pathological mechanism for neurodegenerative disease. In addition, the chapters in this book address a spectrum of novel issues related to CNS development, aging, Alzheimer's disease, as well as to acute CNS injury and to the molecular mechanisms of inter- and intracellular signaling.

What are the challenges for microglial research in the future? To refine the definition of various microglial activation states is one challenge, since there are many different levels of microglial activation. To define the molecular and genetic changes that occur in microglia with aging will become increasingly important as the incidence of Alzheimer's disease continues to grow. To investigate how microglia remodel synaptic connections and how they guide axons during development or during post-injury repair will be important for developing strategies to repair the injured spinal cord. To identify the molecular nature of neuronal signals that control microglial activity will be useful for learning to manipulate microglial function as part of the treatment of injury and disease. To identify how systemic disease affects microglia will remind us neuroscientists that the brain is but one organ in the organism. The list goes on.

I want to express my appreciation to all contributing authors, some of whom are former students, some are colleagues, some are friends, and some are all of the above. It is gratifying to see how everybody's part helped shape the final product. Without them this book would not have been written.

University of Florida
Gainesville, Florida

Wolfgang J. Streit
June 2000

Contents

Preface v
Contributors ix

1. Physiology and Pathophysiology of Microglial Cell Function 1
   Wolfgang J. Streit

2. Roles of Microglia in the Developing Avian Visual System 15
   *Julio Navascués, Miguel A., Cuadros, Ruth Calvente,
   and José L. Marín-Teva*

3. Microglial Ion Channels 36
   Claudia Eder

4. Calcium Signaling in Microglial Cells 58
   Thomas Möller

5. Microglia as a Source and Target of
   Cytokine Activities in the Brain 79
   Uwe-Karsten Hanisch

6. Microglia and Macrophage Responses in Cerebral Ischemia 125
   Guido Stoll, Sebastian Jander, and Michael Schroeter

7. Role of Microglia and Macrophages in Secondary Injury
   of the Traumatized Spinal Cord: Troublemakers or Scapegoats? 152
   Phillip G. Popovich

8. Microglial Response in the Axotomized Facial Motor Nucleus 166
   Gennadij Raivich

9. Neuroprotective Roles of Microglia in the Central Nervous System ... 188
   Kazuyuki Nakajima and Shinichi Kohsaka

10. Influences of Activated Microglia/Brain Macrophages on Spinal Cord Injury and Regeneration 209
    Alexander G. Rabchevsky

11. Opportunities for Axon Repair in the CNS: Use of Microglia and Biopolymer Compositions 227
    Joshua B. Stopek, Wolfgang J. Streit, and Eugene P. Goldberg

12. Beta Amyloid Protein Clearance and Microglial Activation 245
    Sally A. Frautschy, Greg M. Cole, and March D. Ard

13. Microglia and Aging in the Brain 275
    Caleb E. Finch, Todd E. Morgan, Irina Rozovsky, Zhong Xie, Richard Weindruch, and Tomas Prolla

Index 307

Contributors

March D. Ard, Ph.D., Department of Anatomy, University of Mississippi Medical Center, 2500 N. State St., Jackson, MS 39216, USA

Ruth Calvente, Ph.D., Departamento de Biología Celular, Facultad de Ciencias, Universidad de Granada, E-18071, Granada, Spain

Greg M. Cole, Ph.D., UCLA, Departments of Medicine and Neurology, Greater Los Angeles Healthcare System, VA Medical Center 16111 Plummer St., Sepulveda, CA 91343, USA

Miguel A. Cuadros, Ph.D., Departamento de Biología Celular, Facultad de Ciencias, Universidad de Granada, E-18071, Granada, Spain

Claudia Eder, Ph.D., Abteilung Neurophysiologie Insitut für Physiologie der Charité, Humboldt Universitaet zu Berlin, Tucholskystr. 2, D-10117, Berlin Germany
Claudia.eder@charite.de

Caleb E. Finch, Ph.D., Andrus Gerontology Center, School of Gerontology and Department of Biological Sciences, University of Southern California, Los Angeles, CA 90089-0191, USA
cefinch@usc.edu

Sally A. Frautschy, Ph.D., UCLA, Departments of Medicine and Neurology, Greater Los Angeles Healthcare System, VA Medical Center, 16111 Plummer St., Sepulveda, CA 91343, USA
frautsch@ucla.edu

Eugene P. Goldberg, Ph.D., Biomaterials Center, Department of Material Science and Engineering, P.O. Box 116400, RM. 239 MAE, University of Florida College of Engineering, Gainesville, FL 32611, USA

Sebastian Jander, M.D., Department of Neurology, Heinrich-Heine Universität, Düsseldorf, Moorenstr. 5, 40255 Düsseldorf, Germany

Shinichi Kohsaka, Ph.D., M.D., Department of Neurochemistry, National Institute of Neuroscience, 4-1-1 Ogamahigashi, Kodaira Tokyo, 187-8502
kohsaka@ncnp.go.jp

José L. Marín-Teva, Ph.D., INSERM, U.495 Hôpital de la Salpêtrière, 47 Boulevard de l'Hôpital, F-75651 Paris (Cedex 13), France
marintev@ccr.jussieu.fr

Thomas Möller, Ph.D., Department of Neurology, University of Washington, Department of Neurology, Box 356465. 1959 NE Pacific Street, Seattle, WA 98195, USA
moeller@u.washington.edu

Todd E. Morgan, Ph.D., Andrus Gerontology Center, School of Gerontology, University of Southern California, Los Angeles, CA 90089-0191, USA
temorgan@usc.edu

Kazuyuki Nakajima, Ph.D., Institute of Life Science, Soka University 1-236 Tangi-machi, Hachioji, Tokyo 192-8577
nakajima@t.soka.ac.jp

Julio Navascues, Ph.D., Departamento de Biología Celular, Facultad de Ciencias, Universidad de Granada, E-18071, Granada, Spain
navascue@goliat.ugr.es

Phillip G. Popovich, Ph.D., The Ohio State University College of Medicine and Public Health, Department of Molecular Virology, Immunology and Medical Genetics, 333 W. 10th Avenue, 2078 Graves Hall, Columbus, OH 43210, USA
popovich.2@osu.edu

Tomas Prolla, Ph.D., Department of Medical Genetics, University of Wisconsin, Madison, WI 53706, USA

Alexander G. Rabchevsky, Ph.D., 229 Sanders-Brown Center on Aging, University of Kentucky, Lexington, KY 40536-0230, USA
agrab@pop.uky.edu

Gennadij Raivich, Ph.D., M.D., Perinatal Brain Repair Group, Department of Obstetrics and Gynecology, Department of Anatomy, University College London, Gower Street Campus 86-96 Chenies Mews, London WC1E 6HX, UK.
G.Raivich@ucl.ac.uk *and* ravich@neuro.mpg.de

Irina Rozovsky, Ph.D., Andrus Gerontology Center, School of Gerontology, University of Southern California, Los Angeles, CA 90089-0191, USA
rozovsky@mollio.usc.edu

Michael Schroeter, M.D., Department of Neurology, Heinrich-Heine Universität, Düsseldorf, Moorenstr. 5, 40255 Düsseldorf, Germany

Guido Stoll, M.D., Professor of Neurology, Department of Neurology, Julius-Maximilians-Universität Würzburg, Josef-Schneider Str. 11, 97080 Würzburg Germany
guido.stoll@mail.un_wuerzburg.de

Joshua B. Stopek, M.S., Biomaterials Center, Department of Material Science and Engineering, P.O. Box 116400, RM. 239 MAE, University of Florida College of Engineering, Gainesville, FL 32611, USA
jstop@mse.ufl.edu

Wolfgang J. Streit, Ph.D., Department of Neuroscience, Building 59, University of Florida College of Medicine, 100 Newell Drive, P.O. Box 100244, Gainesville, FL 32611, USA
streit@ufbi.ufl.edu

Richard Weindruch, Ph.D., Department of Medicine, University of Wisconsin, VA Hospital (GRECC, 4D), Madison, WI 53705

Zhong Xie B.S., Ph.D., Andrus Gerontology Center, Department of Biological Sciences, University of Southern California, Los Angeles, CA 90089–0191
zhongxie@usc.edu

1

Physiology and Pathophysiology of Microglial Cell Function

WOLFGANG J. STREIT

Introduction

The primary objective of this first chapter is to provide a brief overview and synthesis of the subsequent chapters in this book and to elaborate on some favorite subjects, such as the role of microglia in the normal brain and their role in Alzheimer's disease. Other pathological conditions where microglia are thought to play important roles, such as autoimmune CNS inflammatory disease, experimental allergic encephalomyelitis and multiple sclerosis, or infectious diseases such as HIV, will not be covered specifically in this book, but there are excellent reviews available on these topics (Gonzalez-Scarano and Baltuch (1999); Benveniste et al. (1997)).

A secondary objective is to conceptualize the functional significance of reactive microgliosis and to integrate experimental observations into a biological and philosophical framework. In other words, why do microglia do what they do? There are specific physiological changes that trigger microglial activation and understanding the nature of these will improve our understanding of the causes and consequences of CNS disease and injury. Alzheimer's disease represents an ever-increasing health problem and there is a burgeoning literature implicating microglial cells in disease pathogenesis and, most recently, also in disease resolution, as shown by the vaccination studies in rodents (Schenk et al. (1999); Morgan et al. (2000); Janus et al. (2000)). It is thus likely that continued and improved understanding of microglial cell function will be essential not only for understanding AD etiology, but also for prevention and treatment of the disease. Chapter 12 by Frautschy and colleagues will convince anybody that clearance of amyloid beta protein is a most important function of microglial cells. Below I elaborate on this theme and reinforce the idea that a lack of amyloid clearance and its subsequent accumulation is due to the fact that microglial cells have, in essence, become dysfunctional. Could AD be a disease of prematurely senescent and dysfunctional microglial cells? In Chapter 13, Finch and colleagues summarize what is known about microglia in the aging brain. Although this field, unlike acute CNS injury, has not yet received much attention, it is clear that, with aging, microglia become progressively more activated. This observation must be taken into consideration with regard to the etiology of age-related dementias. Future work in this area will be quite important for a better understanding of AD and other neurodegenerative diseases.

A final, but not necessarily separate, objective of this chapter is to emphasize certain conceptual advances that have been made with regard to microglial biology and this should go hand in hand with the revision of some commonly accepted and widely disseminated facts. For example, the idea that monocytes represent direct precursors for microglia which invade the CNS during development and continue to replace microglia throughout the life of the organisms is no longer tenable (Hurley and Streit (1996); Streit (2001)). Primitive microglial precursors (fetal macrophages) are present in the developing CNS before monocytes are detectable elsewhere in the body (Takahashi et al. (1989); Sorokin et al. (1992)), and studies with bone marrow chimeras have shown that little replacement of the adult microglial population comes from blood-derived cells (Hickey et al. (1992); Lassmann et al. (1993)). So, monocytes that are already terminally differentiated blood macrophages are unlikely to serve as direct microglia precursors. At the same time, it may be stated unequivocally that monocytes and microglia represent closely related cell types that may have a common ancestor during ontogeny. The situation can be seen as being somewhat analogous to the debate over whether man is derived from the chimpanzee.

The concept of activated microglia has been and will continue to be a subject of debate because it represents a complex issue. Activated microglia come in many variants, depending on whether they occur as cultured cells, or *in vivo* in pathological states. For example, the activated microglia that appear in great numbers soon after acute CNS trauma or stroke may be in a different state of activation than the microglia that become activated as a result of aging. Ameboid microglia in the normally developing CNS that look so much like brain macrophages are not activated in same way as microglia-derived brain macrophages in cell culture, or those that appear after traumatic CNS injury (Hurley et al. (1999)). In fact, ameboid microglia may not be activated at all and simply are rounded, like macrophages, because they are immature and have not yet developed ramified processes. Thus, one challenge for the future of microglial research lies in an improved and more refined molecular definition of the various microglial activation states. This is, of course, inextricably linked to the role of microglia as neurotrophic and neurotoxic effectors, a subject that will receive much attention in the chapters that follow.

Microglial functions: from pathology to physiology

The physiological functions of microglia in the normal CNS are largely unknown, and most work has been concerned with their roles in brain pathology. Using primarily rodent and human tissues, detailed histopathological studies on microglial activation have been done in nearly every pathological state that affects the CNS, including some systemic conditions which affect the CNS secondarily (Kreutzberg (1996); Moore and Thanos (1996); Streit and Sparks (1997)). These studies have pointed towards multiple functions of microglia related to their role as phagocytic and antigen presenting cells (immune effector cells), as well as to their ability to execute neurotrophic and neurotoxic activities. The latter, in particular, has caused somewhat of a stir and has created two camps supporting either microglial benevolence or malice. This dichotomy in the perception of microglial cell function, which surfaces throughout this book, represents perhaps the most debated issue in microglial research currently. However, it is reasonable to assume that microglia can act both neurotrophically and neurotoxically. Yet, neurotrophic and neurotoxic actions are probably not being exerted simultaneously by a microglial cell at any given moment, hence the question arises as to what determines when and how microglia assume either a neurotrophic or a neurotoxic state. The answer to this lies in the neurons themselves, and more specifically in the nature of the signaling that occurs between neurons and microglia. Continued research into the mechanisms of how neurons communicate their functional status to microglia will shed light on how microglia respond to neuron-derived signals and whether they do so in a neurotrophic or neurotoxic fashion (Bruce-Keller (1999); Streit et al. (1999, 2000)).

Can we learn about normal physiological functions of microglia by studying them histopathologically in sections from diseased or injured brain tissue? Most definitely. To illustrate an acute microglial reaction, such as would occur following trauma or stroke, an analogy with a simpler biological system is useful. Ants in an ant hill become mobilized very quickly if a portion of their hill is suddenly destroyed. The ants become agitated and they swarm to investigate the insult. Subsequently, they repair and rebuild the damaged site. Analagously, when acute brain damage occurs, microglia become activated and they migrate within the vicinity of the wound to investigate the nature of their now dramatically altered microenvironment. Later, during the post-injury period, they phagocytose debris, produce growth factors and other pro-regenerative molecules to promote tissue repair. In Chapter 6, Stoll and colleagues describe reactive microgliosis after acute ischemic injury. They offer an in-depth histopathological account that is a bit more sophisticated than the ant analogy. Notably, they report robust expression of CD8 surface receptors on activated microglia, which is remarkable because this molecule is traditionally associated with cytotoxic lymphocytes.

Microglia are well equipped for exploring their surroundings as they possess an astonishing variety and abundance of sensors in the form of cell surface receptors and ion channels. Chapter 3 by Eder provides an up-to-date summary of the many types of ion channels identified on microglia. It is clear from her description that ion channels represent critical molecular sensors that can mediate microglial plasticity and enable the cells to discern chemical changes in their microenvironment that inform them about the viability of neurons and other cells. The richness of the surface jungle of microglial receptors is reflected by a prominent microglial glycocalyx, a carbohydrate rich surface coat made up of membrane-associated glyoproteins and glycolipids. Lectin binding studies have consistently revealed the microglial glycocalyx in several different species (Streit (1995)). A glycocalyx with lectin-binding properties is also prominent along the endothelium of the cerebral vasculature, and thus blood vessels and microglia have in common some surface characteristics that may be reminiscent of their mesodermal origins. In terms of physiological function, the presence of a carbohydrate-rich surface coat reflects the cells' enhanced ability to detect and absorb signals from the surrounding milieu.

Signals received by microglia are manifold and include ions, nucleotides, neuropeptides, chemokines/cytokines and neurotransmitters. Appropriate surface receptors facilitate a microglial response to these ligands, which may be present in different concentrations depending on whether the CNS is injured. Some microglial surface receptors can interact with serum-derived components, such as complement, thrombin, or immunoglobulins (Graeber et al. (1988); Ulvestad et al. (1994); Möller et al. (2000)). These molecules, which are largely excluded from the normal CNS, flood the CNS immediately following a physical injury. Because microglia bear complement, thrombin, and Fc receptors, they are able to quickly "mop up" the spillage and prevent binding of these molecules to other CNS cells, which could bring about undesirable effects, such as complement-mediated lysis. Obviously, this constitutes a protective function important for maintaining tissue homeostasis. In the normal CNS, microglial receptors for serum components may thus help keep immobilized small amounts of serum substances that may exude into the parenchyma. In Chapter 4, Möller reviews some of the microglial receptor molecules, including receptors for serum components, purinoceptors, neurotransmitter receptors, and chemokine receptors. His discussion is focused on how these receptors mediate changes in a major signal transduction pathway, free cytoplasmic Ca^{2+} concentration. Following is Chapter 5 by Hanisch, who tackles the huge field of microglial cytokines, chemokines and their receptors. His inclusive review on microglia and cytokine interactions, as well as cytokine-mediated signal transduction cascades, underscores the extraordinary ability of microglia to detect and integrate extracellular signals and, in turn, respond to them by changing their pattern of gene expression.

The microglial response to injury is marked by adjustments in the synthesis of growth factors and cytokines, as well as in the expression of surface receptors in order to optimally cope with the altered state of the tissue. Many of the adjustments in microglial-secreted protein synthesis and gene expression have been measured using both *in vivo* and *in vitro* experimentation, and these measurements have provided clues for a better understanding of microglial cell function. For example, the fact that MHC class II antigens are newly expressed on some populations of activated microglia clearly points towards their function as antigen presenting cells. The fact that acute CNS trauma or stroke within minutes to hours induces production of interleukin-1 (IL-1) and tumor necrosis factor (TNF) in microglia for a short period of time (1–2 days) suggests that these proinflammatory cytokines serve as immediate triggers for autocrine or paracrine production of growth factors and other proregenerative molecules needed for subsequent neuronal rescue and tissue repair (Streit et al. (1998); Rothwell and Luheshi (2000); Fassbender et al. (2000)). *In vivo,* microglial production of IL-1 and TNF is short-lived following acute damage, and production of these cytokines is quickly shut down, suggesting that prolonged expression is not necessary. In fact, should production of these cytokines persist for prolonged periods of time, their presence may be detrimental and contribute to secondary neurodegeneration (Akiyama et al. (2000); Griffin et al. (1989)).

It is well established that microglia produce a variety of molecules with neurotrophic activities. In Chapter 9, Nakajima and Kohsaka focus on the neuroprotective functions of microglia, and these go beyond the production of neurotrophic factors. These authors document for the first time expression of glutamate transporters on activated microglia *in vivo,* suggesting that microglia, like astrocytes, are involved in the removal of excess glutamate from the extracellular space. The picture that emerges in terms of the physiological functions of microglia is one primarily of neuroprotection. One can, in fact, easily envision that microglia are continuosly monitoring neuronal activity during normal brain functioning, intervening whenever necessary to remove excitatory neurotransmitters and/or to supply trophic factors as needed. Histological observations in normal brain show presence of frequent perineuronal microglial satellites (Palacios (1990)), cells that are separated from the neuron by only a few nanometers of extracellular space (Figure 1). This close physical proximity between neurons and microglia suggests that microglia are monitoring neuronal homeostasis and activity. If there are altered neuronal requirements for trophic support during periods of increased activity or stress, perineuronal microglia are perfectly positioned to deliver minute, but targeted, quantities of specific neuronal survival molecules that may be tailored to the individual needs of different neuron populations (see Chapter 9). Thus, microglial satellites may ensure neuronal well-being through constitutive secretion of trophic proteins as a normal physiological function. This idea is strongly supported by observations in the axotomized facial motor nucleus, a well characterized *in vivo* model for studying neuron-microglia interactions during motoneuron regeneration (Kreutzberg, 1996).

Chapter 8 by Raivich summarizes what has been learned about microglial activation using the facial nerve model, primarily in the context of *in vivo* cytokine production. Advances in understanding the functional significance of microglial cytokines have been made possible by producing facial nerve lesions in cytokine knockout animals. Perhaps one of the most intriguing features of the glial response in the facial nucleus after axotomy is the ensheathment of axotomized motoneurons by microglial satellites within a few days after nerve cut. As Raivich points out, the molecular mechanisms that regulate the chemical attraction and adhesion of microglia to the injured nerve cells are complex, but essential for advancing continued understanding of neuron-microglial interactions. Since motoneurons ultimately recover from axotomy and regenerate their axons much of the cytokine activities in the facial nucleus are probably designed to maximize regeneration. One is also reminded of another phenomenon, first reported more than thirty years

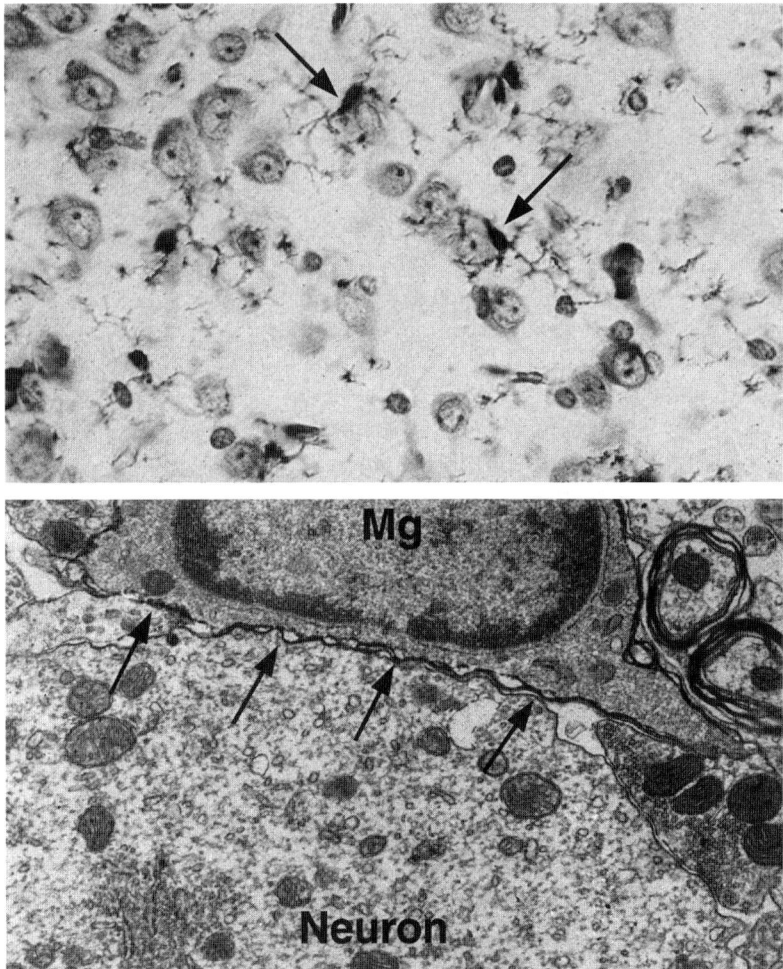

FIGURE 1. Perineuronal microglial satellites as seen by light and electron microscopy. Top panel shows two lectin-labelled microglial satellites partly enclosing neurons with their cytoplasmic extensions (arrows). Rabbit cerebral cortex, cresyl violet counterstain ×350. Bottom panel shows a perineuronal microglial cell (Mg). Arrows point to sites where microglial and neuronal membranes are directly apposed, ×10,000.

ago by Blinzinger and Kreutzberg (1968). These authors observed that, as microglia approach and ensheath axotomized motoneurons, they displace afferent synapses from neuronal cell bodies and dendrites, resulting in a loss of synaptic contacts that may be as high as 80%. Activated microglia are thus involved in lesion-induced synaptic plasticity after motoneuron injury, but possibly also after central lesions (Gehrmann et al. (1991); Schoen and Kreutzberg (1994)), and the question arises whether they particpate in synaptic plasticity and remodelling in the normal CNS. Studies conducted in the developing and adult olfactory system suggest that this is indeed the case (Fiske and Brunjes (2000); Schoen and Kreutzberg (1995)). Microglia are well equipped to participate in synaptic remodelling because they generate a number of protease activities, including matrix metalloproteinase, elastase, and plasminogen activator (Gottschall et al. (1995); Nakajima et al. (1992, 1993, 1996)). They also produce extracellular matrix molecules, such as laminin, thrombospondin and keratan sulfate (Rieske et al.(1989); Rabchevsky and Streit (1997); Chamak et al. (1994); Bertolotto et al. (1993)), as well as neurotrophic factors, which are important in mechanisms controlling synaptic plasticity. Loss of synaptic connections is associated with neurodegenerative diseases, and this loss could be connected with age-related changes in microglial cell function that impair the cells' ability to sustain synaptic connections (see below).

Role of microglia in development and tissue repair

Historically, microglia have received considerable attention with regard to neural development, and earlier work was focused primarily on their ontogeny and their relationship to blood monocytes (Theele and Streit (1993); Ling and Wong (1993)). I consider the debate over the monocytic origin of microglia resolved, since most of the experimental evidence that has become available during the last decade no longer supports a direct monocyte origin. For details of this evidence, the reader is referred to the introductory paragraph of this chapter and to other reviews (Hurley and Streit (1996); Streit et al. (1999); Streit (2001)). What lies ahead in terms of microglia and development are functional issues related to the role of microglia in promoting axon growth and guidance, their ability to regulate apoptosis during development, and also their potential involvement in vasculogenesis and synaptogenesis. In this book, Chapter 2 by Navascués addresses the role of microglia during visual system development. One issue that stands out here is the exquisite migratory ability exhibited by microglia in the developing brain. Perhaps most interesting is their ability to move along radial glial processes, suggesting that microglia are involved in guiding developing neurons along radial glial processes to their final destinations. As with synaptic remodelling, a developmental function for microglia in neuronal guidance seems quite likely, since they do have all the right tools to undertake this task: chemokines and cytokines for guidance and chemoattraction, growth factors for promoting neuronal maturation, extracellular matrix molecules for providing a substrate, and enzymes for tissue remodelling.

From development it is only a small step to tissue repair after injury, and many of the same processes that take place during development may be recapitulated in a CNS lesion site. There are now published studies that have shown beneficial effects of microglia/macrophage transplants in terms of enhancing axonal regrowth, primarily after spinal cord injury, and these are reviewed in Chapter 10 by Rabchevsky. Transplantation studies using microglia/macrophage grafts have potentially far-reaching implications for post-traumatic axon repair, especially after spinal cord injury, where a lack of regeneration in ascending and descending long tracts contributes significantly to functional impairment. In Chapter 11, collaborators from materials science, Stopek and Goldberg, and I consider the future use of synthetic biopolymers as implantable scaffolding devices. The biopolymer compounds under consideration are porous, hydrophilic and bioerodable, and they can be inundated with microglia as biological sources of growth factors and extracellular matrix molecules. Future *in vivo* experiments with such biopolymer-microglia implants may prove them to be useful as bridging devices for guiding regenerating axons across a lesion cavity.

Functional significance of microglial activation

The title of Chapter 7 by Popovich captures the essence of the debate over neuroprotective and neurotoxic functions of microglial cells. Although this issue is of paramount importance for treatment of traumatic CNS injury, I discuss it here only briefly, since Popovich and several of the other authors in this book cover this topic comprehensively. The concept of secondary injury, although widely accepted, is somewhat artificial in that much of what happens immediately after the moment of injury is simply inevitable. While some secondary events may be preventable, it is still highly speculative what exactly is preventable or what should be prevented. Some secondary processes, such as inflammation, are quite beneficial in terms of getting the healing process started. This is where microglia come into the picture, and the key question is how much, if any, secondary damage is caused by activated microglial cells following physical injury. As discussed at length (Streit et al. (1999, 2000)), it is reasonable to assume that microglial activation is largely determined by neuronal signals. The hypothesis is that acute injury causes neurons to generate signals that inform microglia about the neuronal state of vitality and viability. Depending on how severe a degree of neuronal disability is being signalled to microglia after injury, they will take appropriate action and either nurse the injured neurons into regeneration, or kill them if they are not viable. Both types of microglial responses likely occur after acute injury, and both are considered to represent normal physiological and neuroprotective responses (Figure 2). The question arises whether there is a pathological response of microglia, and if so, what causes it. It is conceivable that some processes that are chronic and presumably extrinsic in nature persistently activate microglial cells, and in so doing eventually cause a decline in the physiological ability of microglial cells to maintain homeostasis. This could have detrimental consequences and may lead to bystander damage due to microglial dysfunction. In addition to this *primary* activation of mi-

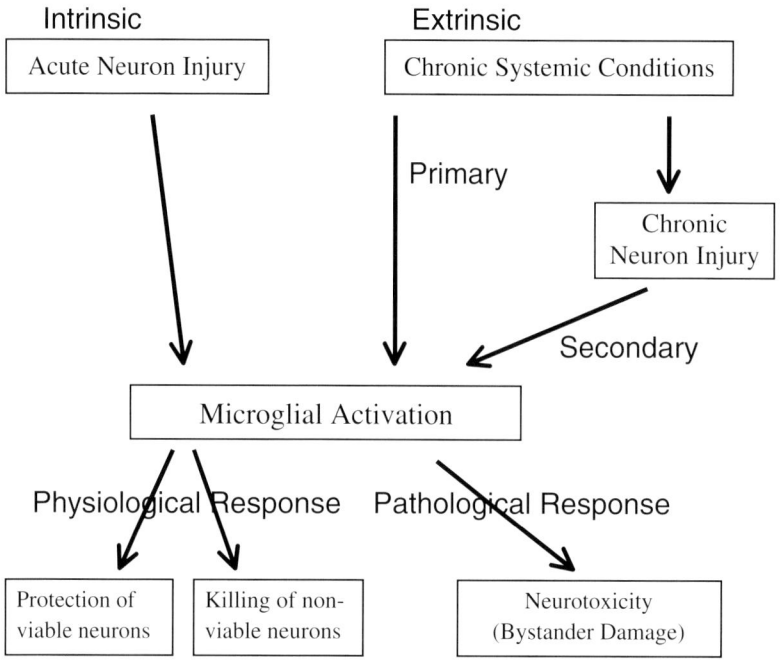

FIGURE 2. Theoretical basis for explaining microglial neuroprotective and neurotoxic activities. The theory states that there are two mechanisms for activating microglia, intrinsic and extrinsic. The intrinsic mechanism involves signaling from acutely injured neurons in the CNS, which trigger normal, physiological microglial activation that is beneficial and designed to ameliorate damage and stress. The extrinsic mechanism involves chronic systemic disease processes producing pathological microglial activation that may lead to neurotoxicity. Extrinisic disease mechanisms may activate microglia directly (primary activation) or indirectly (secondary activation) by way of inducing chronic neuronal injury. It is important to notice that for both primary and secondary activation the stimulus is a chronic one. Abnormal microglial activation occurs in response to any number of (largely unknown) systemic triggers, which owing to their chronic persistence, eventually render microglial cells unable to carry out their normal neuroprotective mission, hence leading to bystander damage or neurotoxicity.

croglia by extrinsic factors, microglia may also be activated secondarily by signals from neurons chronically injured due to systemic disease. One reported example of a systemic condition causing microglial activation is heart disease (Streit and Sparks (1997)). More specifically, in this study we were able to show that high serum cholesterol levels induced experimentally in rabbits were associated with microgliosis and leukocyte extravasation in the brains of these animals. Earlier studies by Sparks et al. (1994) had reported an increase in neuronal intracellular β-amyloid in hypercholesterolemic rabbits. These findings point towards vascular problems as one possible systemic cause for microglial dysfunction. Future research into other systemic diseases with vascular components, such as diabetes or hypertension may offer new insights into this possibility.

Microglia and Alzheimer's disease

Increased microglial activation, as determined by measurement of HLA-DR-immunoreactive microglia, has been reported and taken as evidence to support a causative role of microglia in neurodegeneration as neurotoxin-producing cells (Carpenter et al. (1993)). The idea of microglia as autoaggressive neurotoxic effector cells is also supported by studies in cell culture, which have shown that treatment of microglia with beta amyloid protein elicits either direct neurotoxicity or increased production of potentially neurotoxic compounds (Meda et al. (1995); Giulian et al. (1996); Barger and Harmon (1997); McDonald et al. (1997)). Thus, one prevailing view of AD pathogenesis is that presence of beta amyloid protein in the brain stimulates chronic microglial activation, which results in the production of microglial neurotoxins that bring about neurodegeneration and dementia. Epidemiological studies with nonsteroidal anti-inflammatory agents, as well as the recent beta amyloid vaccination studies, do seem to support this idea. However, it is important to consider the other side of the coin, which portrays microglia as sentinels involved in neuroprotection. Considering AD pathogenesis with this neuroprotective microglial cell function in mind, it is then conceivable that, if for some reason this physiological function of microglia becomes depleted due to systemic disease or aging, presence of dysfunctional microglia in the brain could contribute to AD development. In other words, neurodegeneration may result not from aggression, but from neglect. Impairment of microglial cell function could explain most of the histopathological features of AD, including amyloid deposits, neurofibrillary tangles, and synapse loss. As Frautschy points out in Chapter 12, there appears to be a defect in the ability of microglia to clear away $A\beta$ protein. Coupled with the possibility that there is increased $A\beta$ production, this might explain why $A\beta$ accumulates extracellularly in AD, as with normal aging. Neurodegenerative changes, such as ubiquitination and tangle formation, could result from a decreased ability of microglia to generate growth factors that sustain neuronal viability. Synapse loss, which correlates well with cognitive impairment (Terry et al. (1991)), could be due to a decreased ability of microglia to sustain synaptic connections through growth factor production, or other mechanisms. Moreover, the fact that synapse loss does not correlate with presence of $A\beta$ (Masliah et al. (1993)) could be due to selective microglial dysfunction related either to the cells' ability to support synapses, or to clear away β-amyloid. Another consideration that is important in this context is the fact there is little evidence for leukocyte extravasation in AD (Akiyama et al. (2000)). Experiments in rodents have suggested that presence of activated microglia may be a prerequisite for promoting leukocyte infiltration into the CNS parenchyma (Maehlen et al. (1989)). Thus, if there is widespread microglial activation in AD brain, why is there little or no infiltration of blood-borne cells? Once again, microglial dysfunction could account for this observation, since it may involve reduced production of the microglial chemoattractants necessary for facilitating leukocyte infiltration into the parenchyma.

In summary, it may be helpful for the continued advancement of understanding neurodegenerative disease to consider impairment of microglial cell function as an alternative to their widely perceived role as neurotoxic aggressors. Although the result in both cases is neurodegeneration, determining the way(s) this neurodegeneration comes about is absolulely critical for initiating the best intervention strategies.

Conclusions

It is hoped that readers of this chapter come away with the impression that microglial cells are essential for ensuring proper neuronal functioning, and quick to intervene when neurons become injured or otherwise stressed. Understanding neuron-microglia and microglia-neuron interactions will be essential for understanding CNS physiology and pathology. If one accepts the idea that microglia are sentinels of neuron well-being, it is easy to see how pathological impairment of microglia could have devastating consequences for brain function. The identification of such "microglial diseases" represents a new challenge for years to come.

References

Akiyama H, Barger S, Barnum S, Bradt B, Bauer J, Cole GM, Cooper NR, Eikelenboom P, Emmerling M, Fiebich BL, Finch CE, Frautschy S, Griffin WST, Hampel H, Hull M, Landreth G, Lue LF, Mrak R, Mackenzie M, O'Banion K, Pachter J, Pasinetti G, Plata-Salaman C, Rogers J, Rydel R, Shen Y, Streit W, Strohmeyer R, Tooyoma I, Van Muiswinkel FL, Veerhuis R, Walker D, Webster S, Wegrzyniak B, Wenk G, Wyss-Coray, A (2000) Inflammation and Alzheimer's Disease. *Neurobiol Aging* 21:383–421.

Barger SW, Harmon AD (1997) Microglia activation by Alzheimer amyloid precursor protein and modulation by apolipoprotein E. *Nature* 388:878–881.

Benveniste EN (1997) Role of macrophages/microglia in multiple sclerosis and experimental allergic encephalomyelitis. *J Mol Med* 75:165–173.

Bertolotto A, Caterson B, Canavese G, Migheli A, Schiffer D (1993) Monoclonal antibodies to keratan sulfate immunolocalize ramified microglia in paraffin and cryostat sections of rat brain. *J Histochem Cytochem* 41:481–487.

Blinzinger K, Kreutzberg G (1968) Displacement of synaptic terminals from regenerating motoneurons by microglial cells. *Z Zellforsch* 85:145–157.

Bruce-Keller AJ (1999) Microglial-neuronal interactions in synaptic damage and recovery. *J Neurosci Res* 58:191–201.

Carpenter AF, Carpenter PW, Markesbery WR (1993) Morphometric analysis of microglia in Alzheimer's disease. *J Neuropathol Exp Neurol* 52:601–608.

Chamak B, Morandi V, Mallat M (1994) Brain macrophages stimulate neurite growth and regeneration by secreting thrombospondin. *J Neurosci Res* 38:221–233.

Fassbender K, Schneider S, Bertsch T, Schlueter D, Fatar M, Ragoschke A, Kuhl S, Kischka U, Hennerici M (2000) Temporal profile of release of interleukin-1beta in neurotrauma. *Neurosci Lett* 284:135–138.

Fiske BK, Brunjes PC (2000) Microglial activation in the developing rat olfactory bulb. *Neuroscience* 96:807–815.

Gehrmann J, Schoen SW, Kreutzberg GW (1991) Lesion of the rat entorhinal cortex leads to a rapid microglial activation in the dentate gyrus. *Acta Neuropathol* 82:442–455.

Giulian D, Havercamp LJ, Yu JH, Karshin W, Tom D, Li J, Kirkpatrick J, Kuo LM, Roher AE (1996) Specific domains of beta-amyloid from Alzheimer plaque elicit neuron killing in human microglia. *J Neurosci* 16:6021–6037.

Gonzalez-Scarano F, Baltuch G (1999) Microglia as mediators of inflammatory and degenerative diseases. *Annu Rev Neurosci* 22:219–240.

Gottschall PE, Yu X, Bing B (1995) Increased production of gelatinase B (matrix metalloproteinase-9) and interleukin-6 by activated rat microglia in culture. *J Neurosci Res* 42:335–342.

Graeber MB, Streit WJ, Kreutzberg GW (1988) Axotomy of the rat facial nerve leads to increased CR3 complement receptor expression by activated microglial cells. *J Neurosci Res* 21:18–24.

Griffin WS, Stanley LC, Ling C, White L, Mac-Leod V, Perrot LJ, White CL, Araoz C (1989) Brain interleukin-1 and S-100 immunoreactivity are elevated in Down syndrome and Alzheimer's disease. *Proc Natl Acad Sci USA* 86:7611–7615.

Hickey WF, Vass K, Lassmann H (1992) Bone marrow-derived elements in the central nervous system: an immunohistochemical and ultrastructural survey of rat chimeras. *J Neuropathol Exp Neurol* 51:246–256.

Hurley SD, Streit WJ (1996) Microglia and the mononuclear phagocyte system. In: Topical Issues of Microglial Research, edited by E.A. Ling, C.K. Tan, and C.B.C. Tan, *Singapore Neuroscience Association*, pp 1–19.

Hurley SD, Walter SA, Semple-Rowland SL, Streit WJ (1999) Cytokine transcripts expressed by microglia *in vitro* are not expressed by ameboid microglia of the developing rat central nervous system. *Glia* 25:304–309.

Janus C, Pearson J, McLaurin J, Mathews PM, Jiang Y, Schmidt SD, Chishti MA, Horne P, Heslin D, French J, Mount HTJ, Nixon RA, Mercken M, Bergeron C, Fraser PE, St. George-Hysolop P, Westaway D (2000) Ab peptide immunization reduces behavioral impairment and plaques in a model of Alzheimer's disease. *Nature* 408:979–982.

Kreutzberg GW (1996) Microglia: a sensor for pathological events in the CNS. *Trends Neurosci* 19:312–318.

Lassmann H, Schmied M, Vass K, Hickey WF (1993) Bone marrow derived elements and resident microglia in brain inflammation. *Glia* 7:19–24.

Ling EA, Wong WC. (1993) The origin and nature of ramified and amoeboid microglia: a historical review and current concepts. *Glia* 7:9–18.

Maehlen J, Olsson T, Zachau A, Klareskog L, Kristensson K (1989) Local enhancement of major histocompatibility complex (MHC) class I and II expression and cell infiltration in experimental allergic encephalomyelitis around axotomized motor neurons. *J Neuroimmunol* 23:125–132.

Masliah E, Mallory M, Hansen L, DeTeresa R, Terry RD (1993) Quantitative synaptic alterations in the human neocortex during normal aging. *Neurology* 43:192–197.

McDonald DR, Brunden KR, Landreth F (1997) Amyloid fibrils activate tyrosine kinase-dependent signaling and superoxide production in microglia. *J Neurosci* 17:2284–2294.

Meda L, Cassatella MA, Szendrei GI, Otvos L, Baron P, Villalba M, Ferrari D, Rossi F (1995) Activation of microglial cells by beta-amyloid protein and interferon-gamma. *Nature* 374:647–650.

Möller T, Hanisch UK, Ransom BR (2000) Thrombin-induced activation of cultured rodent microglia. *J Neurochem* 75:1539–1547.

Moore S, Thanos S (1996) The concept of microglia in relation to central nervous system disease and regeneration. *Prog Neurobiol* 48:41–460.

Morgan D, Diamond DM, Gottschall PE, Ugen KE, Dickey C, Hardy J, Duff K, Jantzen P, DiCarlo G, Wilcock D, Connor K, Hatcher J, Hope C, Gordon M, Arendash GW (2000) Ab peptide vaccination prevents memory loss in an animal model of Alzheimer's disease. *Nature* 408:982–985.

Nakajima K, Tsuzaki N, Shimojo M, Hamanoue M, Kohsaka S (1992) Microglia isolated from rat brain secrete a urokinase-type plasminogen activator. *Brain Res* 577:285–292.

Nakajima K, Nagata K, Hamanoue M, Takemoto N, Kohsaka S (1993) Microglia-derived elastase produces a low-molecular-weight plasminogen that enhances neurite outgrowth in rat neocortical explant cultures. *J Neurochem* 61:2155–2163.

Nakajima K, Reddington M, Kohsaka S, Kreutzberg GW (1996) Induction of urokinase-type plasminogen activator in rat facial nucleus by axotomy of the facial nerve. *J Neurochem* 66:2500–2505.

Palacios, G (1990) A double immunocytochemical and histochemical technique for demonstration of cholinergic neurons and microglial cells in basal forebrain and neostriatum of the rat. *Neurosci Lett* 115:13–18.

Rabchevsky AG, Streit WJ (1997) Grafting of cultured microglial cells into the lesioned spinal cord of adult rats enhances neurite outgrowth. *J Neurosci Res* 47:34–48.

Rieske E, Graeber MB, Tetzlaff W, Czlonkowska A, Streit WJ, Kreutzberg GW (1989) Microglia and microglia-derived brain macrophages in culture: generation from axotomized rat facial nuclei, identification and characterization in vitro. *Brain Res* 492:1–14.

Rothwell NJ, Luheshi GN (2000) Interleukin 1 in the brain: biology, pathology and therapeutic target. *Trends Neurosci* 23:618–625.

Schenk D, Barbour R, Dunn W, Gordon G, Grajeda H, Guido T, Hu K, Huang J, Johnson-Wood K, Khan K, Kholodenko D, Lee M, Liao Z, Lieberburg I, Motter R, Mutter L, Soriano F, Shopp G, Vasquez N, Vandevert C, Walker S, Wogulis M, Yednock T, Games D, Seubert P (1999) Immunization with amyloid-β attenuates Alzheimer-disease-like pathology in the PDAPP mouse. *Nature* 400:173–177.

Schoen SW, Kreutzberg GW (1994) Synaptic 59-nucleotidase activity reflects lesion-induced sprouting within the adult rat dentate gyrus. *Exp Neurol* 127:106–118.

Schoen SW, Kreutzberg GW (1995) Evidence that 59-nucleotidase is associated with malleable synapses–an enzyme histochemical investigation of the olfactory bulb of adult rats. *Neuroscience* 65:37–50.

Sorokin SP, Hoyt RF Jr, Blunt DG, McNelly, NA (1992) Macrophage development: II. Early ontogeny of macrophage populations in brain, liver, and lungs of rat embryos as revealed by a lectin marker. *Anat Rec* 232:527–550.

Sparks DL, Scheff SW, Hunsaker III JC, Liu H, Landers T, Gross DR (1994) Induction of Alzheimer-like β-amyloid immunoreactivity in the brains of rabbits with dietary cholesterol. *Exp Neurol* 126:88–94.

Streit WJ (1995) Microglial cells. In: *Neuroglia,* edited by H. Kettenmann and B.R. Ransom. Oxford University Press.

Streit WJ, Sparks DL (1997) Activation of microglia in the brains of humans with heart disease and hypercholesterolemic rabbits. *J Mol Med* 75:130–138.

Streit WJ, Semple-Rowland SL, Hurley SD, Miller RC, Popovich PG, Stokes BT (1998) Cytokine mRNA profiles in contused spinal cord and axotomized facial nucleus suggest a beneficial role for inflammation and gliosis. *Exp Neurol* 152:74–87.

Streit WJ, Walter SA, Pennell NA (1999) Reactive microgliosis. *Prog Neurobiol* 57:563–581.

Streit WJ, Hurley SD, McGraw TS, Semple-Rowland SL (2000) Comparative evaluation of cytokine profiles and reactive gliosis supports a critical role for interleukin-6 in neuron-glia signaling during regeneration. *J Neurosci Res* 61:10–20.

Streit WJ (2001) Microglia and macrophages in the developing CNS. Neurotoxicology, in press.

Takahashi K, Yamamura F, Naito M. (1989) Differentiation, maturation, and proliferation of macrophages in the mouse yolk sac: a light-microscopic, enzyme-cytochemical, immunohistochemical, and ultrastructural study. *J Leukocyte Biol* 45:87–96.

Terry RD, Masliah E, Salmon DP, Butters N, DeTeresa R, Hill R, Hansen LA, Katzman R (1991) Physical basis of cognitive alterations in Alzheimer disease: synapse loss is the major correlate of cognitive impairment. *Ann Neurol* 30:572–580.

Theele DP, Streit WJ. (1993) A chronicle of microglial ontogeny. *Glia* 7:5–8.

Ulvestad E, Williams K, Matre R, Nyland H, Olivier A, Antel J (1994) Fc receptors for IgG on cultured human microglia mediate cytotoxicity and phagocytosis of antibody-coated targets. (primary), and/or *J. Exp. Neurol.* 53:27–36.

2

Roles of Microglia in the Developing Avian Visual System

Julio Navascués[1], Miguel A. Cuadros[1], Ruth Calvente[1], and José L. Marín-Teva[2]

Introduction

Microglia exhibit at least three morphological and functional states during their life cycle: ameboid, ramified and reactive. Ramified microglia, also called resting microglia, are differentiated cells present in the adult CNS that bear thin ramified processes that emerge from the cell body. Ramified microglial cells are considered inactive in the normal adult CNS. They activate in response to CNS insults to become reactive microglia that retract their processes and upregulate the expression of different molecules (reviewed in Streit et al. (1999); Stoll and Jander (1999)).

Ramified microglia derive from the differentiation of ameboid microglia present in the developing CNS (Perry and Gordon (1991); Ling and Wong (1993); Hurley and Streit (1996); Cuadros and Navascués (1998)). Ameboid microglia are characterized by an irregular morphology, with pseudopodia and lamellipodia that emanate either from the cell body, or from short, broad processes (Perry et al. (1985); Ling and Wong (1993); Cuadros et al. (1994, 1997); Navascués et al. (1996); Marín-Teva et al. (1998)). The term ameboid microglia is often inappropriately used to refer to activated-brain macrophages arising from reactive microglia in the adult CNS, since these are morphologically similar to ameboid microglia during development. However, ameboid and reactive microglia represent two distinct functional states of microglia (Streit et al. (1999)). In this chapter, we use the term ameboid microglia exclusively to refer to immature microglia that appear during development.

The developing visual system has been widely used for the study of microglial development because it is particularly well defined, its developmental timing is largely known in different species, and many of its anatomical components are easily accessible to experimental manipulations. This chapter aims to summarize the state of the knowledge about ameboid microglia in different parts of the developing visual system, emphasizing their possible functions during development.

Distribution patterns of microglia in the visual system of adult vertebrates

As in other regions of the brain (Lawson et al. (1990)), microglial cells are present throughout the entire visual system of adult vertebrates, and their normal distribution is influenced by the microenvironment in each part of this system.

Elongated microglial cells with scarce processes, often aligned parallel to the optic axons, are found among the densely packed fiber fascicles in the optic pathways (optic nerves, chiasm and tracts). This distribution pattern has been described in the optic nerve of fish (Dowding et al. (1991); Velasco et al. (1995); Lillo et al. (1998)), amphibians (Goodbrand and Gaze (1991); Naujoks-Manteuffel and Niemann (1994)), birds (Moujahid et al. (1996)), and mammals (Lawson et al. (1994); Reichert and Rotshenker (1996); Wolswijk (1994); Zhang and McKanna (1997)). In the adult retina, most ramified microglial cells are localized within the outer and inner plexiform layers and no microglia are present in the nuclear layers, in which neuronal bodies are densely packed. This layer-specific microglial distribution pattern is similar in all vertebrate species (see Table 3 in Salvador-Silva et al. (2000)). The distribution of microglial cells within the optic tectum of non-mammalian vertebrates is less clearly layer-related than it is within the retina (Cuadros et al. (1994); Dowding et al. (1991); Velasco et al. (1995)), perhaps because, unlike the retinal nuclear layers, all tectal layers contain abundant neuropil.

To summarize, the distribution of microglia in the visual system shows similar features to those observed in the brain as a whole in mammals (Lawson et al. (1990)): microglia tend to be excluded from dense neuronal cell fields, such as retinal nuclear layers, whereas they accumulate in the optic pathways and gray matter zones of abundant neuropil, such as retinal plexiform layers.

Spreading of microglial cells throughout the developing visual system

Ameboid microglia spread throughout the visual system during development in order to attain the normal distribution of ramified microglia in the adult visual system. Cells of the macrophage lineage have been reported to enter the nervous parenchyma in two main waves (Fig. 1): the first at an early stage of visual system development, when retinofugal projections are beginning to develop, and the second at a much more advanced developmental stage, at the time of ganglion cell death and the refinement of connections between the retina and its targets.

2. Roles of Microglia in the Developing Avian Visual System 17

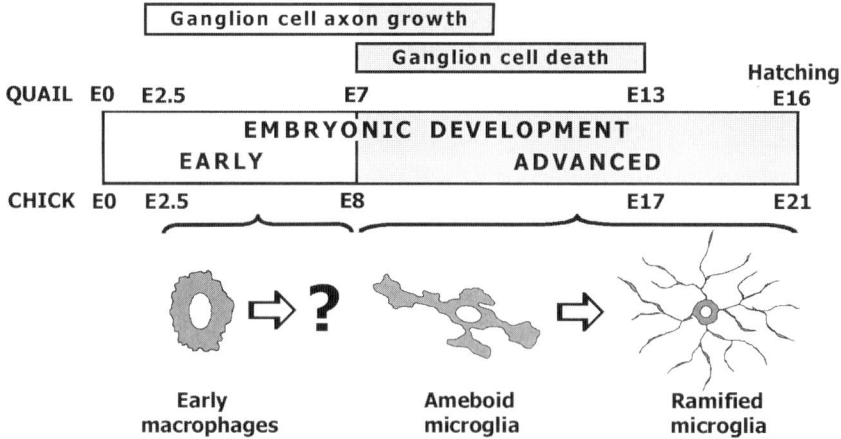

FIGURE 1. Schematic drawing illustrating the existence of two phases in the process of colonization of the avian visual system by cells related to the macrophage lineage during embryonic development. The early embryonic development comprises stages prior to embryonic day 7 (E7) in the quail and E8 in the chick, whereas the advanced embryonic development is considered to occur from these days to hatching. The time ranges of growth of ganglion cell axons in the retina and ganglion cell death are represented at the top to correlate the two developmental phases with these events. Yolk sac-derived early macrophages enter different parts of the visual system during the early development from E2.5 on and their possible fate (death or differentiation into microglia) is unknown (question mark). The appearance of ameboid microglia, which differentiate into ramified microglia, takes place during the advanced development.

Entry of early macrophages

Cells of the macrophage lineage have been observed during early development in the retina (Cuadros et al. (1991); Navascués et al. (1995)), optic nerve (Moujahid et al. (1996)), chiasm (Martín-Partido et al. (1991)) and tectum (Cuadros et al. (1994)) of birds, and in the retina of mammals (Ashwell (1989); Ashwell et al. (1989); Frade and Barde (1998); Knabe et al. (2000)). These early cells of the macrophage lineage belong to a population of yolk sac-derived early macrophages, also known as fetal macrophages (Takahashi et al. (1989); Hurley and Streit (1996)). Their yolk sac origin has been experimentally demonstrated using chick-quail chimeras (Cuadros et al. (1991); Martín-Partido et al. (1991)).

It is not clear whether yolk sac-derived early macrophages in the visual system are microglia precursors. A recent study (Alliot et al. (1999)) supports the origin of microglia from yolk sac-derived progenitors. However, early macrophages are not present in the quail retina at the beginning of the wave of immigration of microglial precursors (Navascués et al. (1995)), suggesting that microglia do not come from early macrophages. Nevertheless, we cannot rule out their possible origin from early macrophages of the optic nerve.

Spreading of microglial progenitors during advanced development: migration and proliferation

The second wave of immigration of macrophage lineage cells into different parts of the visual system takes place at a developmental stage that roughly coincides with the period of retinal ganglion cell death. These immigrating cells are true microglial progenitors, as can be deduced from their progressive transformation during development to become ramified microglia (Cuadros et al. (1994); Navascués et al. (1995)). In our laboratory, we used the retina and the optic tectum of the developing quail as model systems to study the spreading of ameboid microglial cells throughout the nervous parenchyma during development. Contrary to the view that most microglia arise from circulating monocytes traversing the vessel endothelium within the CNS (Perry and Gordon (1991); Ling and Wong (1993); Thanos et al. (1996)), our results support the idea that many non-circulating hematopoietic microglial precursors enter these optic centers at specific points. Indeed, exogenous microglial precursors enter the retina from its central area, occupied by the optic nerve head and the base of the pecten (Navascués et al. (1995)), whereas they enter the optic tectum mainly from the meninges, at a zone localized in the caudal ventromedial tectal surface (Cuadros et al. (1994)). From these entry points, ameboid microglial cells spread throughout the retina or optic tectum by two main mechanisms, migration and proliferation.

The migration of ameboid microglial cells in the retina and optic tectum is not random, but follows a two-phase pattern (Fig. 2; Cuadros et al. (1994); Navascués et al. (1995)). During the first phase, called tangential migration, ameboid microglial cells migrate long distances parallel to the retinal or tectal surface to spread throughout a single layer of each region. In the second phase, microglial cells move perpendicularly to their previous movement to gain access to different depths of the nervous parenchyma. This phase is called radial migration. Recruitment of microglial cells by tangential and radial migration has also been suggested in fish retinas after peripheral cryolesion (Jimeno et al. (1999)).

The tangential migration of ameboid microglial cells takes place on the end-feet of Müller cells (Fig. 2) and along the nerve fiber layer in the retina (Navascués et al. (1995); Marín-Teva et al. (1998)), and through the stratum album centrale in the optic tectum (Cuadros et al. (1994)). The most conspicuous morphological feature of ameboid microglial cells that tangentially migrate along the vitreal part of the developing quail retina is their flattened shape, with extensive lamellipodia emerging either from the cell body or from broad cell processes (Marín-Teva et al. (1998)). These cells have changing morphologies polarized along the line of movement, suggesting a mechanism of migration similar to that of fibroblasts in culture.

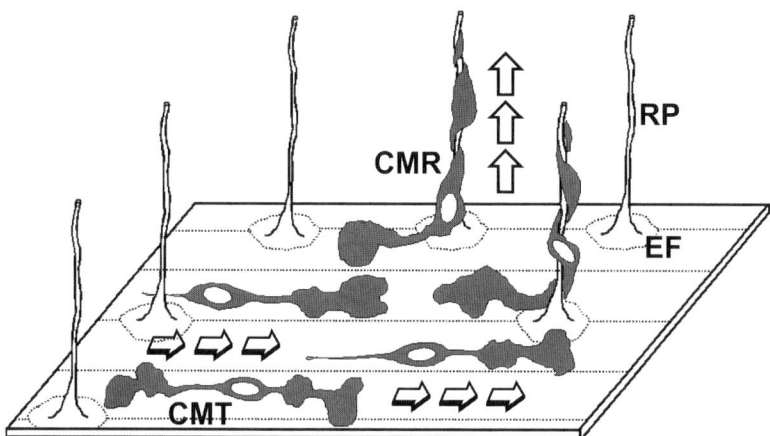

FIGURE. 2. Schematic drawing illustrating the two steps of migration of ameboid microglial cells in the developing retina. Tangential migration (horizontal arrows) occurs parallel to the retinal surface in the vitrealmost part of the retina by movement on Müller cell end-feet (EF), whereas radial migration (vertical arrows) takes place perpendicular to the retinal surface by moving up the Müller cell radial processes (RP). CMT: microglial cell migrating tangentially; CMR: microglial cell migrating radially.

This mechanism consists of the polarized extension of lamellipodia at the leading edge of the cell, strong cell-to-substrate anchorage, forward translocation of the cell body, and retraction of the rear of the cell (Fig. 3A, B). Non-polarized ameboid microglial cells with lamellipodial projections radiating in several directions from the cell body (Fig. 3C) are also found in the developing quail retina, suggesting that they are exploring the surrounding environment to orient their movement (Marín-Teva et al. (1998)). During the orientation phase, lamellipodia may function as devices to explore the substrate and determine the direction of subsequent cell movement.

Many ameboid microglial cells remain attached on isolated retinal sheets containing the inner limiting membrane covered by a carpet of Müller cell end-feet, demonstrating that microglial cells attach to Müller cell end-feet during tangential migration (Marín-Teva et al. (1998, 1999b)). The direction of this migration in non-marginal areas of the quail retina is central-to-peripheral (Navascués et al. (1995); Marín-Teva et al. (1998)), whereas it is circumferential at the retinal margin (Marín-Teva et al. (1999b)). The orientation of the rows of Müller cell processes, which flank grooves paved with end-feet, is different in marginal and non-marginal areas of the retina, and is the same as the direction of tangential migration of microglial cells in each area. This suggests that mechanical guidance is involved in the tangential migration of ameboid microglia throughout the retina. In addition, Müller cell end-feet may also participate in the adhesive guidance of these

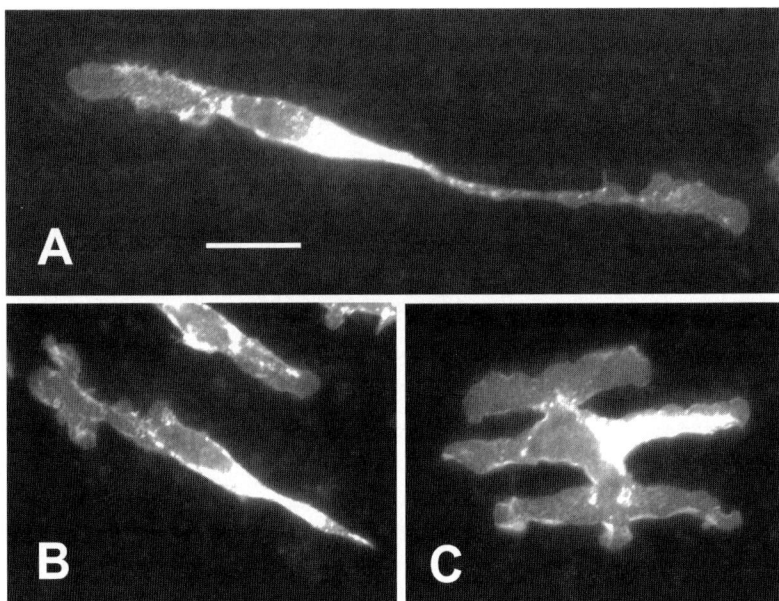

FIGURE 3. Tangential migration of ameboid microglia in the developing quail retina occurs by a mechanism of locomotion including the polarized extension of broad cell processes with lamellipodia (A), forward translocation of the cell body, and retraction of the rear of the cell (B). A non-polarized ameboid microglial cell with lamellipodial projections radiating in several directions from the cell body (C) appears to be exploring the surrounding environment. Scale bar: 20 μm.

cells because adhesion molecules such as N-CAM and laminin have been observed on the Müller cell end-feet of developing chick and quail retina (Halfter et al. (1987); Halfter and Fua (1987)).

Ameboid microglia migrating tangentially in *axon* tracts move through a nonlaminar environment and are rounded instead of flattened in shape. However, they show many similar features to those of microglial cells migrating on Müller cell end-feet in the retina, i.e., they have a cell body sending out broad processes with pseudopodia and lamellipodia (Boya et al. (1991); Ling et al. (1980); Perry et al. (1985); Cuadros et al. (1994, 1997)). Thus, knowledge on microglial migration obtained in the retina can be extended to other parts of the CNS.

The radial migration of ameboid microglial cells has been described in the retina and the optic tectum of the quail embryo. In the optic tectum, ameboid microglia migrate radially from the stratum album centrale towards the pial surface (Cuadros et al. (1994)). In the retina, they move from the vitreal part towards scleral levels (Navascués et al. (1995)). As a consequence, ameboid microglial cells reach the inner and outer plexiform layers in the retina and the different layers in the optic tectum, where they subsequently differentiate. In the developing retina, ameboid microglia appear to use radially oriented processes of Müller glia as a substrate for radial migration (Fig. 2; Navascués et al. (1996)). In other parts of the visual system, such as the optic tectum and visual cortex, radial migration might be supported by other radially oriented structures, such as radial glia and some blood vessels.

The chronology of the radial migration of microglia partially coincides with that of neuronal migration, although the former appears to begin and end later than the latter. For instance, radial migration of neurons in the quail optic tectum takes place mainly between E5 and E10 (Senut and Alvarado-Mallart, 1986) whereas microglial cells migrate radially during the second week of incubation (Cuadros et al. (1994)).

Interestingly, the radial migration of ameboid microglial cells in the quail retina appears to occur in two phases (Marín-Teva et al. (1999c)). First, microglial cells migrate across the nerve fiber layer and the ganglion cell layer to reach the vitreal border of the inner plexiform layer. Second, they migrate across the inner plexiform layer and the inner nuclear layer to reach the outer plexiform layer. This second phase occurs after microglial cells have stopped for several days at the vitreal border of the inner plexiform layer. We do not yet know whether similar cell arrests take place during the radial migration of ameboid microglia in other parts of the CNS.

In addition to cell migration, cell proliferation also contributes to the spreading of microglial precursors throughout different parts of the developing visual system. Ameboid microglia proliferate during normal development in the vertebrate CNS (Ling and Wong (1993)), as shown by the presence of mitotic ameboid microglial cells (Schnitzer (1989); Wu et al. (1996)), the expression of the proliferating cell nuclear antigen in ameboid microglia (Vela-Hernández et al. (1997); Marín-Teva et al. (1999a)), and by autoradiographic studies (Imamoto and Leblond (1978)). In the developing quail retina, ameboid microglial cells enter mitosis while they are migrating (Fig. 4A, B; Marín-Teva et al. (1999a)). As mitosis advances, microglial cells retract their lamellipodia (Fig. 4C, D), which are extended again before the completion of telophase (Fig. 4E, F). Cell division of microglial precursors also occurs in other locations of the developing brain where ameboid microglia are actively migrating, such as the cerebellar white matter (Cuadros et al. (1997)) and the corpus callosum (Kaur and Ling (1991); Wu et al. (1996)). Therefore, proliferation of ameboid microglia while they are migrating appears to be a widespread event that contributes to the dispersion of microglial precursors throughout the nervous parenchyma during normal development.

Distinct populations of microglial progenitors migrate into the developing retina

Two subpopulations of parenchymal microglia have been found in the adult human retina: paravascular and ramified (Provis et al. (1995, 1996)). Paravascular microglia

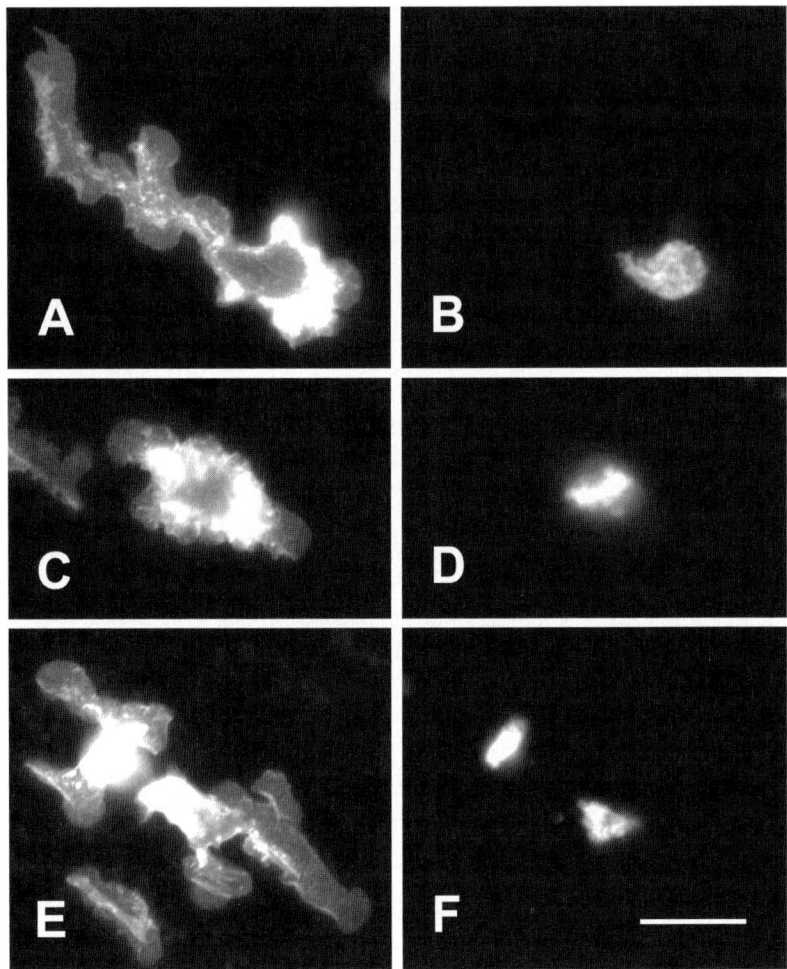

FIGURE. 4. Proliferation of ameboid microglial cells while they are migrating contributes to their spreading throughout the developing quail retina. The left column (A, C, E) shows morphological features of ameboid microglial cells immunostained with the QH1 antibody, and the right one (B, D, F) shows chromosomal material of the above cells stained with Hoechst 33342. A, B: A migrating cell in prophase has not yet retracted its lamellipodia. C, D: A metaphasic cell has retracted its lamellipodia. E, F: A telophasic cell is resuming the emission of lamellipodia. Scale bar: 20 μm.

are cells of macrophage lineage closely associated to the vessel glia limitans, whereas ramified microglia are not associated with vessels and are apparently of dendritic cell lineage. Microglia of dendritic lineage (which are immunoreactive for MHC-I, MHC-II, and CD45 but not for the human macrophage antigen S22) begin to migrate into the retina before vasculogenesis and come from two sources, the optic disc and the retinal margin. By contrast, paravascular microglia of macrophage lineage (which are immunoreactive for the antigen S22) enter the retina later in development, coinciding with vasculogenesis, and come only from the optic disc (Diaz-Araya et al. (1995a, 1995b); Provis et al. (1996)).

Two subpopulations of ameboid microglial cells also migrate into the developing quail retina (Marín-Teva et al. (1999b)). The cells of one subpopulation enter the marginal retina from the ciliary body and appear to be derived from dendritic cells of the ciliary body, whereas the cells of the other microglial subpopulation migrate in a central-to-peripheral direction from the optic nerve head.

Control of migration of microglial progenitors: possible role of neuronal death

The fact that ameboid microglial cells migrate in a consistent manner suggests that migration is controlled by specific factors present in the developing CNS. Different chemokines induce the migration and activation of microglia in the adult CNS (Cross and Woodroofe (1999); Maciejewski-Lenoir et al. (1999)), and their possible influence on the migration of ameboid microglial cells during development remains to be investigated.

Factors released from naturally dying neurons have been proposed to attract ameboid microglial cells in the developing visual system, as supported by the chronologic coincidence of ganglion cell death with the entry of microglial precursors into the retina (Hume et al. (1983); Schnitzer (1989); Perry and Gordon (1991); Pearson et al. (1993)). In fact, a close physical association between dying neurons and microglial cells is frequently observed (Wong and Hughes (1987); Ferrer et al. (1990); Thanos et al. (1996); Egensperger et al. (1996); Moujahid et al. (1996)). However, many microglial precursors enter the mammalian retina well before the time of ganglion cell death, contradicting this hypothesis (Ashwell (1989); Ashwell et al. (1989); Diaz-Araya et al. (1995b)). In the developing quail retina, the time of entry and tangential migration of ameboid microglial cells coincides with that of cell death in the ganglion cell layer, but no spatial colocalization exists between dying ganglion cell bodies and microglial cells migrating tangentially (Marín-Teva et al. (1999c)). This suggests that tangential migration of microglial cells is not directly related to neuronal death in the ganglion cell layer. A similar conclusion has been obtained in the rat retina (Garcia-Valenzuela and Sharma (1999)), in which blood-borne macrophages invade the nerve fiber layer of the retina after optic nerve axotomy, but do not reach the ganglion cell layer despite the existence of massive ganglion cell death. It cannot be ruled out that axons of dying ganglion cells promote migration of microglial cells in contact with them. In summary, it remains to be shown whether ganglion cell death triggers tangential migration of microglia in the developing retina.

A recent study by our group (Marín-Teva et al. (1999c)) has shown that cell death in the inner nuclear layer does not promote the radial migration of ameboid

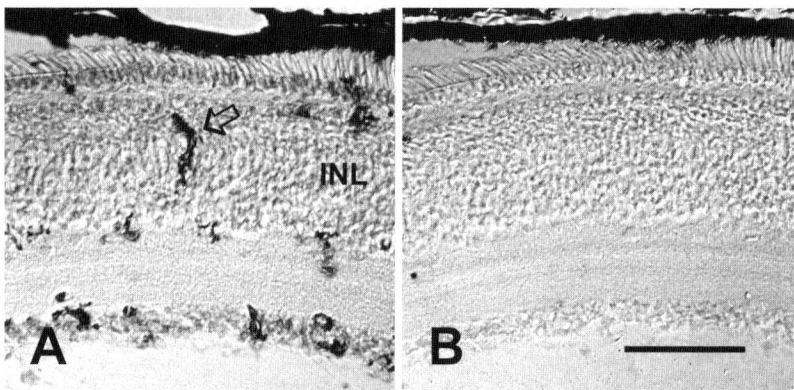

FIGURE 5. Single labeling with either the QH1 antibody (A) or TUNEL (B) in adjacent serial sections of a nasodorsal area of an E15 quail retina. An ameboid microglial cell (arrow in A) is migrating across the inner nuclear layer (INL), whereas no apoptotic fragments are seen in this layer (B). Scale bar: 65 μm.

microglial cells in the quail retina. In fact, microglial cells are not attracted by dying cells in the inner nuclear layer during their arrest for two or three days at the vitreal border of the inner plexiform layer. On the contrary, microglial cells only traverse the inner nuclear layer after cell death has ceased in this layer (Fig. 5).

Other roles of microglial cells in the developing visual system

Is the colonization of the different parts of the vertebrate visual system the only role of ameboid microglial cells during development, or do they carry out other specific functions in the normal development of the optic centers and pathways? A clearly demonstrated function of ameboid microglial cells during the development of the visual system is the phagocytosis of debris from dead neurons, but others, such as the induction of cell death, axonal growth, functional interactions with oligodendrocytes, and vascularization, have been proposed for these cells in other parts of the developing CNS (Ashwell and Bobryshev (1996)) and may also play a part in the developing visual system. The ability of immature resident microglial cells to respond to injury in the developing visual system has yet to be clarified.

Phagocytosis of cell debris

The most evident and widely recognized role of microglial cells during CNS development is the removal of apoptotic cell debris arising from naturally occurring neuronal death (Perry and Gordon (1991); Ferrer et al. (1990); Ashwell and Bobryshev (1996)). This function has been reported in the developing retina (Hume et al. (1983); Wong and Hughes (1987); Ashwell (1989); Ashwell et al. (1989); Pearson et al. (1993); Thanos (1991); Egensperger et al. (1996); Marín-Teva et al. (1999c)) and optic nerve (Moujahid et al. (1996)). In an elegant experiment, Thanos (1991) applied a fluorescent dye in the superior colliculus of newborn rats, i.e., prior to the naturally occurring ganglion cell death. The dye was taken up by retinal ganglion cell terminals and transported retrogradely to their cell bodies in the retina. Thus, ganglion cells were vitally stained with fluorescent labeling. After the programmed natural death of ganglion cells, fluorescent cell debris was engulfed by microglial cells, resulting in the accumulation of the fluorescent dye within the microglial cytoplasm. The very slow catabolism of the dye allowed the long-term labeling of microglial cells, which maintained fluorescence into adulthood. This experimental approach clearly showed that microglial cells take part in the phagocytosis of debris from retinal ganglion cells that undergo naturally occurring cell death during normal development.

In addition to ameboid microglial cells, Müller glial cells also phagocytose debris from apoptotic neurons in the developing retinas of mammals (Penfold and Provis (1986); Provis and Penfold (1988); Egensperger et al. (1996)) and birds (Hughes and LaVelle (1975); Hughes and McLoon (1979); Marín-Teva et al. (1999c); Thanos (1999)). Do Müller cells and ameboid microglial cells phagocytose cell debris simultaneously or successively in the developing retina? In the chick embryo, Müller glia are the only class of cells observed to become phagocytic between embryonic days 9 and 16 (E9-E16) and are replaced in this function by microglial cells from E17 on (Thanos (1999)). However, our results in quail embryos (Marín-Teva et al. (1999c)) suggest that phagocytosis of ganglion cells is carried out simultaneously by ameboid microglia and Müller cells. By contrast, cell debris in the inner nuclear layer are phagocytosed exclusively by Müller cells (Marín-Teva et al. (1999c)). The possible existence of a transfer mechanism of phagocytosed material from Müller to microglial cells warrants more research.

Are all ameboid microglial cells able to phagocytose dead ganglion cell debris during development? Around 26% and 5% of ameboid microglial cells contain apoptotic fragments in the E26 rabbit retina (Ashwell (1989)) and postnatal day 5 rat retina (Ashwell et al. (1989)), respectively. It is possible that all retinal microglia would be able to phagocytose cell debris, but the precise number of microglial cells actively involved in this process at each time point would depend on the number of apoptotic fragments to be phagocytosed. An alternative possibility is that only some ameboid microglial cells have phagocytic capacity. In fact, microglial cells of dendritic cell lineage in the developing human retina have a limited phagocytic capacity in comparison with that of the microglial cell subpopulation of macrophage lineage (Diaz-Araya et al. (1995a); Provis et al. (1996)).

Expression of growth factors and induction of cell death

Microglial cells produce and accumulate neurotrophins, as demonstrated in cultured microglia (Gilad and Gilad (1995)) and in immature microglia *in vivo* (Elkabes et al. (1996)). In turn, the functional behavior of microglia is regulated by neurotrophins (Elkabes et al. (1996); Nakajima et al. (1998)). The differential expression of neurotrophins influences the survival, differentiation and axonal outgrowth of neuronal populations in the developing visual system (Von Bartheld (1998); Frade et al. (1999)). Microglial cells have been suggested to influence cell death in the developing retina. In fact, in a model of light-induced retinal degeneration, neurotrophins derived from microglial cells seem to regulate the release of basic fibroblast growth factor by Müller cells, and this factor in turn promotes photoreceptor survival (Harada et al. (2000)).

Interestingly, macrophages invading the retina during its early development apparently promote cell death through the release of nerve growth factor (NGF) (Frade and Barde (1998, 1999)). NGF induces the death of early differentiating retinal ganglion cells that express the low affinity neurotrophin receptor p75, but not the high affinity receptor *trk*A (Frade et al. (1996); Frade and Barde (1998)). The macrophages that produce NGF are early macrophages (Cuadros et al. (1991)) and their microglial nature (Frade and Barde (1998)) remains to be demonstrated (see above). Evidence of an active role of macrophages in promoting cell death has also been obtained in the regression of the mouse pupillary membrane, a developmentally transient capillary network found in the anterior chamber of the eye (Diez-Roux and Lang (1997)).

Axonal growth

The role of microglia in promoting axonal growth is supported by the enhancement of neurite outgrowth when cultured microglial cells are grafted to the lesioned spinal cord of adult rats (Rabchevsky and Streit (1997)). This role is also likely in the developing CNS because macrophage/microglial cell clusters are frequently localized in regions containing growing axons, or slightly ahead of them (Perry et al. (1985); Cuadros et al. (1993); Ling and Wong (1993); Chamak et al. (1995); Ashwell and Bobryshev (1996); Bass et al. (1998)). Similar observations have been made in the developing optic nerve of birds (Moujahid et al. (1996)). Furthermore, microglial cells are present at the boundary between the inner and outer nuclear layers in the developing mouse retina prior to their separation by the development of axons that leads to the formation of the plexiform layers (Hume et al. (1983)).

Studies on the regeneration of optic axons may shed some light on this issue, although the extrapolation of the results of studies in adulthood to development has to be considered with reservations. Macrophage/microglial cells may favor the regeneration of optic axons, as supported by the different responses of these cells to optic projection injury in lower vertebrates (fish and amphibians), which have a natural ability to regenerate axons in the CNS, and in upper vertebrates, without axonal regeneration in the CNS. The response is rapid and extensive in the visual system of fish (Dowding et al. (1991); Battisti et al. (1995); Velasco et al. (1995); Nona et al. (1998)) and amphibians (Goodbrand and Gaze (1991); Wilson et al. (1992); Naujoks-Manteuffel and Niemann (1994)), whereas it is slower and smaller in mammals (Perry et al. (1987); Hirschberg and Schwartz (1995)). The poor response of macrophage/microglial cells in mammals appears to be due to inherent factors in the CNS microenvironment that inhibit both the migration (Hirschberg and Schwartz (1995)) and activation of recruited macrophages (Lazarov-Spiegler et al. (1998); Leon et al. (2000)).

Macrophage/microglial cells may exert their influence on axonal growth via the secretion of different factors, such as thrombospondin. This extracellular matrix protein is released *in vitro* by macrophages isolated from embryonic rat brain and stimulates the neurite growth of cultured CNS neurons (Chamak et al (1994)). In addition, thrombospondin is expressed by immature microglial cells in developing axon tracts of the rat brain (Chamak et al. (1995)), supporting the notion that microglial cells enhance axonal growth by producing thrombospondin during development.

Interactions with developing oligodendrocytes

Ameboid microglial cells are localized in axon tracts of the developing CNS, where immature oligodendrocytes are also found (Ellison and De Vellis (1995); Wolswijk (1995); Zhang and McKanna (1997)). Moreover, microglial cells interact with oligodendrocytes to stimulate myelinogenesis (reviewed by Ashwell and Bobryshev (1996)) and macrophage-derived cytokines or growth factors may directly or indirectly promote oligodendroglial proliferation and differentiation (Diemel et al. (1998)).

Comparison of the time course and spatial gradients of oligodendrocyte progenitor migration along the developing chick optic nerve (Ono et al. (1997)) and retina (Ono et al. (1998)) with those of ameboid microglia migration in the same parts of the quail embryo (Moujahid et al. (1996); Navascués et al. (1995)) shows that both cell types migrate simultaneously on the same substrates. This spatiotemporal overlapping is compatible with functional interactions between microglial cells and oligodendrocyte progenitors, although this remains to be demonstrated in the developing visual system.

Vascularization

Macrophages are involved in the induction of vascularization (Polverini et al. (1977)). The early vascularization of the chick embryo CNS is related to the presence of early macrophages in close association with growing vessels (Cuadros et al. (1993)). In addition, host microglia invading neural allografts in the rat striatum seem to promote graft neovascularization (Pennell and Streit (1997)). The involvement of microglial cells in CNS vascularization is also suggested by the simultaneous occurrence of vascularization (Hughes et al. (2000)) and the entry of microglial progenitors (Diaz-Araya et al. (1995a, b); Provis et al. (1996)) in the developing human retina. Furthermore, a close topographical association exists between the developing vasculature and cells of macrophage lineage (Penfold et al. (1990)). Recently, an *in vitro* model using retinal explants provided evidence of the participation of microglial cells in retinal vascularization (Knott et al. (1999)).

Microglial cells do not appear to exert a direct action on retinal vascularization, because no differences in the distribution of microglia are detected between avascular and vascular retinas (Ashwell (1989); Ashwell et al. (1989); Boycott and Hopkins (1981); Sanyal and De Ruiter (1985); Navascués et al. (1994)). Retinal vascularization is under the control of the vascular endothelial growth factor (VEGF) secreted by Müller cells and astrocytes (Stone et al. (1995); Provis et al. (1997); Zhang et al. (1999)). Thus, microglial cells may affect retinal vascularization by regulating the release of VEGF by these neuroglial cells. In fact, a diffusible factor released by non-characterized retinal cells in conditions of hypoxia induces astrocytes to produce VEGF (Zhang et al. (1999)).

Response to injury in the developing visual system

Microglia in the adult CNS are actively involved in the response to brain injury (Streit et al. (1999)). Several studies have investigated how immature microglia respond to CNS injury during development (Morioka and Streit (1991); Lawson and Perry (1995); Graeber et al. (1998)), including injury to the developing visual system (Milligan et al. (1991); Cuadros et al. (2000)). Studies on the response of mononuclear phagocytes to retrograde axonal degeneration in the developing and mature dorsal lateral geniculate nucleus (dLGN) of rats (Milligan et al. (1991)) revealed that the response during development is quite different from that in adulthood. First, the mononuclear phagocytes that participate in the response in the developing dLGN are mainly recruited from the blood circulation, whereas the phagocytes that respond to injury in the adult dLGN are activated resident microglia. Second, the response in the developing dLGN is quicker and shorter than in the adult, starting two days after injury, peaking around day four and declining from day five. By contrast, the response in the adult dLGN is first observable seven days after injury and lasts for at least two months (Milligan et al. (1991)).

Recently, we studied the response of macrophage/microglial cells to neuronal death in the developing isthmo-optic nucleus (ION) of avian embryos after colchicine injection in the contralateral eye (Cuadros et al. (2000)). This treatment results in the death of developing ION neurons by the blockage of their retrograde axonal transport from the retina (Fig. 6). The number of macrophage/microglial cells begins to increase around eighteen hours after the injection of colchicine in

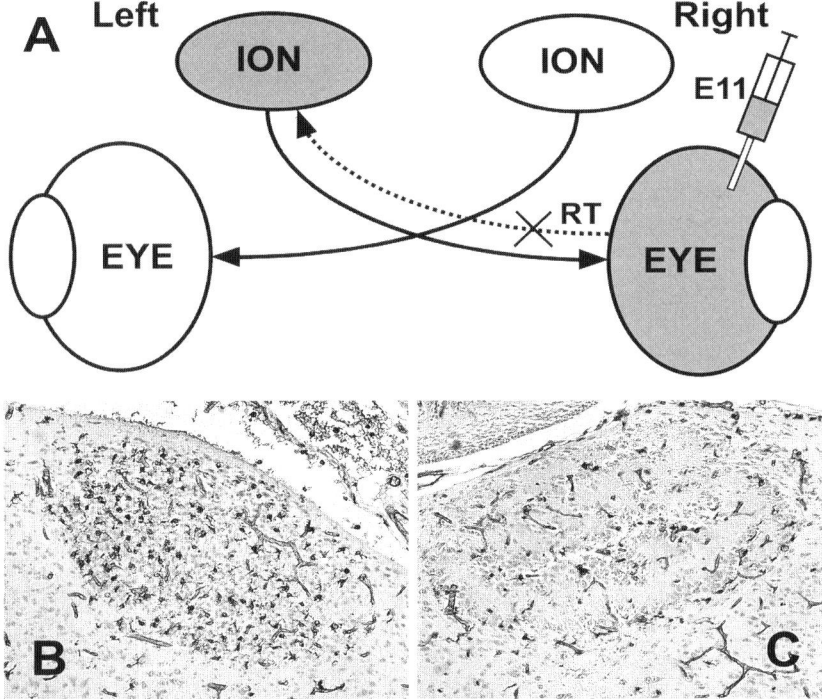

FIGURE 6. Response of macrophage/microglial cells to neuronal death in the developing isthmo-optic nucleus (ION) of the quail embryo. A: Colchicine injection into the right eye at E11 induces massive degeneration of neurons in the contralateral ION by blocking retrograde axonal transport (RT) from the retina. B, C: Distribution of QH1-positive macrophage/microglial cells in the IONs contralateral (B) and ipsilateral (C) to the injected eye, 40 hours post-injection. Numerous QH1-positive macrophage/microglial cells appear in the affected ION (B) in comparison with the control ION (C).

the contralateral eye, peaks at two days post-injection, and progressively declines thereafter, coinciding with the disappearance of ION neurons. The origin of the macrophage/microglial cells could not be determined because the antibodies used in our study do not distinguish blood-recruited macrophages from resident microglia. Thus, more research is needed to determine whether immature resident microglia are able to respond to injury in the developing visual system.

Concluding Remarks

The developing visual system of vertebrates has been used as a model to study the migration of ameboid microglial cells and their involvement in the phagocytosis of naturally dying neurons. Many issues have yet to be resolved, in particular, the mechanisms that control microglial migration and the possible cooperation of microglia with Müller glia in the clearance of cell debris. Studies in the visual system have also demonstrated differences in the microglial response to injury between developing and adult animals. Additional functions of immature microglia in the developing visual system have been suggested by studies in other parts of the CNS, and these warrant further investigation.

Acknowledgements

This paper was supported by grant PM97-0178 from the DGICYT of the Spanish Ministry of Education and Culture. Thanks are due to Richard Davies for improving the English style of the chapter.

References

Alliot FI, Godin I, Pessac B (1999). Microglia derive from progenitors, originating from the yolk sac, and which proliferate in the brain. *Dev Brain Res* 117:145-152.

Ashwell K (1989). Development of microglia in the albino rabbit retina. *J Comp Neurol* 287:286-301.

Ashwell KWS, Holländer H, Streit W, Stone J (1989). The appearance and distribution of microglia in the developing retina of the rat. *Vis Neurosci* 2:437-448.

Ashwell KWS, Bobryshev YV (1996). The developmental role of microglia. In: *Topical Issues in Microglial Research* (Ling EA, Tan CK, Tan CBC, eds.), pp 65-82. Singapore: Singapore Neuroscience Association.

Bass WT, Singer GA, Liuzzi FJ (1998). Transient lectin binding by white matter tract border zone microglia in the foetal rabbit brain. *Histochem J* 30:657-666.

Battisti WP, Wang J, Bozek K, Murray M (1995). Macrophages, microglia, and astrocytes are rapidly activated after crush injury of the goldfish optic nerve: a light and electron microscopic analysis. *J Comp Neurol* 354:306-320.

Boya J, Calvo JL, Carbonell AL, Borregon A (1991). A lectin histochemistry study on the development of rat microglial cells. *J Anat* 175:229-236.

Boycott BB, Hopkins JM (1981). Microglia in the retina of monkey and other mammals: its distinction from other types of glia and horizontal cells. *Neuroscience* 6:679-688.

Chamak B, Morandi V, Mallat M (1994). Brain macrophages stimulate neurite growth and regeneration by secreting thrombospondin. *J Neurosci Res* 38:221-233.

Chamak B, Dobbertin A, Mallat M (1995). Immunohistochemical detection of thrombospondin in microglia in the developing rat brain. *Neuroscience* 69:177-187.

Cross AK, Woodroofe MN (1999). Chemokines induce migration and changes in actin polymerization in adult rat brain microglia and a human fetal microglial cell line in vitro. *J Neurosci Res* 55:17-23.

Cuadros MA, García-Martín M, Martin C, Ríos A (1991). Haemopoietic phagocytes in the early differentiating avian retina. *J Anat* 177:145–158.

Cuadros MA, Martin C, Coltey P, Almendros A, Navascués J (1993). First appearance, distribution, and origin of macrophages in the early development of the avian central ner-

vous system. *J Comp Neurol* 330:113–129.
Cuadros MA, Moujahid A, Quesada A, Navascués J (1994). Development of microglia in the quail optic tectum. *J Comp Neurol* 348:207–224.
Cuadros MA, Rodríguez-Ruiz J, Calvente R, Almendros A, Marín-Teva JL, Navascués J (1997). Microglia development in the quail cerebellum. *J Comp Neurol* 389:390–401.
Cuadros MA, Navascués J (1998). The origin and differentiation of microglial cells during development. *Prog Neurobiol* 56:173–189.
Cuadros MA, Martín D, Pérez-Mendoza D, Navascués J, Clarke PGH (2000). Response of macrophage/microglial cells to experimental neuronal degeneration in the avian isthmo-optic nucleus during development. *J Comp Neurol* 423:659–669.
Diaz-Araya CM, Provis JM, Penfold PL (1995a). Ontogeny and cellular expression of MHC and leucocyte antigens in human retina. *Glia* 15:458–470.
Diaz-Araya CM, Provis JM, Penfold PL, Billson FA (1995b). Development of microglial topography in human retina. *J Comp Neurol* 363:53–68.
Diemel LT, Copelman CA, Cuzner ML (1998). Macrophages in CNS remyelination: friend or foe? *Neurochem Res* 23:341–347.
Diez-Roux G, Lang RA (1997). Macrophages induce apoptosis in normal cells in vivo. *Development* 124:3633–3638.
Dowding AJ, Maggs A, Scholes J (1991). Diversity amongst the microglia in growing and regenerating fish CNS: immunohistochemical characterization using FL.1, an anti-macrophage monoclonal antibody. *Glia* 4:345–364.
Egensperger R, Maslim J, Bisti S, Holländer H, Stone J (1996). Fate of DNA from retinal cells dying during development: uptake by microglia and macroglia (Müller cells). *Dev Brain Res* 97:1–8.
Elkabes S, DiCicco-Bloom EM, Black IB (1996). Brain microglia/macrophages express neurotrophins that selectively regulate microglial proliferation and function. *J Neurosci* 16:2508–2521.
Ellison JA, De Vellis J (1995). Amoeboid microglia expressing GD3 ganglioside are concentrated in regions of oligodendrogenesis during development of the rat corpus callosum. *Glia* 14:123–132.
Ferrer I, Bernet E, Soriano E, Del Rio T, Fonseca M (1990). Naturally occurring cell death in the cerebral cortex of the rat and removal of dead cells by transitory phagocytes. *Neuroscience* 39:451–458.
Frade JM, Rodríguez-Tébar A, Barde YA (1996). Induction of cell death by endogenous growth factor through its p75 receptor. *Nature* 383:166–168.
Frade JM, Barde YA (1998). Microglia-derived nerve growth factor causes cell death in the developing retina. *Neuron* 20:35–41.
Frade JM, Barde YA (1999). Genetic evidence for cell death mediated by nerve growth factor and the neurotrophin receptor p75 in the developing mouse retina and spinal cord. *Development* 126:683–690.
Frade JM, Bovolenta P, Rodríguez-Tébar A (1999). Neurotrophins and other growth factors in the generation of retinal neurons. *Microsc Res Tech* 45:243–251.
Garcia-Valenzuela E, Sharma SC (1999). Laminar restriction of retinal macrophagic response to optic nerve axotomy in the rat. *J Neurobiol* 40:55–66.
Gilad GM, Gilad VH (1995). Chemotaxis and accumulation of nerve growth factor by microglia and macrophages. *J Neurosci Res* 41:594–602.
Goodbrand IA, Gaze R (1991). Microglia in tadpoles of *Xenopus laevis:* normal distribution and the response to optic nerve injury. *Anat Embryol* 184:71–82.

Graeber MB, Lopez-Redondo F, Ikoma E, Ishikawa M, Imai Y, Nakajima K, Kreutzberg GW, Kohsaka S (1998). The microglia/macrophage response in the neonatal rat facial nucleus following axotomy. *Brain Res* 813:241–253.

Halfter W, Fua CS (1987). Immunohistochemical localization of laminin, neural cell adhesion molecule, collagen type IV and T-61 antigen in the embryonic retina of the Japanese quail by in vivo injection of antibodies. *Cell Tiss Res* 249:487–496.

Halfter W, Reckhaus W, Kröger S (1987). Nondirected axonal growth on basal lamina from avian embryonic neural retina. *J Neurosci* 7:3712–3722.

Harada T, Harada C, Nakayama N, Okuyama S, Yoshida K, Kohsaka S, Matsuda H, Wada K (2000). Modification of glial-neuronal cell interactions prevents photoreceptor apoptosis during light-induced retinal degeneration. *Neuron* 26:533–541.

Hirschberg DL, Schwartz M (1995). Macrophage recruitment to acutely injured central nervous system is inhibited by a resident factor: a basis for an immune-brain barrier. *J Neuroimmunol* 61:89–96.

Hughes S, Yang HJ, Chan-Ling T (2000). Vascularization of the human fetal retina: roles of vasculogenesis and angiogenesis. *Invest Ophthalmol Vis Sci* 41:1217–1228.

Hughes WF, LaVelle A (1975). The effects of early tectal lesions on development in the retinal ganglion cell layer of chick embryos. *J Comp Neurol* 163:265–283.

Hughes WF, McLoon SC (1979). Ganglion cell death during normal retinal development in the chick: comparisons with cell death induced by early target field destruction. *Exp Neurol* 66:587–601.

Hume DA, Perry VH, Gordon S (1983). Immunohistochemical localization of a macrophage specific antigen in developing mouse retina: phagocytosis of dying neurons and differentiation of microglial cells to form a regular array in the plexiform layers. *J Cell Biol* 97:253–257.

Hurley SD, Streit WJ (1996). Microglia and the mononuclear phagocyte system. In: *Topical Issues in Microglial Research* (Ling EA, Tan CK, Tan CBC, eds.), pp 1–19. Singapore: Singapore Neuroscience Association.

Imamoto K, Leblond CP (1978). Radioautographic investigation of gliogenesis in the corpus callosum of young rats. II. Origin of microglial cells. *J Comp Neurol* 180:134–164.

Jimeno D, Velasco A, Lillo C, Lara JM, Aijón J (1999). Response of microglial cells after a cryolesion in the peripheral proliferative retina of tench. *Brain Res* 816:175–189.

Kaur C, Ling EA (1991). Study of the transformation of amoeboid microglial cells into microglia labelled with the isolectin Griffonia simplicifolia in postnatal rats. *Acta Anat* 142:118–125.

Knabe W, Süss M, Kuhn HJ (2000). The patterns of cell death and of macrophages in the developing forebrain of the tree shrew *Tupaia belangeri*. *Anat Embryol* 201:157–168.

Knott RM, Robertson M, Muckersie E, Folefac VA, Fairhurst FE, Wileman SM, Forrester JV (1999). A model system for the study of human retinal angiogenesis: activation of monocytes and endothelial cells and the association with the expression of the monocarboxylate transporter type 1 (MCT-1). *Diabetologia* 42:870–877.

Lawson LJ, Perry VH, Dri P, Gordon S (1990). Heterogeneity in the distribution and morphology of microglia in the normal adult mouse brain. *Neuroscience* 39:151–170.

Lawson LJ, Frost L, Risbridger J, Fearn S, Perry VH (1994). Quantification of the mononuclear phagocyte response to Wallerian degeneration of the optic nerve. *J Neurocytol* 23:729–744.

Lawson LJ, Perry VH (1995). The unique characteristics of inflammatory responses in mouse brain are acquired during postnatal development. *Eur J Neurosci* 7:1584–1595.

Lazarov-Spiegler O, Solomon AS, Schwartz M (1998). Peripheral nerve-stimulated macrophages simulate a peripheral nerve-like regenerative response in rat transected optic nerve. *Glia* 24:329–337.

Leon S, Yin Y, Nguyen J, Irwin N, Benowitz LI (2000). Lens injury stimulates axon regeneration in the mature rat optic nerve. *J Neurosci* 20:4615–4626.

Lillo C, Velasco A, Jimeno D, Lara JM, Aijón J (1998). Ultrastructural organization of the optic nerve of the tench (Cyprinidae, Teleostei). *J Neurocytol* 27:593–604.

Ling EA, Penney D, Leblond CP (1980). Use of carbon labelling to demonstrate the role of blood monocytes as precursors of the amoeboid cells present in the corpus callosum of postnatal rats. *J Comp Neurol* 193:631–657.

Ling EA, Wong WC (1993). The origin and nature of ramified and amoeboid microglia: a historical review and current concepts. *Glia* 7:9–18.

Maciejewski-Lenoir D, Chen SZ, Feng LL, Maki R, Bacon KB (1999). Characterization of fractalkine in rat brain cells: migratory and activation signals for CX3CR-1-expressing microglia. *J Immunol* 163:1628–1635.

Marín-Teva JL, Almendros A, Calvente R, Cuadros MA, Navascués J (1998). Tangential migration of ameboid microglia in the developing quail retina: mechanism of migration and migratory behavior. *Glia* 22:31–52.

Marín-Teva JL, Almendros A, Calvente R, Cuadros MA, Navascués J (1999a). Proliferation of actively migrating ameboid microglia in the developing quail retina. *Anat Embryol* 200:289–300.

Marín-Teva JL, Calvente R, Cuadros MA, Almendros A, Navascués J (1999b). Circumferential migration of ameboid microglia in the margin of the developing quail retina. *Glia* 27:226–238.

Marín-Teva JL, Cuadros MA, Calvente R, Almendros A, Navascués J (1999c). Naturally occurring cell death and migration of microglial precursors in the quail retina during normal development. *J Comp Neurol* 412:255–275.

Martín-Partido G, Cuadros MA, Martin C, Coltey P, Navascués J (1991). Macrophage-like cells invading the suboptic necrotic centres of the avian embryo diencephalon originate from haemopoietic precursors. *J Neurocytol* 20:962–968.

Milligan CE, Levitt P, Cunningham TJ (1991). Brain macrophages and microglia respond differently to lesions of the developing and adult visual system. *J Comp Neurol* 314:136–146.

Morioka T, Streit WJ (1991). Expression of immunomolecules on microglial cells following neonatal sciatic nerve axotomy. *J Neuroimmunol* 35:21–30.

Moujahid A, Navascués J, Marín-Teva JL, Cuadros MA (1996). Macrophages during avian optic nerve development: relationship to cell death and differentiation into microglia. *Anat Embryol* 193:131–144.

Nakajima K, Kikuchi Y, Ikoma E, Honda S, Ishikawa M, Liu Y, Kohsaka S (1998). Neurotrophins regulate the function of cultured microglia. *Glia* 24:272–289.

Naujoks-Manteuffel C, Niemann U (1994). Microglial cells in the brain of *Pleurodeles waltl* (Urodela, Salamandridae) after Wallerian degeneration in the primary visual system using *Bandeiraea simplicifolia* isolectin B4-cytochemistry. *Glia* 10:101–113.

Navascués J, Moujahid A, Quesada A, Cuadros MA (1994). Microglia in the avian retina: immunocytochemical demonstration in the adult quail. *J Comp Neurol* 350:171–186.

Navascués J, Moujahid A, Almendros A, Marín-Teva JL, Cuadros MA (1995). Origin of microglia in the quail retina: central-to-peripheral and vitreal-to-scleral migration of microglial precursors during development. *J Comp Neurol* 354:209–228.

Navascués J, Cuadros MA, Almendros A (1996). Development of microglia: evidence from studies in the avian central nervous system. In: *Topical Issues in Microglia Research* (Ling EA, Tan CK, Tan CBC, eds.), pp 43–64. Singapore: Singapore Neuroscience Association.

Nona SN, Thomlinson AM, Stafford CA (1998). Temporary colonization of the site of lesion by macrophages is a prelude to the arrival of regenerated axons in injured goldfish optic nerve. *J Neurocytol* 27:791–803.

Ono K, Yasui Y, Rutishauser U, Miller RH (1997). Focal ventricular origin and migration of oligodendrocyte precursors into the chick optic nerve. *Neuron* 19:283–292.

Ono K, Tsumori T, Kishi T, Yokota S, Yasui Y (1998). Developmental appearance of oligodendrocytes in the embryonic chick retina. *J Comp Neurol* 398:309–322.

Pearson HE, Payne BR, Cunningham TJ (1993). Microglial invasion and activation in response to naturally occurring neuronal degeneration in the ganglion cell layer of the postnatal cat retina. *Dev Brain Res* 76:249–255.

Penfold PL, Provis JM (1986). Cell death in the development of the human retina: phagocytosis of pyknotic and apoptotic bodies by retinal cells. *Graefe's Arch Clin Exp Ophthalmol* 224:549–553.

Penfold PL, Provis JM, Madigan MC, Van Driel D, Billson FA (1990). Angiogenesis in normal human retinal development: the involvement of astrocytes and macrophages. *Graefe's Arch Clin Exp Ophthalmol* 228:255–263.

Pennell NA, Streit WJ (1997). Colonization of neural allografts by host microglial cells: relationship to graft neovascularization. *Cell Transplant* 6:221–230.

Perry VH, Hume DA, Gordon S (1985). Immunohistochemical localization of macrophages and microglia in the adult and developing mouse brain. *Neuroscience* 15:313–326.

Perry VH, Brown MC, Gordon S (1987). The macrophage response to central and peripheral nerve injury. A possible role for macrophages in regeneration. *J Exp Med* 165:1218–1223.

Perry VH, Gordon S (1991). Macrophages and the nervous system. *Int Rev Cytol* 125:203–244.

Polverini PJ, Cotran RS, Gimbrone MA, Unanue ER (1977). Activated macrophages induce vascular proliferation. *Nature* 269:804–806.

Provis JM, Penfold PL (1988). Cell death and the elimination of retinal axons during development. *Progr Neurobiol* 31:331–347.

Provis JM, Penfold PL, Edwards AJ, Van Driel D (1995). Human retinal microglia: expression of immune markers and relationship to the glia limitans. *Glia* 14:243–256.

Provis JM, Diaz CM, Penfold PL (1996). Microglia in human retina: a heterogeneous population with distinct ontogenies. *Persp Dev Neurobiol* 3:213–222.

Provis JM, Leech J, Diaz CM, Penfold PL, Stone J, Keshet E (1997). Development of the human retinal vasculature: cellular relations and VEGF expression. *Exp Eye Res* 65:555–568.

Rabchevsky AG, Streit WJ (1997). Grafting of cultured microglial cells into the lesioned spinal cord of adult rats enhances neurite outgrowth. *J Neurosci Res* 47:34–48.

Reichert F, Rotshenker S (1996). Deficient activation of microglia during optic nerve degeneration. *J Neuroimmunol* 70:153–161.

Salvador-Silva M, Vidal-Sanz M, Villegas-Pérez MP (2000). Microglial cells in the retina of *Carassius auratus*: effects of optic nerve crush. *J Comp Neurol* 417:431–447.

Sanyal S, De Ruiter A (1985). Inosine diphosphatase as a histochemical marker of retinal microvasculature, with special reference to transformation of microglia. *Cell Tiss Res* 241:291–297.

Schnitzer J (1989). Enzyme-histochemical demonstration of microglial cells in the adult and postnatal rabbit retina. *J Comp Neurol* 282:249–263.

Senut MC, Alvarado-Mallart RM (1986). Development of the retinotectal system in normal quail embryos: cytoarchitectonic development and optic fiber innervation. Dev Brain Res 29:123–140.

Stoll G, Jander S (1999). The role of microglia and macrophages in the pathophysiology of the CNS. *Progr Neurobiol* 58:233–247.

Stone J, Itin A, Alon T, Pe'er J, Gnessin H, Chan-Ling T, Keshet E (1995). Development of retinal vasculature is mediated by hypoxia-induced vascular endothelial growth factor (VEGF) expression by neuroglia. *J Neurosci* 15:4738–4747.

Streit WJ, Walter SA, Pennell NA (1999). Reactive microgliosis. *Progr Neurobiol* 57: 563–581.

Takahashi K, Yamamura F, Naito M (1989). Differentiation, maturation, and proliferation of macrophages in the mouse yolk sac: a light-microscopic, enzyme-cytochemical, immunohistochemical, and ultrastructural study. *J Leukoc Biol* 45:87–96.

Thanos S (1991). The relationship of microglial cells to dying neurons during natural neuronal cell death and axotomy-induced degeneration of the rat retina. *Eur J Neurosci* 3:1189–1207.

Thanos S (1999). Genesis, neurotrophin responsiveness, and apoptosis of a pronounced direct connection between the two eyes of the chick embryo: a natural error or a meaningful developmental event? *J Neurosci* 19:3900–3917.

Thanos S, Moore S, Hong YM (1996). Retinal Microglia. *Progr Ret Eye Res* 15:331–361.

Vela-Hernández JM, Dalmau I, González B, Castellano B (1997). Abnormal expression of the proliferating cell nuclear antigen (PCNA) in the spinal cord of the hypomyelinated Jimpy mutant mice. *Brain Res* 747:130–139.

Velasco A, Caminos E, Vecino E, Lara JM, Aijón J (1995). Microglia in normal and regenerating visual pathways of the tench (Tinca tinca L., 1758; Teleost): a study with tomato lectin. *Brain Res* 705:315–324.

Von Bartheld CS (1998). Neurotrophins in the developing and regenerating visual system. *Histol Histopathol* 13:437–459.

Wilson MA, Gaze R, Goodbrand IA, Taylor JSH (1992). Regeneration in the *Xenopus* tadpole is preceded by a massive macrophage/microglial response. *Anat Embryol* 186:75–89.

Wolswijk G (1994). G_{D3}^+ cells in the adult rat optic nerve are ramified microglia rather than O-2A^{adult} progenitor cells. *Glia* 10:244-249.

Wolswijk G (1995). Strongly G_{D3}^+ cells in the developing and adult rat cerebellum belong to the microglial lineage rather than to the oligodendrocyte lineage. *Glia* 13:13–26.

Wong ROL, Hughes A (1987). Role of cell death in the topogenesis of neuronal distributions in the developing cat retinal ganglion cell layer. *J Comp Neurol* 262:496–511.

Wu CH, Wen CY, Shieh JY, Ling EA (1996). Use of lectin as a tool for the study of microglial cells: expression and regulation of lectin receptors in normal development and under experimental conditions. In: *Topical Issues in Microglia Research* (Ling EA, Tan CK, Tan CBC, eds.), pp. 83–104. Singapore: Singapore Neuroscience Association.

Zhang MZ, McKanna JA (1997). Gliogenesis in postnatal rat optic nerve: LC1+ microglia and S100-β+ astrocytes. *Dev Brain Res* 101:27–36.

Zhang Y, Porat RM, Alon T, Keshet E, Stone J (1999). Tissue oxygen levels control astrocyte movement and differentiation in developing retina. *Dev Brain Res* 118:135–145.

3

Microglial Ion Channels

CLAUDIA EDER

Introduction

In all living cells, ion channels are required to regulate the membrane potential and intracellular ion concentrations. Functional ion channels allow movements of cations or anions across the membrane, which subsequently may influence a variety of cellular processes, such as proliferation, excitability, migration, apoptosis, secretion and others. Ion channels are specialized membrane proteins that span the plasma membrane. They form hydrophilic pores through which ions flow from one side of the membrane to the other down their electrochemical gradient.

Microglial ion channels have been studied using the patch clamp technique (Hamill et al. (1981)). This method allows recordings of either ion currents from the whole cell, or single channel currents from patches of the cell membrane. Measurements of ion currents in the whole-cell configuration are shown in Figures 1, 3 and 4, and examples of single-channel currents are illustrated in Figure 2. Single-channel currents can be measured in patches of the membrane in intact cells (cell-attached mode), or in excised patches that have been pulled from the cell with either the external surface (outside-out mode) or the cytoplasmic surface (inside-out mode) of the membrane facing the bath solution. The sum of single-channel currents produces the whole-cell current of a cell. In order to carry out whole-cell recordings, it is necessary to rupture the membrane under the pipette following seal formation. Thus, whole-cell recordings permit diffusional exchange of the pipette solution with cytoplasmic constituents. In some cases, the perforated patch-clamp method has been applied to achieve electrical access to the cell while minimizing cytoplasmic dialysis. In perforated-patch experiments, pore-forming antibiotic molecules, such as nystatin or amphotericin B, are added to the pipette solution. After seal formation, the antibiotics form small pores in the membrane patch that had been trapped within the pipette. These pores are exclusively selective for monovalent ions, and they lower the resistance of the patch enough to allow voltage-clamp of the whole cell membrane. Because these pores do not allow larger molecules such as proteins to pass, in perforated-patch recordings second messenger mechanisms within the cell remain intact. In general, membrane currents of a cell are measured in the voltage clamp mode of the patch-clamp technique where the membrane potential is controlled, and the current activated at a certain potential is monitored. Ion currents are studied using either voltage pulse or voltage ramp protocols. Voltage pulses are usually applied from a negative holding potential to a certain potential for a duration of a few milliseconds or seconds, as shown for the examples in Fig. 1 and Fig. 3. This approach allows investigations of current kinetics, such as time-dependent activation and inactivation behavior. Application of voltage ramps means that the membrane voltage is steadily increased from a negative to a positive potential (or in opposite direction) within a few milliseconds or seconds, as shown in Fig. 2 and Fig. 4. The advantage of voltage ramp protocols is that activation threshold and reversal potential of currents can easily be estimated and changes in current amplitude can be visualized over the whole voltage range. The disadvantage is that time-dependent gating (opening and closing kinetics) will distort and complicate the data.

Ion channels can be distinguished based on their ion selectivity, conductances, gating properties, kinetics, and pharmacology. Based on the ion selectivity, i.e., the nature of the ions that are allowed to pass through the open channel, microglial ion channels can be classified into K^+, H^+, Na^+, Ca^{2+} and Cl^- channels. The single-channel conductance is a measure of the rate at which ions pass through the channel. The unitary conductance of a channel can reach from a few fS, e.g. as determined for proton channels, up to hundreds of pS, e.g. as determined for voltage-dependent Ca^{2+}-activated K^+ channels. These conductances correspond to $\sim 10^4$–10^8 ions per second permeating individual ion channels. The process of opening and closing a channel is termed gating. Channel opening and closing may be influenced by the actions of certain intracellular messenger molecules or ions, such as G-proteins or calcium ions. Several ion channels in microglia, such as H^+, Na^+ or delayed rectifier K^+ channels, show voltage-dependent gating, i.e., the rate at which these channels open and close depends on the membrane potential. Important characteristics of the kinetic behavior of voltage-dependent ion channels are the rates of activation and inactivation. The increase in the open probability of a channel induced by voltage changes is termed activation. Following their activation at a given voltage, some channels progressively close, i.e., the channels undergo inactivation. Inactivated channels are refractory to a second stimulus, and must first recover from inactivation before they can reopen.

Within the central nervous system, substantial changes in extracellular pH and ion concentrations occur during neuronal activity. For example, the concentration of potassium ions in the extracellular space increases during neuronal activity and can be largely augmented under pathological conditions, such as inflammation, anoxia, epilepsy or spreading depression (Heinemann et al. (1986); Somjen et al. (1992); Heinemann and Eder (1997); Streit et al. (1999)). Furthermore, in many regions of the central nervous system, a rapid extracellular alkalinization is seen at the onset of neuronal activity, which is often followed by a more slowly developing extracellular acidification (Chesler (1990); Eder and DeCoursey (2001)). One has to keep in

TABLE 1: Ion channels in microglia

Channel type	Species	Gene	Pharmacology	Regulation / modulation
Inward rectifier K^+ channel (~30 pS)	rat; mouse; bovine; human	Kir2.1	Ba^{2+}; Cs^+; TEA; quinine; Na^+	G-protein activation (TNF, GTP-γS, EGF, C5a); PKC activation; internal acidification; increase in $[Ca^{2+}]_i$; GM-CSF; LPS; IFN-γ
Delayed rectifier K^+ channel (13 pS)	rat; mouse; human	Kv1.3	CTX; KTX; NTX; MTX; 4-AP; Cd^{2+}; Zn^{2+}; Ba^{2+}; TEA	increase in $[Ca^{2+}]_i$; PKC activation; tyrosine kinase phosphorylation; arachidonic acid; prostaglandin E_2; changes in intracellular and extracellular pH; LPS; IFN-γ; GM-CSF; TGF-β; astrocytic factor(s)
Voltage-dependent Ca^{2+}-activated K^+ channel (>200 pS)	bovine; human	Slo (?)	TEA	
Voltage-independent Ca^{2+}-activated K^+ channel	mouse	IK1	CTX; La^{3+}; Ba^{2+}; Cd^{2+}; TEA	
G-protein-activated K^+ channel	mouse	?	4-AP; TEA	
H^+ channel	mouse; rat;	NOH-1	Zn^{2+}; La^{3+}; Ni^{2+};	cytochalasin D; colchicine; LPS;
	human	(?) $gp91^{phox}$ (?)	Cd^{2+}; Co^{2+}; Ba^{2+}; 4-AP; TEA	astrocytic factor(s)
Na^+ channel	rat; human	?	TTX	astrocytic factor(s)
Ca^{2+}-release activated Ca^{2+} channel	rat	trp (?)		PKA activation; PKC activation
Voltage-dependent Cl^- channel (280-325 pS)	bovine; human	?	Ba^{2+}	
Voltage-independent stretch-activated Cl^- channel	rat; mouse	?	DIDS; SITS; NPPB; IAA-94; flufenamic acid	

Abbreviations: CTX, charybdotoxin; KTX, kaliotoxin; NTX, noxiustoxin; MTX, margatoxin; TEA, tetraethylammonium; 4-AP, 4-aminopyridine; TTX, tetrodotoxin; DIDS, 4,4'-diisothiocyanatostilbene-2,2'-disulfonic acid; SITS, 4-acetamino-4'-isothiocyanatostilbene-2,2'-disulfonic acid; NPPB, 5-nitro-2-(3-phenylpropylamino)benzoic acid; IAA-94, 6,7-dichloro-2-cyclopentyl-2,3-dihydro-2-methyl-1-oxo-1H-inden-5-yl(oxy)acetic acid; GM-CSF, granulocyte/macrophage colony-stimulating factor; IFN-γ, interferon-γ; LPS, lipopolysaccharide; TNF, tumor necrosis factor; EGF, epidermal-growth factor; C5a, complement factor C5a; PKC, protein kinase C; $[Ca^{2+}]i$, intracellular free Ca^{2+} concentration; TGF-β, transforming-growth factor-β.

mind that changes in the extracellular milieu can dramatically alter the activity of microglial ion channels and subsequently induce changes in microglial functioning.

Potassium channels

The majority of studies on microglial electrophysiology has focused on K^+ channels. During the past ten years, patch clamp studies revealed an unexpected diversity of K^+ channels in microglia, namely inward rectifier, delayed rectifier, Ca^{2+}-activated and G-protein-activated K^+ channels.

Potassium channels are involved in the regulation of the resting membrane potential of microglial cells. Functional inward rectifier K^+ channels maintain a highly negative resting membrane potential. In microglia with prominent inward rectifier currents, resting membrane potentials of about -70 mV were determined (Nörenberg et al. (1994a); Fischer et al. (1995); Chung et al. (1999); Eder et al. (1999)), whereas significantly more positive values were measured in microglia lacking these channels (Fischer et al. (1995)). It has also been demonstrated that blockage of inward rectifier channels with Ba^{2+} results in strong membrane depolarization (Visentin et al. (1995); Chung et al. (1999)).

In the absence, or during blockage, of inward rectifier channels, delayed rectifier outward K^+ channels are responsible for setting the resting membrane potential in microglial cells. Moreover, delayed rectifier channels may contribute to repolarization of the cell membrane in situations that shift the membrane potential above the delayed rectifier current activation threshold. For example, they may be involved in the reestablishment of the negative membrane potential during membrane potential oscillations that had been observed in lipopolysaccharide-activated microglia (Nörenberg et al. (1994a)). Since delayed rectifier channels activate with depolarization, they have a much greater capacity than inward rectifier channels to resist depolarizing influences.

A negative resting membrane potential might be a prerequisite condition for initiating several microglial functions, such as phagocytosis, antigen presentation, motility, cytokine secretion and others. Pharmacological lesion experiments suggest that functional inward rectifier and delayed rectifier channels may play a role in microglial proliferation (Schlichter et al. (1996); Kotecha and Schlichter (1999)). Moreover, inhibition of lipopolysaccharide-induced production of interleukin-1β was observed in cultured microglia upon blockage of delayed rectifier channels with 4-aminopyridine (Caggiano and Kraig (1998)). It is also possible that delayed rectifier channels of microglia play an important role in volume regulation, as has been reported for other immune cells (Deutsch and Chen (1993)). It has been shown for lymphocytes that the regulatory volume decrease following cell swelling involves ion efflux through delayed rectifier K^+ channels and swelling-activated Cl^- channels (Garber and Cahalan (1997)).

Many physiologically active substances increase the concentration of intracellular calcium ions ($[Ca^{2+}]_i$). In microglia, augmentation of $[Ca^{2+}]_i$ causes inhibition of both inward rectifier and delayed rectifier K^+ channels (Nörenberg et al. (1994a); Ilschner et al. (1995)). Under these conditions, however, membrane depolarization is prevented by opening of Ca^{2+}-activated K^+ channels (McLarnon et al. (1995, 1997); Eder et al. (1997)). The hyperpolarized membrane potential is important for maintaining a large driving force for Ca^{2+} influx through Ca^{2+}-release-activated Ca^{2+} (CRAC) channels, which modulate the cytosolic Ca^{2+} concentration and subsequently regulate gene expression of the cells (Lewis and Cahalan (1995); DeCoursey and Grinstein (1999)).

The importance of G-protein-activated K^+ channels in microglial functioning remains to be investigated. These channels may maintain the hyperpolarized membrane potential of microglial cells upon activation of G-protein signaling pathways, for example following exposure to complement factor C5a, tumor-necrosis factor or epidermal-growth factor (Ilschner et al. (1995, 1996)).

The expression pattern of microglial K^+ channels depends on environmental stimuli. Dramatic changes in the expression levels of inward rectifier and delayed rectifier K^+ channels occur in microglia during processes of activation and deactivation, whereas Ca^{2+}-activated and G-protein-activated K^+ channels appear to be stably expressed in the membrane of microglial cells. Prominent inward rectifier K^+ currents have been demonstrated in a large number of *in vitro* and *in situ* recordings of rat, mouse, bovine and human microglia, whereas delayed rectifier outward K^+ currents were not detected in the majority of unstimulated microglia of cell culture or brain sections (Kettenmann et al. (1990); Korotzer and Cotman (1992); Nörenberg et al. (1992); Brockhaus et al. (1993); Sievers et al. (1994); Eder et al. (1995a); Fischer et al. (1995); Visentin et al. (1995); McLarnon et al. (1995, 1997); Schlichter et al. (1996)). Treatment of microglia with proinflammatory cytokines, such as interferon-γ or granulocyte/macrophage colony-stimulating factor, cause downregulation of inward rectifier channels, which is accompanied by simultaneous upregulation of delayed rectifier channels (Eder et al. (1995); Fischer et al. (1995)). Similar changes in the K^+ channel expression pattern were observed upon microglial activation with cell wall components of gram-negative or gram-positive bacteria (Nörenberg et al. (1992, 1994a); Draheim et al. (1999)). In microglia of rat brain slices, delayed rectifier channels were expressed following facial nerve axotomy (Boucsein et al. (2000)). Although upregulation of delayed rectifier channels was always seen upon microglial activation, expression of these outward K^+ channels cannot be used as a marker for microglial activation. Recent studies have demonstrated that delayed rectifier channels are also upregulated during processes of microglial deactivation. Microglial cells deactivated by astrocyte-conditioned medium, transforming growth factor (TGF)-β1 or TGF-β2 express delayed rectifier channels of similar properties to those of cytokine- or lipopolysaccharide-activated microglia (Schmidtmayer et al. (1994); Eder et al. (1996, 1997); Schilling et al. (2000)). Thus, upregulation of delayed recifier K^+ channels indicates the transformation of microglia from one functional state into another, rather than a defined functional state of the cells.

Inward rectifier K^+ channels

Inward rectifier K^+ channels open upon membrane hyperpolarization. These channels strongly depend on the concentration of external potassium ions ($[K^+]_o$), i.e., elevation of $[K^+]_o$ shifts the reversal potential of the currents to more positive potentials and increases the chord conductance. The time-dependent decay of microglial inward rectifier currents seen in recordings using normal Ringer's solution is due to a time- and voltage-dependent inhibition by external sodium ions. Inward rectifier currents do not show time-dependent inactivation after elimination of Na^+ from the extracellular solution. Microglial inward rectifier currents are effectively inhibited by extracellular barium ions. As shown in Fig. 1, Ba^{2+} causes a time- and voltage-dependent current blockage. Inward rectifier currents are also inhibited by Cs^+ or quinine. In single-channel recordings using symmetrical isotonic $[K^+]$ solutions, a conductance of about 30 pS was determined for microglial inward rectifier channels (Kettenmann et al. (1990); Nörenberg et al. (1994a); Eder et al. (1995a); Schilling et al. (2000)).

It has been demonstrated in reverse transcriptase-polymerase chain reaction (RT-PCR) analyses that Kir2.1 mRNA is present in activated and deactivated microglia (Küst et al. (1999); Schilling et al. (2000)). The Kir2.1 channel cloned from mouse macrophage cDNA (Kubo et al. (1993)) exhibits nearly identical characteristics to the microglial inward rectifier channel, including strong dependence on $[K^+]_o$ and sensitivity to external Na^+, Ba^{2+} and Cs^+. Thus, it is most likely that Kir2.1 encodes microglial inward rectifier channels. In addition to Kir2.1 mRNA, ROMK1 mRNA was found in rat microglia (Küst et al. (1999)). However, there is no evidence that ROMK1 channels are expressed in the membrane of microglia. Inward rectifier currents of microglia are not reduced in the presence of the specific ROMK1 channel inhibitor δ-dendrotoxin (Schilling et al. (2000)).

Inward rectifier channels in microglia are directly modulated by phorbol esters, G-proteins, intracellular Ca^{2+}, or changes in the intracellular pH. Phosphorylation of protein kinase C appears to modulate opening of inward rectifier channels. However, the effects of phorbol esters on microglial inward rectifier channels are controversial: whereas inward rectifier currents of human microglia were inhibited (Yoo et al. (1996)), increase in inward rectifier current density was observed in rat microglial cells following treatment with phorbol 12-myristate 13-acetate (Visentin and Levi (1997)). The controversial effects of protein kinase C activators may be explained by differences in the initial phosphorylation state of the channel, or by differences in the expression or activation of protein kinase C isoforms. Activation of G-proteins induced in microglia by internal perfusion with GTP-γS, by receptor activation with tumor necrosis factor or epidermal-growth factor, or by stimulation with complement factor C5a was found to inhibit inward rectifier currents (Ilschner et al. (1995, 1996)). Microglial inward rectifier currents are also reduced by elevating the concentration of intracellular Ca^{2+} (Ilschner et al. (1995)) or by acidification of the intracellular milieu to pH values of less than 7.0 (Eder et al. (1995a)).

Delayed rectifier K1 Channels

Delayed rectifier channels have been characterized in microglia activated by lipopolysaccharide (Nörenberg et al. (1994a)) and granulocyte/macrophage colony-stimulating factor (Eder et al. (1995b)) or deactivated by transforming growth factor β1 (Schilling et al. (2000)). Delayed rectifier currents activate at potentials positive to -40 mV, and increase in amplitude at more depolarizing potentials. The currents exhibit sigmoidal voltage-dependent activation kinetics and can be fitted to a Hodgkin-Huxley-type n^4j model. Inactivation of delayed rectifier currents can be fitted by a single exponential with time constants of about 600 ms at potentials *above* 0 mV. Delayed rectifier currents in microglia recover slowly from inactivation and exhibit pronounced use dependence during repeated pulse application at high frequency (Nörenberg et al. (1994a); Eder et al. (1995b); Schilling et al. (2000)).

Single delayed rectifier channel currents of a TGF-β1-treated microglial cell are shown in Fig. 2. The single delayed rectifier channel conductance was 25–30 pS using symmetrical high K^+ solutions, and 13 pS in patches studied with Ringer's solution (Schilling et al. (2000)).

The most potent blockers of microglial delayed rectifier channels are several scorpion toxins, including charybdotoxin, kaliotoxin, margatoxin or noxiustoxin (Eder et al. (1995b, 1996), Schlichter et al. (1996); Schilling et al. (2000)). These peptide toxins inhibit delayed rectifier currents in microglia at nanomolar concentrations. Microglial delayed rectifier currents are also sensitive to external 4-aminopyridine and tetraethylammonium. Inorganic divalent cations, such as Ba^{2+}, Cd^{2+} or Zn^{2+}, reduce delayed rectifier current amplitude and shift the conductance-voltage curve to more positive potentials (Nörenberg et al. (1994a); Eder et al. (1995b)).

Microglial delayed rectifier channels closely resemble cloned Kv1.3 channels (Douglass et al. (1990); Grissmer et al. (1990)). Using RT-PCR, the presence of Kv1.3 mRNA has been demonstrated in LPS-activated, as well as in TGF-β-deactivated, microglial cells (Nörenberg et al. (1993); Schilling et al. (2000)), suggesting that Kv1.3 codes for microglial delayed rectifier channels. In addition to Kv1.3, RT-PCR investigations on rat microglia revealed mRNA expression of Kv1.2, Kv1.5 and Kv1.6 (Küst et al. (1999); Kotecha and Schlichter (1999)). More experiments are required to identify conditions under which microglial cells are capable of expressing the various types of Kv channels. It has been proposed that rat microglia obtained from tissue prints of hippocampal slices express Kv1.5 channels (Kotecha and Schlichter (1999)).

A variety of substances is known to influence the activity of delayed rectifier channels. Microglial delayed rectifier channels are regulated by protein and tyrosine kinases, by changes in the concentration of intracellular free Ca^{2+}, by pH changes, and in the presence of arachidonic acid or prostaglandin E_2. Similar to studies on inward rectifier channels (see above), controversial effects of protein kinase C activators have been described for microglial delayed rectifier channels. In studies on human microglial cells, activation of protein kinase C with phorbol 12-myristate 13-acetate led to an enhancement of outward K^+ currents (Yoo et al. (1996)), whereas delayed rectifier currents in rat microglia decreased upon treatment with the phorbol ester (Visentin and Levi (1997)). Activation of endogenous src-family tyrosine kinases causes inhibition of delayed rectifier currents in microglia (Cayabyab et al. (2000)). Inhibition of delayed rectifier currents observed upon oxygen/glucose deprivation can be related to tyrosine phosphorylation of the Kv1.3 protein. Furthermore, decrease in delayed rectifier current amplitude was observed in microglia in patch-clamp measurements with increased concentration of internal free Ca^{2+} ($[Ca^{2+}]_i$), or after elevation of $[Ca^{2+}]_i$ following application of the Ca^{2+} ionophore A23187 (Nörenberg et al. (1994a)). Extracellular acidification results in a shift of steady-state activation and inactivation curves for microglial delayed rectifier currents in depolarizing direction, whereas alkalinization induces the opposite effect. Furthermore, delayed rectifier currents inactivate more slowly upon extracellular acidification than upon alkalinization. Variations in the internal pH alter current amplitude, but do not affect the kinetics, of delayed rectifier currents in microglia. Intracellular acidosis reduces delayed rectifier currents, whereas alkalosis increases the currents (Eder and Heinemann (1996)).

Complex changes in delayed rectifier channel activity were also observed in microglia following treatment with arachidonic acid. Decrease in peak amplitudes, as well as acceleration of the current onset and inactivation rates, were induced by arachidonic acid. In addition, steady-state activation and inactivation curves of the channels were shifted to more negative potentials. These changes were explained by interactions of arachidonic acid with membrane lipids and by direct channel blockage (Visentin and Levi (1998)). Prostaglandin E_2, a member of the eicosanoid family, that is produced from arachidonic acid has been shown to inhibit the lipopolysaccharide-induced upregulation of delayed rectifier channels in cultured rat microglia (Caggiano and Kraig (1998)).

Ca^{2+}-activated K^+ channels

Increases in the concentration of intracellular calcium ions lead to activation of Ca^{2+}-activated K^+ channels. Microglial Ca^{2+}-activated K^+ channels can be characterized

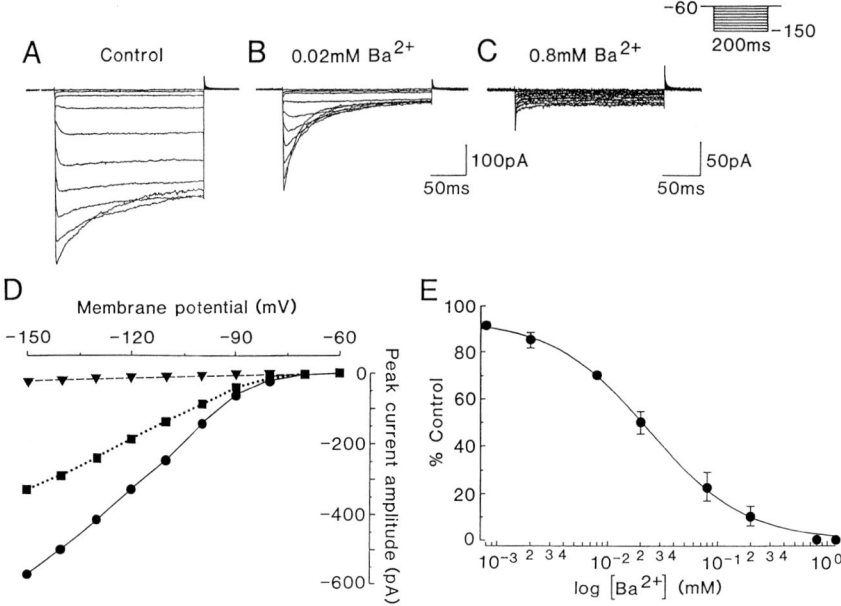

FIGURE 1: Inward rectifier currents in murine microglia. Extracellularly applied Ba^{2+} inhibits inward rectifier currents in a voltage- and time-dependent manner. Cells were hyperpolarized for 200 ms from the holding potential of -60 mV to potentials between -70 and -150 mV in 10 mV increments. Current recordings were made in the absence (A) or in the presence of 0.02 mM Ba^{2+} (B) or 0.8 mM Ba^{2+} (C). The corresponding current-voltage curves of peak amplitudes are shown in D (● control; ■ 0.02 mM Ba^{2+}; ▼ 0.8 mM Ba^{2+}). The concentration-response curve for Ba^{2+} is illustrated in E. (Reproduced from Eder et al. (1995a).)

as voltage-dependent and voltage-independent channels. The voltage-dependent Ca^{2+}-activated K^+ channels observed in bovine and human microglia (McLarnon et al. (1995, 1997)) have large single channel conductance of >200 pS, and resemble BK-type channels in other cell preparations. These channels are regulated by both Ca^{2+} and voltage. Increasing $[Ca^{2+}]_i$ shifts the voltage dependence of channel opening to more negative potentials, tending to open more channels at any given membrane potential. The voltage-dependent Ca^{2+}-activated K^+ channels of microglia are effectively inhibited by external tetraethylammonium. The molecular nature of these microglial channels has not been determined. Based on their characteristics, voltage-dependent Ca^{2+}-activated K^+ channels of microglia closely resemble the cloned *Slo* channels (Atkinson et al. (1991)).

Voltage-independent Ca^{2+}-activated K^+ channels of intermediate single channel conductance have been detected in murine microglia (Eder et al. (1997)). In contrast to the BK-type currents, voltage-independent Ca^{2+}-activated K^+ currents

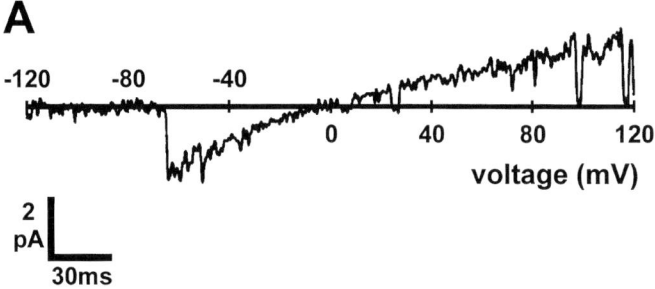

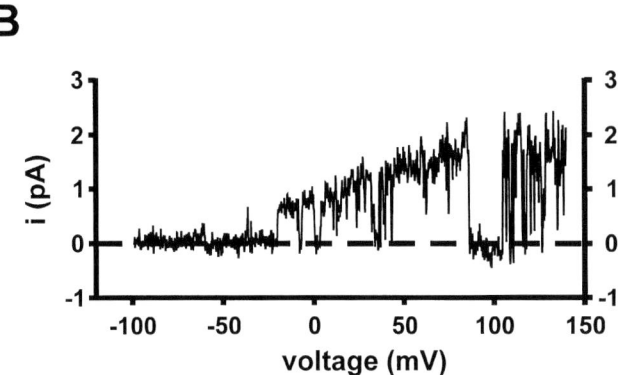

FIGURE 2: Single delayed rectifier channel currents. A: Current recorded during a single voltage ramp in an inside-out patch from a TGF-β1-treated microglial cell. Pipette and bath contained K$^+$-Ringer solution. B: Single delayed rectifier channel current in a cell-attached patch from a TGF-β-treated microglial cell with Ringer's solution in the pipette. The bath contained K$^+$-Ringer's solution, which was assumed to clamp the resting membrane potential near 0 mV. V$_{hold}$ = −80 mV, ramp rate 0.8 V/s. (Reproduced from Schilling et al. (2000).)

exhibit only weak sensitivity to extracellular tetraethylammonium. Charybdotoxin inhibits voltage-independent Ca^{2+}-activated K$^+$ currents in murine microglia at nanomolar concentrations, whereas kaliotoxin and apamin fail to block the currents. In addition, currents are reduced by several inorganic polyvalent cations, including La^{3+}, Ba^{2+} or Cd^{2+} (Eder et al. (1997)).

With respect to their pharmacological profile and their voltage-independent gating, Ca^{2+}-activated K$^+$ channels of murine microglia closely resemble intermediate conductance Ca^{2+}-activated K$^+$ channels found in other macrophages (Gallin (1991); DeCoursey and Grinstein (1999)). The recently cloned channel gene IK1 (Ishii et al. (1997); also known as SK4 (Joiner et al. (1997))) encodes for microglial voltage-independent Ca^{2+}-activated K$^+$ channels as demonstrated by RT-PCR analyses (Repp, Dreyer and Eder, unpublished observations).

G-protein-activated K^+ channels

Outwardly rectifying K^+ currents were observed in microglia after extracellular application of complement factor C5a, tumor necrosis factor or epidermal growth factor, and in the presence of intracellular GTP-γS (Ilschner et al. (1995, 1996)). Upon inhibition of G-proteins by preincubation of microglial cells with pertussis toxin, by application of extracellular N-ethylmaleimide, or by intracellular perfusion with GDP-βS, these outward K^+ currents could not be induced. Microglial G-protein-activated K^+ currents do not show apparent time-dependent activation or inactivation, and are inhibited by extracellular 4-aminopyridine (Ilschner et al. (1995)). Inhibitors of Ca^{2+}-activated K^+ channels, such as charybdotoxin or iberiotoxin, do not affect microglial G-protein-activated K^+ currents (Schilling and Eder, unpublished observations).

HERG-like K^+ channels

An additional type of K^+ channels, namely HERG-like K^+ channels, has been described in a rat microglial cell line (Zhou et al. (1998)). The presence of HERG-like channels in primary microglia cultures, or in microglia of brain sections, has not been reported so far. In the MLS-9 cell line, inward HERG-like currents were seen upon membrane hyperpolarization, when cells were studied in high K^+-containing solution. In contrast to inward rectifier currents, HERG-like currents exhibited pronounced time-dependent inactivation and were inhibited by the specific HERG channel blocker E-4031. External Cs^+ or Ba^{2+} had little effect on the HERG-like currents of MLS-9 cells (Zhou et al. (1998)).

Proton channels

In response to brain injury, ischemia, or inflammation, activated microglial cells generate superoxide anions, and subsequently other oxygen free radicals (reviewed in Colton and Gilbert (1993); Love (1999)). The potential for stimulated microglia to generate oxygen radicals during the respiratory burst may also have implications in several degenerative neurological diseases. Voltage-gated proton channels in the plasma membrane of microglia most likely serve to extrude acid during the respiratory burst. In addition, H^+ efflux compensates for the positive charge introduced into the cell by NADPH oxidase activity (Henderson et al. (1987)), thus limiting changes in membrane potential during the respiratory burst. The importance of this role in phagocytes is demonstrated by the inhibition of superoxide anion release by H^+ channel block by Cd^{2+} or Zn^{2+} (Henderson et al. (1988); Lowenthal and Levy (1999)).

It can also be assumed that in microglia outward H^+ currents contribute to the repolarization of the cell membrane in situations that shifts the membrane potential above the H^+ current activation threshold. For example, strong membrane depolarization is induced by increases in $[K^+]_o$ during spreading depression or following ion fluxes through ATP receptor channels.

Similar to observations made for microglial potassium channels (see above), proton channel expression changes during processes of microglial activation and deactivation (Klee et al. (1999)). Exposure of microglia to lipopolysaccharide or to astrocyte-conditioned medium decreases the microglial H^+ current density. The mechanisms leading to reduced H^+ channel expression have not fully been understood.

Activation of H^+ currents strongly depends on voltage, intracellular pH and extracellular pH (reviewed in DeCoursey and Cherny (1994)). Shifts in activation threshold and conductance/voltage curve of H^+ currents to more hyperpolarizing potentials, as well as augmentation of H^+ current amplitude, are seen upon decreases in intracellular pH, or upon increases in extracellular pH (Fig. 3). Within the central nervous system, neuronal activity is accompanied by substantial changes in extracellular pH. Changes in pH occur under pathological situations, such as ischemia and epilepsy, but also in the normally functioning brain (reviewed in Chesler (1990); Chesler and Kaila (1992); Eder and DeCoursey (in press)). Such activity-dependent pH changes of the external milieu most likely influence the activity of microglial proton channels. Voltage-gated H^+ channels open only when there is an outward pH gradient, so that they only carry outward current in the steady-state. In microglia, time-dependent activation of H^+ currents can be fitted by a single exponential.

Microglial H^+ currents are inhibited by several inorganic polyvalent cations in the micromolar concentration range. The following order of blocking activity has been determined: $Zn^{2+} > La^{3+} > Ni^{2+} > Cd^{2+} > Co^{2+} > Ba^{2+}$. Simultaneous with the reduction in current amplitude, a shift in the conductance/voltage curve of H^+ currents to more positive potentials is seen upon application of polyvalent cations. H^+ currents in microglia are also inhibited by superfusion with the K^+ channel blockers 4-aminopyridine or tetraethylammonium (Eder et al. (1995a)).

H^+ channels in microglia are regulated by cytoskeletal disruptive agents (Klee et al. (1998)). Decrease in current density and increase in activation time constant of H^+ currents were observed after exposure of microglia to cytochalasin D or colchicine. It is possible that cytoskeletal reorganization processes induce alterations in H^+ current density and activation time constant in microglia upon deactivation with astrocyte-conditioned media, or upon activation with lipopolysaccharide (Klee et al. (1999)).

The molecular nature of microglial voltage-gated proton channels has not been identified. Henderson and colleagues (1995) identified the $gp91^{phox}$ glycoprotein, one of the membrane-bound components of the NADPH oxidase, as comprising a proton channel. Banfi and colleagues (2000) proposed recently that the gene NOH-1 may encode voltage-gated proton channels of phagocytes.

Sodium channels

Isolated cultured microglial cells only rarely express voltage-gated Na^+ channels. In contrast, microglia cocultured with astrocytes exhibit large Na^+ currents in response to depolarizing voltage commands (Korotzer and Cotman (1992); Sievers et al. (1994); Schmidtmayer et al. (1994)). Expression of Na^+ channels is not induced by astrocyte-conditioned medium, suggesting that insoluble astrocytic factors induce upregulation of Na^+ channels in microglia (Eder et al. (1999)). Although upregulation of Na^+ channels is accompanied by transformation of microglia from ameboid into ramified morphology, Na^+ channels are not involved in microglial ramification. Inhibition of the channels with tetrodotoxin failed to influence ramification of microglia (Eder et al. (1998)). The functional role of Na^+ channels in microglia remains to be elucidated. Activation of Na^+ channels leads to rapid and transient membrane depolarization of microglial cells. Depolarization of the cell membrane might be necessary to trigger signaling cascades for microglial activation and a subsequent immune response. It has been demonstrated, for example, that membrane depolarization induces conformational changes of the MHC class I molecule (Bene et al. (1997)), which are required for sucessful antigen presentation (Catipovic et al. (1994)).

Na^+ currents activate at depolarizing potentials positive to -40 mV, and reach current maximum at about 0 mV. They exhibit rapid time-dependent activation and inactivation. Tetrodotoxin inhibits voltage-gated Na^+ channels of microglia at nanomolar concentrations (Korotzer and Cotman (1992); Nörenberg et al. (1994b); Schmidtmayer et al. (1994)).

Calcium channels

Ca^{2+}-release-activated Ca^{2+} channels

Regulation of the free intracellular Ca^{2+} concentration plays a major role in physiological signal transduction. A variety of Ca^{2+}-dependent processes regulate cell functions, and many of the essential enzymes in signaling cascades are Ca^{2+}-dependent. In immune cells, Ca^{2+} influx through Ca^{2+}-release-activated Ca^{2+} channels (CRAC channels) provides the trigger for activation of signaling cascades leading to gene expression and cell activation, or apoptosis (Lewis and Cahalan (1995)).

In patch-clamp recordings, CRAC channels of microglia have been activated either by passive Ca^{2+} store depletion with the Ca^{2+} chelator BAPTA, or by store depletion in response to added inositol 1,4,5-triphosphate (Nörenberg et al. (1997); Hahn et al. (2000)). In the whole-cell configuration, CRAC currents develop slowly after a delay of a few seconds. The activation time course can be fitted monoexponentially. CRAC currents in microglia exhibit strong inward rectification. As

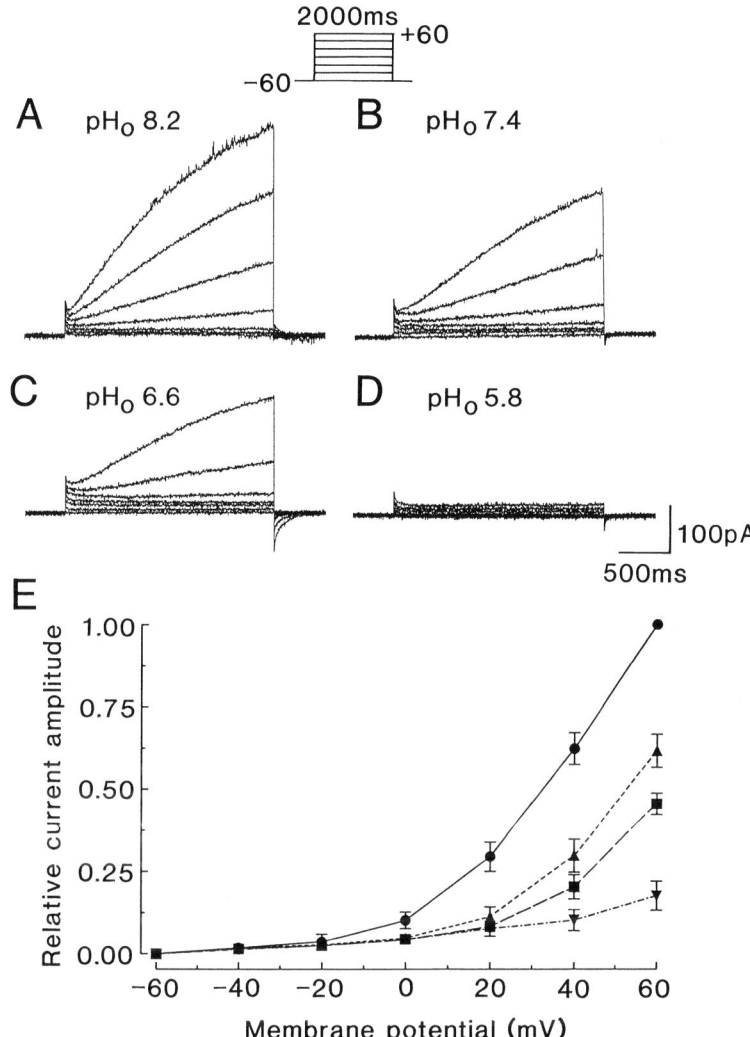

FIGURE 3: Voltage-gated proton currents. Families of H^+ currents recorded from the same microglial cell at several pH_o. Identical families of voltage pulses were applied in each solution (see inset). At higher pH_o, H^+ channels open at more negative voltages; conversely, low pH_o inhibits the current. In (E) the currents at the end of the pulses in the four solutions (A–D) are plotted: A, pH_o 8.2 (●), B, pH_o 7.4 (▼), C, pH_o 6.6 (■), D, pH_o 5.8 (◆). (Reproduced from Eder et al. (1995a).)

first demonstrated in Jurkat T lymphocytes (Kerschbaum and Cahalan (1999)), the use of sodium ions as a current carrier enabled single-channel recordings of CRAC channels in microglia. A single-channel conductance of 42 pS was determined for microglial CRAC channels carrying Na^+ (Hahn et al. (2000)).

Modulation of microglial CRAC channel activity has been observed by activators of protein kinase A and protein kinase C. Whereas CRAC channels were enhanced by the protein kinase A activator dibutyryl-cAMP, current reduction was seen after activation of protein kinase C by phorbol 12,13-dibutyrate (Hahn et al. (2000)).

L-type voltage-gated Ca^{2+} channels

Based on observations that microglial activity is modulated by several inhibitors of voltage-gated Ca^{2+} channels, including nifedipine, verapamil and diltiazem (Spranger et al. (1998); Silei et al. (1999)), it had been proposed that L-type voltage-gated Ca^{2+} channels are expressed in the membrane of microglial cells. However, the inhibitory effects of nifedipine, verapamil and diltiazem on microglial proliferation and respiratory-burst generation (Colton et al. (1994); Spranger et al. (1998); Silei et al. (1999)) can also be attributed to their action on microglial delayed rectifier channels. Kv1.3 potassium channels are inhibited by several Ca^{2+} channel inhibitors, including nifedipine, verapamil and diltiazem (DeCoursey et al. (1985); Grissmer et al. (1994); Rauer and Grissmer (1996)), suggesting that these inhibitors cannot be used to demonstrate functional roles of voltage-gated Ca^{2+} channels in microglial activity. In the majority of studies on microglial ion channels, voltage-gated Ca^{2+} channels have not been detected. The presence of L-type Ca^{2+} channels in microglia has only been reported in one paper on cultured rat microglia (Colton et al. (1994)). Ca^{2+} currents were seen in a small percentage of microglial cells after exposure to the Ca^{2+} channel opener BAY K 8644. These currents activated at potentials of about -35 mV, and showed a U-shaped activation curve.

Chloride channels

Single channel and whole cell recordings revealed the existence of two types of chloride channels in microglia, namely voltage-independent and voltage-dependent Cl^- channels. Functional roles of Cl^- channels in microglia can be related to changes in resting membrane potential of the cells. Activation of Cl^- channels results in a shift of the resting membrane potential to more positive values, i.e., near the Cl^- reversal potential. In immune cells, including macrophages and lymphocytes, chloride channels significantly affect volume regulation (Garber and Cahalan (1997); DeCoursey and Grinstein (1999)). Opening of Cl^- channels provides the initial trigger for the regulatory volume decrease. Activation of Cl^- channels following cell swelling results in membrane depolarization to potentials above the activation threshold of delayed rectifier channels. Thus, Cl^-

and K$^+$ channels become simultaneously active and allow Cl$^-$ and K$^+$ to leave the cell. Subsequently, water follows the osmolytes, and cellular volume returns to normal.

In microglia, functional Cl$^-$ channels are also required for the induction of ramification. Microglial cells undergo dramatic morphological changes, namely transformation from ameboid into ramified morphology, following exposure to astrocyte-conditioned medium (Eder et al. (1997, 1999)). Addition of the Cl$^-$ channel blockers DIDS, SITS, NPPB, or flufenamic acid to the astrocyte-conditioned medium reversibly inhibited microglial ramification (Eder et al. (1998)).

It has also been reported that proliferation of microglia induced by macrophage colony-stimulating factor was significantly reduced in the presence of flufenamic acid, NPPB or IAA-94. The half-maximal inhibition of microglial proliferation by the Cl$^-$ channel inhibitors occurred at micromolar concentrations similar to those required to block Cl$^-$ currents (Schlichter et al. (1996)).

Voltage-independent Cl$^-$ channels

Voltage-independent Cl$^-$ channels have been identified in whole-cell measurements of rat (Visentin et al. (1995); Schlichter et al. (1996)) and mouse (Eder et al. (1998)) microglia. These channels are activated either by cell swelling following exposure of microglia to hypo-osmotic external solution, or by mechanically induced stretching of the cell membrane (Schlichter et al. (1996); Eder et al. (1998)). Cl$^-$ channels activate slowly, i.e., within a few minutes, after the initial membrane stretch. In whole-cell recordings they undergo rundown, which is prevented by addition of ATP to the pipette solution. Stretch-activated Cl$^-$ currents exhibit outward rectification and do not show time- or voltage-dependent gating. Lowering the concentration of extracellular Cl$^-$ shifts the reversal potential of the current in depolarizing direction as shown in Fig. 4B. The well-known Cl$^-$ channel blockers DIDS and SITS inhibit microglial Cl$^-$ currents in a voltage- and time-dependent manner. Currents are also blocked by NPPB, IAA-94, or flufenamic acid in the micromolar concentration range, but are unaffected by extracellular application of barium or lanthanum.

Cloned Cl$^-$ channels invoked in cell volume regulation include the ClC-2 channel (Gründer et al. (1992)), the ClC-3 channel (Duan et al. (1997)) and pICln (Gschwentner et al. (1995)). It is possible that one of those genes encodes voltage-independent Cl$^-$ channels in microglia. The molecular identity of microglial Cl$^-$ channels remains to be determined.

Voltage-dependent Cl$^-$ channels

Voltage-dependent Cl$^-$ channels of large unitary conductance (280–325 pS) were detected in inside-out patches of bovine and human microglial cells (McLarnon et

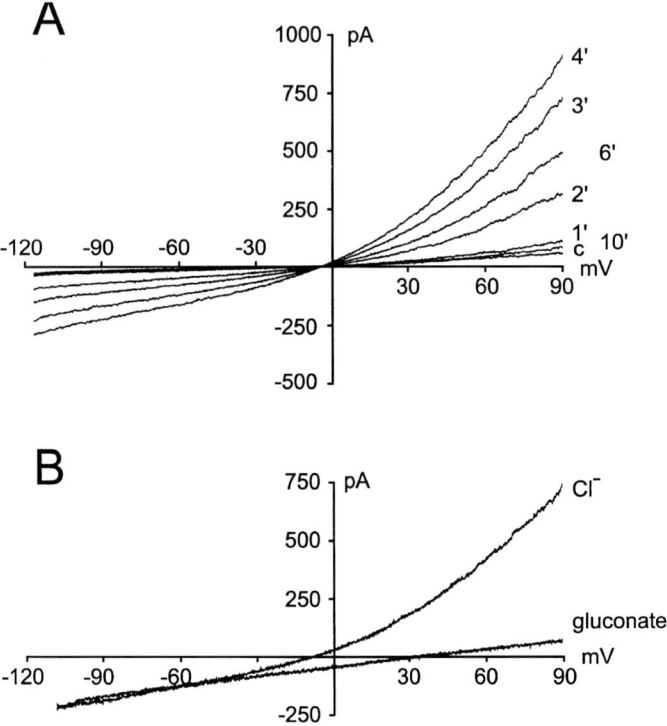

FIGURE 4: Stretch-activated Cl⁻ currents. A- Current recordings under control (before stretching of the membrane) and at several minutes (as indicated) after application of a membrane stretch using a K$^+$-free intracellular solution. B- Current recording in the presence of 142 mM Cl⁻ (Cl⁻) or after lowering concentration of extracellular Cl⁻ to 12 mM by equimolar substitution with gluconate (gluconate). (Reproduced from Eder et al. (1998).) Copyright 1998 by the Society for Neuroscience.

al. (1995, 1997)). It appears that in microglia voltage-dependent Cl⁻ channels are virtually silent in whole-cell measurements, and only become activated after excision of the patch. Similar observations have been made in other macrophages (Gallin (1991)).

Conclusions

When comparing microglial ion channels with those found in macroglial cells (reviewed in Sontheimer (1994)) and those described in macrophages of peripheral tissue (reviewed in Gallin (1991); DeCoursey and Grinstein (1999)), it can be concluded that expression, properties and regulation of microglial ion channels are nearly identical to those of other macrophages. In contrast, ion channels of microglia dramatically differ from those of macroglial cells, i.e., astrocytes, oligodendrocytes and Schwann cells, with respect to their biophysical and pharmacological properties.

Acknowledgements

I am very grateful to Thomas E. DeCoursey for many helpful discussions of, and comments on, this article. Experiments in the author's laboratory were supported by grant SFB 507/C3 of the Deutsche Forschungsgemeinschaft.

References

Atkinson NS, Robertson GA, Ganetzky B (1991). A component of calcium-activated potassium channels encoded by the Drosophila slo locus. *Science* 253:551–555.

Banfi B, Maturana A, Jaconi S, Arnaudeau S, Laforge T, Sinha B, Ligeti E, Demaurex N, Krause KH (2000). A mammalian H^+ channel generated through alternative splicing of the NADPH oxidase homolog NOH-1. *Science* 287:138–142.

Bene L, Szöllösi J, Balazs M, Matyus L, Gaspar R, Ameloot M, Dale RE, Damjanovich S (1997). Major histocompatibility complex class I protein conformation altered by transmembrane potential changes. *Cytometry* 27:353–357.

Boucsein C, Kettenmann H, Nolte C (2000). Electrophysiological properties of microglial cells in normal and pathologic rat brain slices. *Eur J Neurosci* 12:2049–2058.

Brockhaus J, Ilschner S, Banati RB, Kettenmann H (1993). Membrane properties of ameboid microglial cells in the corpus callosum slice from early postnatal mice. *J Neurosci* 13:4412–4421.

Brown H, Kozlowski R, Perry H (1998). The importance of ion channels for macrophage and microglial activation in vitro. *Glia* 22:94–97.

Caggiano AO, Kraig RP (1998). Prostaglandin E2 and 4-aminopyridine prevent the lipopolysaccharide-induced outwardly rectifying potassium current and interleukin-1β production in cultured rat microglia. *J Neurochem* 70:2357–2368.

Catipovic B, Talluri G, Oh J, Wei T, Su XM, Johansen TE, Edidin M, Schneck JP (1994). Analysis of the structure of empty and peptide-loaded major histocompatibility complex molecules at the cell surface. *J Exp Med* 180:1753–1761.

Cayabyab FS, Khanna R, Jones OT, Schlichter LC (2000). Suppression of the rat microglia Kv1.3 current by src-family tyrosine kinases and oxygen/glucose deprivation. *Eur J Neurosci* 12:1949–1960.

Chesler M (1990). The regulation and modulation of pH in the nervous system. *Prog Neurobiol* 34:401–427.

Chesler M, Kaila K (1992). Modulation of pH by neuronal activity. *Trends Neurosci* 15:396–402.

Chung S, Jung W, Lee MY (1999). Inward and outward rectifying potassium currents set membrane potentials in activated rat microglia. *Neurosci Lett* 262:121–124.

Colton CA, Gilbert DL (1993). Microglia, an in vivo source of reactive oxygen species in the brain. *Adv Neurol* 59:321–326.

Colton CA, Jia M, Li MX, Gilbert DL (1994). K$^+$ modulation of microglial superoxide production: involvement of voltage-gated Ca^{2+} channels. *Am J Physiol* 266:C1650–C1655.

DeCoursey TE, Chandy KG, Gupta S, Cahalan MD (1985). Voltage-dependent ion channels in T-lymphocytes. *J Neuroimmunol* 10:71–95.

DeCoursey TE, Cherny VV (1994). Voltage-activated hydrogen ion currents. *J Membr Biol* 141:203–223.

DeCoursey TE, Grinstein S (1999). Ion channels and carriers in leukocytes. In: *Inflammation: Basic principles and clinical correlates*. JI Gallin and R Snyderman, eds. Lippincott Williams & Wilkins, New York, pp. 639–659.

Draheim HJ, Prinz M, Weber JR, Weiser T, Kettenmann H, Hanisch UK (1999). Induction of potassium channels in mouse brain microglia: cells acquire responsiveness to pneumococcal cell wall components during late development. *Neuroscience* 89:1379–1390.

Deutsch C, Chen LQ (1993). Heterologous expression of specific K$^+$ channels in T lymphocytes: functional consequences for volume regulation. *Proc Natl Acad Sci USA* 90:10036–10040.

Douglass J, Osborne PB, Cai YC, Wilkinson M, Christie MJ, Adelman JP (1990). Characterization and functional expression of a rat genomic DNA clone encoding a lymphocyte potassium channel. *J Immunol* 144:4841–4850.

Duan D, Winter C, Cowley S, Hume JR, Horowitz B (1997). Molecular identification of a volume-regulated chloride channel. *Nature* 390:417–421.

Eder C, DeCoursey TE (2001). Voltage-gated proton channels in microglia. *Prog Neurobiol*, 64:277–305.

Eder C, Fischer HG, Hadding U, Heinemann U (1995a). Properties of voltage-gated currents of microglia developed with macrophage colony-stimulating factor. *Pflügers Arch* 430:526–533.

Eder C, Fischer HG, Hadding U, Heinemann U (1995b). Properties of voltage-gated potassium currents of microglia differentiated with granulocyte/macrophage colony-stimulating factor. *J Membr Biol* 147:137–146.

Eder C, Heinemann U (1996). Proton modulation of outward K$^+$ currents in interferon-γ-activated microglia. *Neurosci Lett* 206:101–104.

Eder C, Klee R, Heinemann U (1996). Blockage of voltage-gated outward K$^+$ currents of ramified murine microglia by scorpion peptide toxins. *Neurosci Lett* 219:29–32, 1996.

Eder C, Klee R, Heinemann U (1997a). Distinct soluble astrocytic factors induce expression of outward K$^+$ currents and ramification of brain macrophages. *Neurosci Lett* 226:147–150.

Eder C, Klee R, Heinemann U (1997b). Pharmacological properties of Ca^{2+}-activated K$^+$ currents of ramified murine brain macrophages. *Naunyn Schmiedeberg's Arch Pharmacol* 356:233–239.

Eder C, Klee R, Heinemann U (1998). Involvement of stretch-activated Cl$^-$ channels in ramification of murine microglia. *J Neurosci* 18:7127–7137.

Eder C, Schilling T, Heinemann U, Haas D, Hailer N, Nitsch R (1999). Morphological, immunophenotypical and electrophysiological properties of resting microglia in vitro. *Eur J Neurosci* 11:4251–4261.

Fischer HG, Eder C, Hadding U, Heinemann U (1995). Cytokine-dependent K$^+$ channel profile of microglia at immunologically defined functional states. *Neuroscience* 64:183–191.

Gallin EK (1991). Ion channels in leukocytes. *Physiol Rev* 71:775–811.
Garber SS, Cahalan MD (1997). Volume-regulated anion channels and the control of a simple cell behavior. *Cell Physiol Biochem* 7:229–241.
Grissmer S, Dethlefs B, Wasmuth JJ, Goldin AL, Gutman GA, Cahalan MD, Chandy KG (1990). Expression and chromosomal localization of a lymphocyte K^+ channel gene. *Proc Natl Acad Sci USA* 87:9411–9415.
Grissmer S, Nguyen AN, Aiyar J, Hanson DC, Mather RJ, Gutman GA, Karmilowicz MJ, Auperin DD, Chandy KG (1994). Pharmacological characterization of five cloned voltage-gated K^+ channels, types Kv1.1, 1.2, 1.3, 1.5, and 3.1, stably expressed in mammalian cell lines. *Mol Pharmacol* 45:1227–1234.
Gründer S, Thiemann A, Pusch M, Jentsch TJ (1992). Regions involved in the opening of CIC-2 chloride channel by voltage and cell volume. *Nature* 360:759–762.
Gschwentner M, Nagl UO, Woll E, Schmarda A, Ritter M, Paulmichl M (1995). Antisense oligonucleotides suppress cell-volume-induced activation of chloride channels. *Pflügers Arch* 430:464–470.
Hahn J, Jung W, Kim N, Uhm DY, Chung S (2000). Characterization and regulation of rat microglial Ca^{2+} release-activated Ca^{2+} (CRAC) channel by protein kinases. *Glia* 31:118–124.
Hamill OP, Marty A, Neher E, Sakmann B, Sigworth FJ (1981). Improved patch-clamp techniques for high-resolution current recording from cells and cell-free membrane patches. *Pflügers Arch* 391, 85–100.
Heinemann U, Eder C (1997). Control of neuronal excitability. In: *Epilepsy: A comprehensive textbook.* (Eds.: Engel J Jr., Pedley TA), pp. 237–250, Lippincott-Raven Publishers, Philadelphia.
Heinemann U, Konnerth A, Pumain R, Wadman WJ (1986). Extracellular calcium and potassium concentration changes in chronic epileptic brain tissue. In: *Advances in Neurology/Basic mechanism of the epilepsies* (Eds.: Delgado-Escueta AV, Ward AA, Woodbury DM, Porter RJ), pp. 641–661, Raven Press, New York.
Henderson LM, Chappell JB, Jones OTG (1987). The superoxide-generating NADPH oxidase of human neutrophils is electrogenic and associated with an H^+ channel. *Biochem J* 246:325–329.
Henderson LM, Chappell JB, Jones OTG (1988). Superoxide generation by the electrogenic NADPH oxidase of human neutrophils is limited by the movement of a compensating charge. *Biochem J* 255:285–290.
Henderson LM, Banting G, Chappell JB (1995). The arachidonate-activable [sic], NADPH oxidase-associated H^+ channel: evidence that gp91-phox functions as an essential part of the channel. *J Biol Chem* 270:5909–5916.
Ilschner S, Nolte C, Kettenmann H (1996). Complement factor C5a and epidermal growth factor trigger the activation of outward potassium currents in cultured murine microglia. *Neuroscience* 73:1109–1120.
Ilschner S, Ohlemeyer C, Gimpl G, Kettenmann H (1995). Modulation of potassium currents in cultured murine microglial cells by receptor activation and intracellular pathways. *Neuroscience* 66:983–1000.
Ishii TM, Silvia C, Hirschberg B, Bond CT, Adelman JP, Maylie J (1997). A human intermediate conductance calcium-activated potassium channel. *Proc Natl Acad Sci USA* 94:11651–11656.
Joiner WJ, Wang LY, Tang MD, Kaczmarek LK (1997). hSK4, a member of a novel subfamily of calcium-activated potassium channels. *Proc Natl Acad Sci USA* 94:11013–11018.

Kerschbaum HH, Cahalan MD (1999). Single-channel recording of a store-operated Ca^{2+} channel in Jurkat T lymphocytes. *Science* 283:836–839.

Kettenmann H, Hoppe D, Gottmann K, Banati R, Kreutzberg GW (1990). Cultured microglial cells have a distinct pattern of membrane channels different from peritoneal macrophages. *J Neurosci Res* 26:278–287.

Klee R, Heinemann U, Eder C (1998). Changes in proton currents in murine microglia induced by cytoskeletal disruptive agents. *Neurosci Lett* 247:191–194.

Klee R, Heinemann U, Eder C (1999). Voltage-gated proton currents in microglia of distinct morphology and functional state. *Neuroscience* 91:1415–1424.

Korotzer AR, Cotman CW (1992). Voltage-gated currents expressed by rat microglia in culture. *Glia* 6:81–88.

Kotecha SA, Schlichter LC (1999). A Kv1.5 to Kv1.3 switch in endogenous hippocampal microglia and a role in proliferation. *J Neurosci* 19:10680–10693.

Kubo Y, Baldwin TJ, Jan YN, Jan LY (1993). Primary structure and functional expression of a mouse inward rectifier potassium channel. *Nature* 362:127–133.

Küst BM, Biber K, van Calker D, Gebicke-Haerter PJ (1999). Regulation of K^+ channel mRNA expression by stimulation of adenosine A_{2a}-receptors in cultured rat microglia. *Glia* 25:120–130.

Lewis RS, Cahalan MD (1995). Potassium and calcium channels in lymphocytes. *Annu Rev Immunol* 13:623–653.

Love S (1999). Oxidative stress in brain ischemia. *Brain Pathol* 9:119–131.

Lowenthal A, Levy R (1999). Essential requirement of cytosolic phospholipase A_2 for activation of the H^+ channel in phagocyte-like cells. *J Biol Chem* 274:21603–21608.

McLarnon JG, Sawyer D, Kim SU (1995). Cation and anion unitary ion channel currents in cultured bovine microglia. *Brain Res* 693:8–20.

McLarnon JG, Xu R, Lee YB, Kim SU (1997). Ion channels of human microglia in culture. *Neuroscience* 78:1217–1228.

Nörenberg W, Appel K, Bauer J, Gebicke-Haerter PJ, Illes P (1993). Expression of an outwardly rectifying K^+ channel in rat microglia cultivated on teflon. *Neurosci Lett* 160:69–72.

Nörenberg W, Cordes A, Blöhbaum G, Fröhlich R, Illes P (1997). Coexistence of purino- and pyrimidinoceptors on activated rat microglial cells. *Brit J Pharmacol* 121:1087–1098.

Nörenberg W, Gebicke-Haerter PJ, Illes P (1992). Inflammatory stimuli induce a new K^+ outward current in cultured rat microglia. *Neurosci Lett* 147:171–174.

Nörenberg W, Gebicke-Haerter PJ, Illes P (1994a). Voltage-dependent potassium channels in activated rat microglia. *J Physiol* 475:15–32.

Nörenberg W, Illes P, Gebicke-Haerter PJ (1994b). Sodium channel in isolated human brain macrophages (microglia). *Glia* 10:165–172.

Rauer H, Grissmer S (1996). Evidence for an internal phenylalkylamine action on the voltage-gated potassium channel Kv1.3. *Mol Pharmacol* 50:1625–1634.

Schilling T, Quandt FN, Cherny VV, Zhou W, Heinemann U, DeCoursey TE, Eder C (2000). Upregulation of Kv1.3 K^+ channels in microglia deactivated by TGF-β. *Am J Physiol (Cell Physiol)* 279:C1123–C1134.

Schlichter LC, Sakellaropoulos G, Ballyk B, Pennefather PS, Phipps DJ (1996). Properties of K^+ and Cl^- channels and their involvement in proliferation of rat microglial cells. *Glia* 17:225–236.

Schmidtmayer J, Jacobsen C, Miksch G, Sievers J (1994). Blood monocytes and spleen macrophages differentiate into microglia-like cells on monolayers of astrocytes: Membrane currents. *Glia* 12:259–267.

Sievers J, Schmidtmayer J, Parwaresch R (1994). Blood monocytes and spleen macrophages differentiate into microglia-like cells when cultured on astrocytes. *Anat Anz* 176:45–51.

Silei V, Fabrizi C, Venturini G, Salmona M, Bugiani O, Tagliavini F, Lauro GM (1999). Activation of microglial cells by PrP and beta-amyloid fragments raises intracellular calcium through L-type voltage sensitive calcium channels. *Brain Res* 818:168–170.

Sontheimer H (1994). Voltage-dependent ion channels in glial cells. *Glia* 11:156–172.

Spranger M, Kiprianova I, Krempien S, Schwab S (1998). Reoxygenation increases the release of reactive oxygen intermediates in murine microglia. *J Cereb Blood Flow Metab* 18:670–674.

Somjen GG, Aitken PG, Czeh GL, Herreras O, Jing J, Young JN (1992). Mechanism of spreading depression: a review of recent findings and a hypothesis. *Can J Physiol and Pharmacol* 70:S248–254.

Streit WJ, Walter SA, Pennell NA (1999). Reactive microgliosis. *Prog Neurobiol* 57:563–581.

Visentin S, Agresti C, Patrizio M, Levi G (1995). Ion channels in rat microglia and their different sensitivity to lipopolysaccharide and interferon-γ. *J Neurosci Res* 42:439–451.

Visentin S, Levi G (1997). Protein kinase C involvement in the resting and interferon-γ-induced K^+ channel profile of microglial cells. *J Neurosci Res* 47:233–241.

Visentin S, Levi G (1998). Arachidonic acid-induced inhibition of microglial outward-rectifying K^+ current. *Glia* 22:1–10.

Yoo AS, McLarnon JG, Xu RL, Lee YB, Krieger C, Kim SU (1996). Effects of phorbol ester on intracellular Ca^{2+} and membrane currents in cultured human microglia. *Neurosci Lett* 218:37–40.

Zhou W, Cayabyab FS, Pennefather PS, Schlichter LC, DeCoursey TE (1998). HERG-like K^+ channels in microglia. *J Gen Physiol* 111:781–794.

4

Calcium Signaling in Microglial Cells

THOMAS MÖLLER

Receptor-mediated changes in the free cytoplasmic Ca^{2+} concentration ($[Ca^{2+}]_c$) represent one of the major signal transduction pathways by which information from extracellular signals is transferred to intracellular sites. The signal is conveyed by the magnitude, duration and location of the changes in $[Ca^{2+}]_c$, and is usually initiated by the binding of an extracellular signaling molecule / ligand to its plasma membrane receptor. This chapter will provide a brief overview of the basic mechanisms of Ca^{2+} signaling, describe the available data about microglial $[Ca^{2+}]_c$ signals, and discuss current challenges and future directions of this emerging field.

1. Basic mechanisms of Ca^{2+} signaling

The field of Ca^{2+} signaling has recently been covered in detail by review articles of exceptional quality (Kostyuk and Verkhratsky (1994); Bootman and Berridge (1995); Clapham (1995); Berridge (1997); Deitmer et al. (1998); Verkhratsky et al. (1998); Berridge et al. (1999)). This chapter will provide an overview of the basic mechanisms of Ca^{2+} signaling as they pertain to microglial cells. Readers interested in a more in-depth review of $[Ca^{2+}]_c$ signaling may wish to consult the literature mentioned above. Microglial $[Ca^{2+}]_c$, like that of any other eukaryotic cell, is tightly regulated. Microglial $[Ca^{2+}]_c$ signaling, as referred to in this chapter, is defined as transitory change in $[Ca^{2+}]_c$ in a time range from milliseconds to minutes. From a "resting" $[Ca^{2+}]_c$ of about 50–150 nM, $[Ca^{2+}]_c$ usually rises up to hundreds of nM, sometimes even several μM within milliseconds, followed by an often much slower decay. Ca^{2+} signals are generated by the interplay of four main processes (Fig. 1): 1.) Ca^{2+} influx from the extracellular space across the plasma membrane into the cytoplasm; 2.) Ca^{2+} release from internal stores into the cytosol. Both mechanisms contribute to the rising phase of a Ca^{2+} signal, whereas 3.) Ca^{2+} extrusion from the cytoplasm across the plasma membrane and 4.) Ca^{2+} sequestration into intracellular Ca^{2+} stores play a role in the recovery

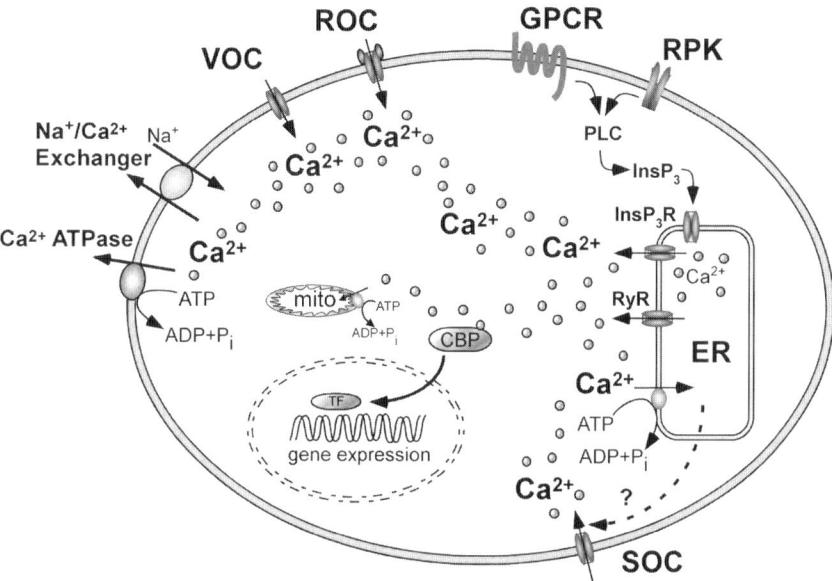

FIGURE. 1. Mechanisms of Ca^{2+} signaling
Simplified scheme of cellular Ca^{2+} signaling.
Abbreviations as follows: ADP, adenosine diphosphate; ATP, adenosine trisphosphate; CBP, Ca^{2+}-binding protein; ER, endoplasmic reticulum; GPCR, G-protein-coupled receptor; $InsP_3$, inositol 1,4,5-trisphosphate; $InsP_3R$, inositol 1,4,5-trisphosphate receptor; Mito, mitochondrion; P_i, inorganic phosphate; PLC, phospholipase C; ROC, receptor-operated channel; RPK, receptor protein kinase; RyR, ryanodine receptor; SOC, store-operated channel; TF transcription factor; VOC, voltage-operated channel; ?, unknown signal.

phase of a Ca^{2+} signal. Ca^{2+} influx, as well as Ca^{2+} release, are gated passive Ca^{2+} redistributions, whereas Ca^{2+} extrusion and Ca^{2+} sequestration are active, energy-dependent transport processes.

1.1. Ca^{2+} influx

Conceptually, Ca^{2+} influx via the microglial plasma membrane can occur by three different mechanisms: 1.) Voltage-operated Ca^{2+} channels (VOCs), lead to $[Ca^{2+}]_c$ increase upon depolarization of the plasma membrane; 2.) Ca^{2+}-permeable receptor-operated channels (ROCs) are triggered by ligand binding; 3.) Ca^{2+}-permeable store-operated channels (SOCs) are opened upon depletion of intracellular Ca^{2+} stores, and are supposed to aid in replenishing Ca^{2+}-depleted stores.

1.2. Ca^{2+} release

Two families of intracellular channels releasing calcium from internal stores have been described: the inositol 1,4,5-trisphosphate ($InsP_3$) receptors ($InsP_3Rs$) and the ryanodine receptors (RyRs). $InsP_3Rs$ link plasma membrane receptor activation with the release of calcium from intracellular stores (Berridge (1993); Patel et al. (1999)). In brief, G-protein-coupled receptors (GPCRs), or receptor protein kinases (RPKs), activate specific isoforms of phospholipase C (PLC). PLC converts membrane-bound phosphatidylinositol (4,5)-bisphosphate into $InsP_3$ and diacylglycerol (DAG). $InsP_3$ acts as a second messenger by binding to $InsP_3Rs$ at the endoplasmic reticulum (ER) and triggering the release of calcium from these internal stores. RyRs play an important role in "excitable" cells (i.e. neurons and muscle cells), where these "Ca^{2+}-induced calcium release" (CICR) channels augment the $[Ca^{2+}]_c$ signal triggered by membrane depolarization and concomitant activation of VOCs.

1.3. Ca^{2+} extrusion

The steep $[Ca^{2+}]$ gradient across the plasma membrane of all eukaryotic cells is maintained by ATP-dependent Ca^{2+} pumps in the plasma membrane. These pumps maintain a $\sim 2 \times 10^4$-fold lower $[Ca^{2+}]_c$ than $[Ca^{2+}]_o$ ($\sim$100nM vs. $\sim$2mM) and enable cells to use temporal and spatial changes in $[Ca^{2+}]_c$ for signal transduction. Under conditions of high Ca^{2+} load, the Ca^{2+} pumps are aided by Na^+/Ca^{2+} exchangers, which passively use the Na^+ gradient generated by Na^+ pumps to extrude Ca^{2+} from the cytoplasm.

1.4. Ca^{2+} sequestration

In addition to Ca^{2+} extrusion from the cytoplasm to the extracellular space, cells pump and sequester cytoplasmic Ca^{2+} into the intracellular stores of the ER or mitochondria. The "sarco/endoplasmic calcium ATPase" (SERCA) located on the ER is responsible for the refilling of the ER Ca^{2+} stores, which play a crucial role in G-protein/$InsP_3$-mediated $[Ca^{2+}]_c$ signaling. Mitochondria serve as a Ca^{2+} sink/buffer during times of high $[Ca^{2+}]_c$, and seem to play a role in the regulation of SOCs.

1.5. Generation of $[Ca^{2+}]_c$ signals

The interplay of the four aforementioned mechanisms enables cells to generate $[Ca^{2+}]_c$ signals, which may vary in magnitude, as well as in temporal and spatial properties. For purposes of this chapter, we shall distinguish between four types

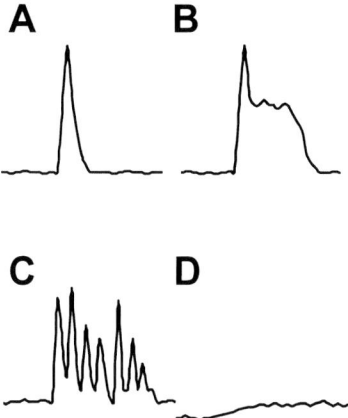

FIGURE. 2. Forms of $[Ca^{2+}]_c$ signals
A.) Transient $[Ca^{2+}]_c$ signal with a fast initial peak and a fast decay component, generated by the release of Ca^{2+} from internal stores and/or by the transient opening of plasma membrane channels. B.) Transient $[Ca^{2+}]_c$ signal composed of an initial peak and a prolonged plateau phase, which usually occurs as a result of an initial Ca^{2+} release followed by a Ca^{2+} influx via ROCs or SOCs. C) Oscillatory $[Ca^{2+}]_c$ signal generated by a complex interplay between Ca^{2+} release from internal stores and the opening of SOCs. D.) Slow-rising $[Ca^{2+}]_c$ signals, which most likely result from activation of a very low plasma membrane conductance, or modulation of Ca^{2+} extrusion mechanisms. All graphs display $[Ca^{2+}]_c$ vs. time A–C in seconds, D minutes.

of $[Ca^{2+}]_c$ signals (Fig. 2): 1.) A transient $[Ca^{2+}]_c$ signal with a fast initial peak and a fast decay component, which is generated by the release of Ca^{2+} from internal stores and/or by the transient opening of plasma membrane channels and a concomitant Ca^{2+} influx (Fig. 2A); 2.) A transient $[Ca^{2+}]_c$ signal comprising an initial peak and a prolonged plateau phase, which usually occurs as a result of an initial Ca^{2+} release followed by a Ca^{2+} influx via ROCs or SOCs (Fig. 2B). In the case of release/ROC, the cell expresses both metabotropic and ionotropic receptors for the same substance (e.g. ATP). In the case of the release/SOC model, the initial release of calcium from internal stores would lead to their depletion. By an as yet unknown mechanism, the depleted stores signal the SOCs to open, leading to Ca^{2+} influx underlying the observed plateau phase. This mechanism, which will aid in replenishing the depleted internal stores, was named "capacitative Ca^{2+} influx" and is a common feature of unexcitable cells (Putney (1986, 1990); Penner et al. (1993); Parekh and Penner (1997); Putney and McKay (1999)). The plateau phase can be of variable length, from a few seconds up to several minutes, and seems to play a crucial role in downstream events, such as modulation of substance release, cell motility and gene expression (Parekh and Penner (1997); Pettit and Fay (1998); Barritt (1999)). 3.) Oscillatory $[Ca^{2+}]_c$ signals, which are generated by a complex interplay between Ca^{2+} release from internal stores and the opening of SOCs (Fig. 2C). 4.) Very slow rising $[Ca^{2+}]_c$ signals, which most likely are due to activation of a very low plasma membrane conductance, or modulation of Ca^{2+} extrusion mechanisms (Fig. 2D).

2. Receptor-mediated $[Ca^{2+}]_c$ signals in microglia cells

2.1. P2 receptors

P2 receptors, formerly named purinoceptors, are activated by purines and pyrimidines (ATP, UTP, ADP, etc.) (Abbracchio and Burnstock (1998); Ralevic and Burnstock (1998)). Two distinct gene families encode P2 receptors, with P2X receptors being ligand-gated ionic channels, whereas P2Y receptors are metabotropic G-protein-coupled receptors. The distinct $P2X_{(1-7)}$ differ in their ion selectivity, gating properties and sensitivity to purino nucleotides. Five mammalian P2Y families ($P2Y_1$, $P2Y_2$, $P2Y_4$, $P2Y_6$ and $P2Y_{11}$) have been identified so far, showing different sensitivities to purino and pyrimidine nucleotides (for a review of P2 receptors, see Abbracchio and Burnstock (1998); Ralevic and Burnstock (1998)). In the CNS, purines and pyrimidines are known as neurotransmitters, but can also be released from injured cells.

P2 receptor-mediated $[Ca^{2+}]_c$ signals are the best so far investigated on microglia. Initial data reported the gating of a Ca^{2+}-permeable P2X receptor on cultured mouse microglial cells and showed an ATP-induced $[Ca^{2+}]_c$ signal by fluorescent Ca^{2+} imaging (Walz et al. (1993)). Interestingly, the $[Ca^{2+}]_c$ signal was sensitive to $[Ca^{2+}]_o$, indicating a purely ionotropic response. Nevertheless, later observations could show $[Ca^{2+}]_c$ signals persisting in Ca^{2+}-free buffer, or elicited by metabotropic P2 agonists in mouse, rat and human microglial cells (Ferrari et al. (1996); Toescu et al. (1998); McLarnon et al. (1999b); Möller et al. (2000b); Morigiwa et al. (2000); Wang et al. (2000)). Additionally, functional P2 receptors have also been shown in microglial cells *in situ* (Möller et al. (2000b)). Even though not necessarily dealing with $[Ca^{2+}]_c$ signaling, several reports have shown that P2 agonists regulate the induction of immediate early genes and transcription factors (Priller et al. (1995); Ferrari et al. (1996); Ferrari et al. (1997); Di Virgilio et al. (1999); Ferrari et al. (1999); Sanz and Di Virgilio (2000)). The role of $[Ca^{2+}]_c$ signals for these events is unclear, whereas P2 receptor-induced secretion of plasminogen and TNF-α has been shown to be Ca^{2+}-dependent (Inoue et al. (1998); Hide et al. (2000)). Although recent observations reported the regulation of P2 receptors by microglial activation with LPS or hypoxia (Möller et al. (2000b); Morigiwa et al. (2000)), the question which P2 receptors are actually expressed in microglial cells remains unanswered. Early identification of receptor subtypes was based solely on pharmacological tools. Unfortunately, these data are difficult to interpret, because the pharmacological profiles of the postulated receptors do not match the cloned P2 receptors (for review: Ralevic and Burnstock (1998)). Nevertheless, current data support the functional expression of (unidentified) members of the ionotropic P2X family (Walz et al. (1993); Ferrari et al. (1996); Nörenberg et al. (1997); McLarnon et al. (1999b); Visentin et al. (1999); Möller et al. (2000b)) as well as (unidentified) subtypes of metabotropic P2Y receptors (Ferrari et al. (1996); Toescu et al. (1998); McLarnon et al. (1999b); Möller et al. (2000b); Morigiwa et al. (2000)). Thus, molecular data are needed to end the ongoing discussion respecting which P2 receptor subtypes are actually expressed in microglial cells.

2.2. Complement receptors

The complement system is a key component in innate immunity and plays an important role in host defense (Morgan (2000)). Interestingly, complement expression and activation has also been shown in the CNS and is believed to play an important role in neurological diseases such as Alzheimer's disease (AD) or multiple sclerosis (MS) (Morgan and Gasque (1996); Morgan et al. (1997); Mukherjee and Pasinetti (2000); Thomas et al. (2000)). The complement anaphylatoxins C3a and C5a are two proinflammtory polypeptides released locally at sites of complement activation. The expression of receptors for C5a (C5aR) was first shown in human microglia (Lacy et al. (1995)). Later data showed the ability of C3a and C5a to induce $[Ca^{2+}]_c$ signals in cultured mouse microglial cells (Nolte et al. (1996); Möller et al. (1997a)). Remarkably, C5a was the first agonist shown to elicit $[Ca^{2+}]_c$ signals in mouse microglial cells *in situ* (Möller et al. (1997a)). Recent data showed the upregulation of microglial C3aR and C5aR in a model of focal cerebral ischemia, suggesting a role for these receptors outside neurodegenerative diseases (Van Beek et al. (2000)).

2.3. Chemokine receptors

Chemokines are chemoattractant cytokines that are intimately involved in inflammatory processes (Gale and McColl (1999)). In the CNS, these molecules are thought to play an important role in the pathogenesis of neuroinflammatory diseases that range from multiple sclerosis and stroke to HIV encephalitis (Sanders et al. (1998); Asensio and Campbell (1999)). Yet, chemokines and their receptors have a widespread distribution in the CNS, and recent data indicate that chemokines might play a signaling role in the developing and normal adult CNS (Asensio and Campbell (1999)). Chemokines and their G-protein coupled receptors are grouped into four subfamilies (α–δ) depending on certain cysteine motifs in their primary amino acid structure. Most of the chemokines bind to several receptors, and nearly all the receptors bind more than one chemokine. The details are highly complex and cannot be covered here. Several recent reviews have provided excellent overview of the area and should be consulted by the interested reader (Sanders et al. (1998); Asensio and Campbell (1999); Gale and McColl (1999); Murdoch and Finn (2000); Rossi and Zlotnik (2000); Zlotnik and Yoshie (2000)).

The interest in microglial chemokine receptors was sparked by the finding that some chemokine receptors act as coreceptors for HIV infection, and that microglial cells are the main CNS cell type that can be productively infected with HIV (Alkhatib et al. (1996); Doranz et al. (1996); Oberlin et al. (1996); He et al. (1997); Berger et al. (1999)). Subsequently, microglial cells were shown to express the chemokine receptors CCR1, CCR2, CCR3, CCR5, CXCR4 and CX_3CR1, but it should be noticed that some authors report divergent subtype expression (Albright et al. (1999); Boddeke et al. (1999b); Jiang (1998); Hegg et al. (2000)). Microglial $[Ca^{2+}]_c$ signals have been reported for many chemokines as they overlap in their receptor specificity. Chemokine-induced $[Ca^{2+}]_c$ signals where shown for RANTES, which generally acts on CCR1, CCR3, CCR4 and CCR5 (Albright et al. (1999); Boddeke et al. (1999b); Hegg et al. (2000); Sheng et al. (2000)), MIP1-α and MIP-1β acting on CCR1, CCR5 and CCR9 (Albright et al. (1999); Boddeke et al. (1999b)), MCP-1 known to activate CCR2 and CCR9 (Boddeke et al. (1999b)), eotaxin via CCR3, CCR9 and CXCR3 (but not seen by Albright et al. (1999); Hegg et al. (2000)), SDF-1α on CXCR4 (Tanabe et al. (1997); Albright et al. (1999)) and fractalkine acting via CX_3CR1 (Harrison et al. (1998); Boddeke et al. (1999a); Maciejewski-Lenoir et al. (1999)). Most of the reports used the chemokine-induced $[Ca^{2+}]_c$ signals only as a readout for functional receptors, not investigating the involved mechanism in detail. Nevertheless, it was shown that the RANTES and SDF-1α-induced $[Ca^{2+}]_c$ signals were pertussis-toxin-sensitive (Tanabe et al. (1997); Albright et al. (1999)). So far, the best-investigated responses are the $[Ca^{2+}]_c$ signals elicited in human microglia by eotaxin and RANTES (Hegg et al. (2000)). The RANTES-induced signal was dependent on $[Ca^{2+}]_o$ and blocked by La^{3+}, a broad-spectrum Ca^{2+} channel blocker. Eotaxin and RANTES-triggered $[Ca^{2+}]_c$ signals were sensitive to nimodipine, a voltage operated calcium channel blocker, indicating a strong influx component. Interestingly, the HIV protein Tat, which is released from infected cells, has also been shown to elicit $[Ca^{2+}]_c$ signals in cultured human microglial cells, most likely mediated by CCR3, whereas gp41 and gp120 were not able to induce $[Ca^{2+}]_c$ signals in the same preparations (Hegg et al. (2000); Sheng et al. (2000)).

2.4. Endothelin receptors

Endothelins are a family of three vaso-active peptides, mainly produced by endothelial cells (Masaki et al. (1994); Rubanyi and Polokoff (1994)). The expression of endothelins, as well as their two G-protein-coupled receptors, has been shown in the CNS, and endothelin might be released from injured endothelial cells or astrocytes during ischemia or trauma (Pollock (1995)). MacCumber (1990); Goldman (1991)). Initial data implicated the expression of the ET_B receptor subtype in microglial cells in an ischemic injury model (Yamashita et al. (1994)). More recent data showed the functional expression of ET_B and endothelin-induced $[Ca^{2+}]_c$ signals in cultured mouse and human microglial cells (Möller et al. (1997b); McLarnon et al. (1999a)). In mouse microglial cells, the $[Ca^{2+}]_c$ signals were mainly caused by Ca^{2+} released from internal stores (Möller et al. (1997b)), whereas in human microglia the $[Ca^{2+}]_c$ signals showed also a strong influx component, most likely mediated by SOCs (McLarnon et al. (1999a)).

2.5. Glutamate receptors

Glutamate is the most abundant excitatory neurotransmitter in the CNS. It exerts its action through several families of ionotropic and metabotropic recptors (Nakanishi et al. (1998); Ozawa et al. (1998)). Early $[Ca^{2+}]_c$ signaling experiments in microglial cells could not detect $[Ca^{2+}]_c$ signals elicited by glutamate, or the metabotropic glutamate receptor agonist 1S,3R-ACPD (*trans*-(1S,3R)-1-ammno-1,3-cyclopentadicarboxylic acid) (Möller, unpublished observations; Whittemore et al. (1993)). Nevertheless, recent data showed metabotropic glutamate receptor expression and 1S,3R-ACPD-induced $[Ca^{2+}]_c$ signals (Biber et al. (1999)), as well as glutamate-induced ion currents via the AMPA/kaninate subtype of glutamate receptors (GluR1–GluR4) (Noda et al. (2000)). Nevertheless, a significant Ca^{2+} influx via these channels seems unlikely because the data suggested the expression of the GluR2 subunit, which renders AMPA/kainate receptors Ca^{2+}-impermeable. This was confirmed by electrophysiological experiments, which showed only a very small current upon replacing high $[Na^+]_o$ by high $[Ca^{2+}]_o$ (Noda et al. (2000)).

2.6. Platelet-activating factor receptors

Platelet-activating factor (PAF, 1-*O*-alkyl-2-acetyl-*sn*-glycero-3-phosphocholine) is a lipid mediator originally characterized as a potent activator of platelets, and was later shown to have diverse biological effects on various cells and tissues, including CNS (Zimmerman et al. (1996); Prescott et al. (2000)). The first report of PAF inducing $[Ca^{2+}]_c$ signals in microglial cells appeared in 1995 (Righi et al. (1995)). This paper mainly characterized the PAF responses of a microglial cell line, but also showed some evidence of PAF-induced $[Ca^{2+}]_c$ signals in mouse primary microglial cells. Later data confirmed the expression of PAF receptors on rat microglia *in situ* and demonstrated PAF-induced Ca^{2+}-dependent arachidonic acid (AA) release from cultured microglia of the same species (Mori et al. (1996)). The PAF-induced $[Ca^{2+}]_c$ signals commonly consisted of an initial release from internal stores, followed by a long-lasting plateau. This plateau component was dependent on the $[Ca^{2+}]_o$ and was most likely due to the activation of SOCs. Interestingly, in human microglial cells, the peak of the $[Ca^{2+}]_c$ signal induced by application of PAF was greatly reduced (down to ~20%) in Ca^{2+}-free medium, indicating a very strong role of SOC-mediated Ca^{2+} influx (Wang et al. (1999)). It seems noteworthy that PAF is the agonist inducing the most prominent plateau phase of all reported agonists so far. This prolonged, elevated $[Ca^{2+}]_c$ has been shown to be crucial for the observed AA release (Mori et al. (1996)), but might also induce other changes in cell phenotype, as discussed above.

2.7. Thrombin receptors

Thrombin is a multifunctional serine-protease, best known for its important role in the blood coagulation cascade. In addition, thrombin was shown to induce a wide spectrum of responses, including $[Ca^{2+}]_c$ signals in leukocytes and CNS cells (Grand et al. (1996); Coughlin (2000)). Recent data showed that thrombin induced $[Ca^{2+}]_c$ signals in cultured rat microglial cells (Möller et al. (2000a)). The $[Ca^{2+}]_c$ signals were due to a release from internal stores, and most likely initiated by the activation of a protease-activated receptor (PAR). Interestingly, under experimental conditions that led to the activation of SOCs, the SOC-mediated Ca^{2+} influx was greatly reduced by the application of thrombin (Möller et al. (2000a)). This might be due to thrombin-induced depolarization, and concomitant reduction in the Ca^{2+} driving force, or to a direct interaction with the SOCs.

2.8. Cholinergic receptors

Acetylcholine (ACh) is a well established neurotransmitter in both the central and peripheral nervous systems of mammals, and elicits its effects via muscarinic (metabotropic) and nicotinic (ionotropic) ACh receptors. Carbachol (CCh), used as an agonist of metabotropic ACh receptors, was shown early on to induce $[Ca^{2+}]_c$ signals in cultured rat microglial cells (Whittemore et al. (1993)). The observed $[Ca^{2+}]_c$ signal was most likely owed to a release of Ca^{2+} from internal stores as it persisted in Ca^{2+}-free buffer. A similar conclusion was drawn from experiments carried out on human microglial cells, which reported $[Ca^{2+}]_c$ signals elicited by ACh and CCh (Zhang et al. (1998)).

2.9. Other receptors

Some agonist-induced $[Ca^{2+}]_c$ signals have been reported as single observations, or were used as controls for other experiments. For example, histamine-mediated $[Ca^{2+}]_c$ signals where used as a control in a paper originally reporting LPS-induced $[Ca^{2+}]_c$ signals (Bader et al. (1994)). Nevertheless, $[Ca^{2+}]_c$ signals where only found in a subpopulation of microglial cells, and no further characterization of the histamine-induced signals was made. Similarly, macrophage-colony stimulating factor (M-CSF), a known mitogen for microglial cells, has been shown to induce $[Ca^{2+}]_c$ oscillations (Ohsawa et al. (2000)) and an article describing carbachol-induced $[Ca^{2+}]_c$ signals in cultured rat microglia reported $[Ca^{2+}]_c$ signals elicited by norepinephrine without further investigation of the underlying mechanisms (Whittemore et al. (1993)). Interestingly, in this paper, the authors report the inability of ATP and 1*S*,3*R*-ACPD to induce $[Ca^{2+}]_c$ signals, a result in conflict with data obtained by others (Walz et al. (1993); Toescu et al. (1998); Biber et al. (1999); McLarnon et al. (1999b); Möller et al. (2000b)).

3. Calcium-Permeable Ion Channels on Microglial Cells

As with any other eukaryotic cell type, microglial cells express a wide variety of ion channels (C. Eder, Chapter 3, this book; Eder, 1998). These include Ca^{2+}-permeable voltage-operated channels (VOCs), receptor-operated channels (ROCs) and store-operated channels (SOCs). In contrast to the mechanisms relying on the release of calcium from internal stores, these plasma membrane channels give cells access to an "unlimited" source of Ca^{2+}.

3.1. Voltage-operated channels

The data on VOCs on microglia are very limited. A first report described a very small current developing after treating cultured rat microglial cells with Bay-K-8644, a VOC facilitator (Colton et al. (1994)). Nevertheless, others could not record any $[Ca^{2+}]_c$ increases using high $[K^+]_o$, which should have been sufficient to open these channels (Möller, unpublished observations; McLarnon et al. (1999b)). This might suggest that these channels may not play an important role under normal conditions, or that their expression might vary depending on culture conditions. Additional, but indirect, evidence was provided by the findings that both chemokines and β-amyloid (Aβ) (25–35) and prion protein PrP (106–126) triggered $[Ca^{2+}]_c$ signals. These signals were sensitive to the VOC blockers verapamil, nifedipine, diltiazem or nimodipine, suggesting the possible involment of VOCs (Silei et al. (1999); Hegg et al. (2000)). Nevertheless, the expression of VOCs in microglial cells has not been addressed directly and, since the VOC blockers used might interfere with other pathways, the nature and function of VOCs in microglial cells remains elusive.

TABLE 1. Summary of agonists inducing Ca^{2+} signals in microglial cells.

Agonist	Species	Comment	References
1S,3R-ACPD			Biber et al. (1999)
		No effect of 1S,3R-ACPD	Whittemore et al. (1993)
Carbachol	rat		Whittemore et al. (1993)
	human		Zhang et al. (1998)
Chromogranin A	rat		Taupenot et al. (1996); Ciesielski-Treska et al. (1998)
Complement fragments C3a/C5a	mouse	C5a	Nolte et al. (1996)
	mouse *in situ*	C3a and C5a	Möller et al. (1997a)
Endothelin	mouse		Möller et al. (1997b)
	human		McLarnon et al. (1999a)
Eotaxin	human		Hegg et al. (2000)
		No effect of eotaxin	Albright et al. (1999)
Fractalkine	rat		Harrison et al. (1998)
			Maciejewski-Lenoir et al. (1999)
			Boddeke et al. (1999a)
Histamine	Rat		Bader et al. (1994)
Lipopolysaccharide (LPS)	Rat		Bader et al. (1994)
Macophage-colony stimulating factor (M-CSF)	mouse (MG5)		Ohsawa et al. (2000)
MCP-1	rat		Boddeke et al. (1999b)
MIP 1-α/MIP-1β	human		Albright et al. (1999)
	rat		Boddeke et al. (1999b)
Norepinephrine	rat		Whittemore et al. (1993)
P2 agoinists	mouse (N9)		Ferrari et al. (1996)
	mouse		Toescu et al. (1998)
	mouse *in situ*		Möller et al. (2000b)
	rat		Morigiwa et al. (2000)
			Hide et al. (2000)
			Inoue et al. (1998)
	human		Wang et al. (2000)
			McLarnon et al. (1999b)
Platelet-activating factor (PAF)	mouse (BV-2)		Righi et al. (1995)
	rat		Mori et al. (1996)
	human		Wang et al. (1999)
RANTES	human		Albright et al. (1999)
			Hegg et al. (2000)
			Sheng et al. (2000)
	rat		Boddeke et al. (1999b)
SDF-1α	human		Albright et al. (1999)
	mouse (N9)		Tanabe et al. (1997)
Thrombin	mouse		Möller et al. (2000a)

3.2. Receptor-operated channels

The only Ca^{2+}-permeable ROCs known on microglia so far are the receptors of the P2X family, which were described earlier in this chapter. Interestingly, the only other described ROCs on microglia, the glutamate receptors of the AMPA/kainate type (GluR1–4, see 2.3), have a very low Ca^{2+} permeability, which seems to be undetectable under normal ionic conditions.

3.3. Store-operated channels

Ca^{2+} influx via SOCs plays an important role in all unexcitable cells (Putney (1986, 1990); Parekh and Penner (1997); Barritt (1999); Putney and McKay (1999)). This so-called "capacitative Ca^{2+} influx" was first postulated on purely theoretical considerations (Putney (1986)), but thereafter physical correlates of this mechanism have been described (Hoth (1992); Zitt (1996); Parekh (1997)). The first indirect evidence for functional SOCs in microglial cells was reported by $[Ca^{2+}]_c$-imaging experiments investigating complement fragments (Möller et al. (1997a)). The authors reported that the plateau phase of the complement fragment-induced $[Ca^{2+}]_c$ signal was dependent on $[Ca^{2+}]_o$. A similar plateau, induced by pharmacological depletion of the internal stores, showed the same dependency on $[Ca^{2+}]_o$, indicating the gating of SOCs. Subsequently, electrophysiological experiments demonstrated a current induced by depletion of the internal stores, which showed similar dependency on $[Ca^{2+}]_o$ (Nörenberg et al. (1997)). Additional Ca^{2+}-imaging experiments revealed that the SOC mechanism triggered by depletion of internal Ca^{2+} stores was operational for more than 20 minutes (Toescu et al. (1998)), and a recent report investigated the electrophysiological properties of the SOC mechanism in more detail (Hahn et al. (2000)). Interestingly, SOCs seem to play a prominent role in human microglia, where several reports showed the strong dependency of metabotropic agonist-induced $[Ca^{2+}]_c$ signals on external $[Ca^{2+}]_o$ (Zhang et al. (1998); McLarnon et al. (1999a); Wang et al. (1999); Goghari et al. (2000); Wang et al. (2000)). Nevertheless, no molecular data on the expression of specific SOCs in microglial cells are yet available. Because these channels might play an important role in substance release, cell motility and gene expression, more investigations are warranted.

4. Other $[Ca^{2+}]_c$ Signals in Microglial Cells

There is a variety of reports that either showed $[Ca^{2+}]_c$ signals induced by substances, where the corresponding receptor is not known, or where the application of a substance led only to a very slow rise in $[Ca^{2+}]_c$, thus suggesting a mechanism other than $InsP_3$ mediated release or fast-gated plasma membrane channels.

4.1. Amyloid β

The Amyloid β(Aβ) peptide is the major constituent of senile plaques, the pathologic hallmark of Alzheimer's disease (AD). Several receptors have been postulated to transduce the effects of Aβ, including the scavenger receptor (El Khoury et al. (1996); Paresce et al. (1996)), the receptor for advanced glycation end products (RAGE) (Yan et al. (1996)) and the serpin-enzyme complex receptor (Boland et al. (1996)). Nevertheless, not much is known about the ability of any of these receptors to induce $[Ca^{2+}]_c$ signals, and none of these receptors alone is able to account for all the effects seen with microglia after stimulation with Aβ. This suggests that Aβ may act on more than one receptor, rendering the various results reported on microglial $[Ca^{2+}]_c$ difficult to interpret. Aβ (25–35) was shown to increase microglial $[Ca^{2+}]_c$ within 5 minutes to 1 hour in a mainly non-reversible fashion (Korotzer et al. (1995)). The same paper showed an even slower (6 hour) increase in microglial $[Ca^{2+}]_c$ incubated with Aβ (1–42). A later paper reported the sensitivity of the Aβ-(25–35)-induced $[Ca^{2+}]_c$ rise to the VOC blockers verapamil, nifedipine and diltiazem (Silei et al. (1999)). It seems likely that the Aβ-induced changes in the microglial phenotype might lead to perturbations of Ca^{2+} extrusion mechanisms, or to an increased Ca^{2+} permeability of the plasma membrane. Interestingly, Aβ (1–40) and Aβ (1–42) are also capable of inserting into the cell membrane and forming Ca^{2+}-permeable channels (Rhee et al. (1998); Lin et al. (1999)).

4.2. Prion proteins

The main characteristic of prion diseases such as bovine spongiform encephalopathy (BSE) in cattle, scrapie in sheep, or Creutzfeldt-Jakob disease (CJD) in humans, is the accumulation of altered isoforms of prion protein (PrP). As with Aβ, the "receptor" for prion proteins (PrPs) remains unknown. A synthetic PrP peptide of the amino acids 106 to 126 induced a slow $[Ca^{2+}]_c$ signal, which was irreversible in about 40% of the tested microglial cells (Herms et al. (1997)). Later reports showed the time dependence of the PrP-(106–129)-induced $[Ca^{2+}]_c$ rise (Silei et al. (2000)), and indicated the involvement of VOCs (Silei et al. (1999)). Again, these signals are most likely due to perturbations of Ca^{2+} extrusion mechanisms, or to an increased Ca^{2+} permeability of the plasma membrane. In addition, as with Aβ fragments, PrP (106–126) was shown to form Ca^{2+}-permeable channels by itself (Lin et al. (1997)).

4.3. Others

Chromogranin A (CGA), a secretory protein released from neuroendocrine cells has been shown to elicit several responses in microglia, including $[Ca^{2+}]_c$ signals (Taupenot et al. (1996); Ciesielski-Treska et al. (1998)). Currently, no receptor is known for CGA, and it may be that CGA induced signals via an as yet unidentified chemokine receptor, as shown for the related Chromogranin C (Kong et al. (1998)).

Lipopolysaccharide (LPS), also known as endotoxin, is shed by gram-negative bacteria and is a potent activator of immune cells, including microglia (for review: Gehrmann et al. (1995); Kreutzberg (1996)). Even though there is ongoing discussion about the nature of the LPS receptor (Fenton and Golenbock, (1998); Beutler, (2000)), the effects of LPS on microglial cells are well described (for review: Gehrmann et al. (1995); Kreutzberg (1996)). Besides the established induction of cytokines, etc., one publication reported $[Ca^{2+}]_c$ signals elicited by LPS (Bader et al. (1994)).

Interleukin-1β (IL-1β) is a proinflammatory cytokine. It is well established that microglial cells, in addition to secreting IL-1β, also respond to this immunomodulator (Gehrmann et al. (1995)). A recent report showed the induction of a slow $[Ca^{2+}]_c$ signal after application of IL-1β (Goghari et al. (2000)). This signal was partially dependent on $[Ca^{2+}]_o$, indicating an influx component via an unknown mechanism, as well as a possible modulation of internal release or Ca^{2+} extrusion.

Ammonium (NH_3/NH_4^+), which can accumulate in the CNS after liver failure, was shown to induce $InsP_3$ $[Ca^{2+}]_c$ signals in cultured mouse microglial cells (Minelli et al. (2000)). Interestingly, the authors suggest direct interactions of NH_3/NH_4^+ with PLC. Phorbol myristate acetate (PMA) causes the translocation of protein kinase C (PKC) from the cytosol to the membrane, and has been widely used to investigate the effects of PKC activation. In cultured human microglial cells, PMA induced a slow $[Ca^{2+}]_c$ signal that depended completely on $[Ca^{2+}]_o$ and was irreversible within 25 minutes. It seems likely that PMA leads to PKC-dependent phosphorylation events, which ultimately lead to the opening of an as yet unknown plasma membrane conductance.

5. Ca^{2+} stores, Ca^{2+} buffers and Ca^{2+} binding proteins

In contrast to neurons and other glial cells such as astrocytes, not much is known about the exact nature of microglial Ca^{2+} stores and buffers. It seems safe to assume that, as in any other non-excitable cell, the Ca^{2+} stores of the ER and the mitochondria play a major role. We know from many reports that blockage of SERCA by the drugs thapsigargin or CPA impairs the $[Ca^{2+}]_c$ signaling of many agonists (e.g. Righi et al. (1995); Möller et al. (1997a); Toescu et al. (1998); Minelli et al. (2000); Wang et al. (2000)). It also appears that microglial Ca^{2+} stores are leaky. Comparatively short times in Ca^{2+}-free buffer (less than 10 minutes), or agonist-induced depletion of the Ca^{2+} stores in Ca^{2+}-free solution, lead to a very strong rebound Ca^{2+} entry after reintroduction of Ca^{2+} (Möller et al. (1997a); Toescu et al. (1998); McLarnon et al. (1999a); Wang et al. (1999)). This rebound Ca^{2+} entry is most likely due to an influx via SOCs, indicating the preceding depletion of internal stores. On the other hand, the only report interfering pharmacologically with the mitochondrial Ca^{2+} stores could not report any involvement of mitochondria in $[Ca^{2+}]_c$ signals under their experimental conditions (Toescu et al. (1998)). At this time, more data are needed to understand the intricacies of microglial Ca^{2+} stores.

Even less is known about the function of microglial Ca^{2+} binding proteins, which are responsible for the transduction of $[Ca^{2+}]_c$ signals to more downstream events such as gene expression or substance release. The upregulation of calmodulin has been reported in microglial cells in the hippocampus of kainic acid-treated mice (Sola et al. (1997)) and several reports suggest the involvement of the Ca^{2+} binding protein Iba1 in microglial activation (Imai et al. (1996); Ito et al. (1998); Ohsawa et al. (2000)). Here too, more research seems indicated to determine how microglial $[Ca^{2+}]_c$ signals specifically influence downstream regulation of microglial activities.

6. Challenges and Future Directions

Ca^{2+}-imaging has proven useful for the discovery of many receptors on microglial cells. One of the challenges of the work reviewed here seems to lie in the discrepancies reported. Why do some authors see ATP-induced signals, but others do not? What P2 receptors are actually expressed? Do VOCs play a role in $[Ca^{2+}]_c$ signaling? The answer to all these questions may lie buried in the material and methods of the papers. Some authors use cell lines, others primary cells, some use serum, others work serum-free and sometimes an additional control experiment might have helped to better refine the data. Nevertheless, a considerable body of data on microglial calcium already exists, and it will only grow in the future. Newly applied techniques, such as knock-out or transgenic animals, as well as multiphoton confocal microscopy, will help to investigate microglial behavior in more intact tissues and bring us closer to an understanding of the role of $[Ca^{2+}]_c$ signaling in microglial activation.

7. References

Abbracchio MP, Burnstock G (1998). Purinergic signalling: pathophysiological roles. *Jpn J Pharmacol* 78:113–145.

Albright AV, Shieh JT, Itoh T, Lee B, Pleasure D, O'Connor MJ, Doms RW, Gonzalez-Scarano F (1999). Microglia express CCR5, CXCR4, and CCR3, but of these, CCR5 is

the principal coreceptor for human immunodeficiency virus type 1 dementia isolates. *J Virol* 73:205–213.
Alkhatib G, Combadiere C, Broder CC, Feng Y, Kennedy PE, Murphy PM, Berger EA (1996). CC CKR5: a RANTES, MIP-1alpha, MIP-1beta receptor as a fusion cofactor for macrophage-tropic HIV-1. *Science* 272:1955–1958.
Asensio VC, Campbell IL (1999). Chemokines in the CNS: plurifunctional mediators in diverse states. *Trends Neurosci* 22:504–512.
Bader MF, Taupenot L, Ulrich G, Aunis D, Ciesielski-Treska J (1994). Bacterial endotoxin induces [Ca^{2+}]i transients and changes the organization of actin in microglia. *Glia* 11:336–344.
Barritt GJ (1999). Receptor-activated Ca^{2+} inflow in animal cells: a variety of pathways tailored to meet different intracellular Ca^{2+} signalling requirements. *Biochem J* 337:153–169.
Berger EA, Murphy PM, Farber JM (1999). Chemokine receptors as HIV-1 coreceptors: roles in viral entry, tropism, and disease. *Annu Rev Immunol* 17:657–700.
Berridge M, Lipp P, Bootman M (1999). Calcium signalling. *Curr Biol* 9:R157–159.
Berridge MJ (1993). Inositol trisphosphate and calcium signalling. *Nature* 361:315–325.
Berridge MJ (1997). Elementary and global aspects of calcium signalling. *J Physiol* (Lond) 499:291–306.
Beutler B (2000). Tlr4: central component of the sole mammalian LPS sensor. *Curr Opin Immunol* 12:20–26.
Biber K, Laurie DJ, Berthele A, Sommer B, Tolle TR, Gebicke-Harter PJ, van Calker D, Boddeke HW (1999). Expression and signaling of group I metabotropic glutamate receptors in astrocytes and microglia. *J Neurochem* 72:1671–1680.
Boddeke EW, Meigel I, Frentzel S, Biber K, Renn LQ, Gebicke-Harter P (1999a). Functional expression of the fractalkine (CX3C) receptor and its regulation by lipopolysaccharide in rat microglia. *Eur J Pharmacol* 374:309–313.
Boddeke EW, Meigel I, Frentzel S, Gourmala NG, Harrison JK, Buttini M, Spleiss O, Gebicke-Harter P (1999b). Cultured rat microglia express functional beta-chemokine receptors. *J Neuroimmunol* 98:176–184.
Boland K, Behrens M, Choi D, Manias K, Perlmutter DH (1996). The serpin-enzyme complex receptor recognizes soluble, nontoxic amyloid-beta peptide but not aggregated, cytotoxic amyloid-beta peptide. *J Biol Chem* 271:18032–18044.
Bootman MD, Berridge MJ (1995). The elemental principles of calcium signaling. *Cell* 83:675–678.
Ciesielski-Treska J, Ulrich G, Taupenot L, Chasserot-Golaz S, Corti A, Aunis D, Bader MF (1998). Chromogranin A induces a neurotoxic phenotype in brain microglial cells. *J Biol Chem* 273:14339–14346.
Clapham DE (1995). Calcium signaling. *Cell* 80:259–268.
Colton CA, Jia M, Li MX, Gilbert DL (1994). K+ modulation of microglial superoxide production: involvement of voltage-gated Ca2+ channels. *Am J Physiol* 266:C1650–1655.
Coughlin SR (2000). Thrombin signalling and protease-activated receptors. *Nature* 407:258–264.
Deitmer JW, Verkhratsky AJ, Lohr C (1998). Calcium signalling in glial cells. *Cell Calcium* 24:405–416.
Di Virgilio F, Sanz JM, Chiozzi P, Falzoni S (1999). The P2Z/P2X7 receptor of microglial cells: a novel immunomodulatory receptor. *Prog Brain Res* 120:355–368.

Doranz BJ, Rucker J, Yi Y, Smyth RJ, Samson M, Peiper SC, Parmentier M, Collman RG, Doms RW (1996). A dual-tropic primary HIV-1 isolate that uses fusin and the beta-chemokine receptors CKR-5, CKR-3, and CKR-2b as fusion cofactors. *Cell* 85:1149–1158.

Eder C (1998). Ion channels in microglia (brain macrophages). *Am J Physiol* 275:C327–342.

El Khoury J, Hickman SE, Thomas CA, Cao L, Silverstein SC, Loike JD (1996). Scavenger receptor-mediated adhesion of microglia to beta-amyloid fibrils [see comments]. *Nature* 382:716–719.

Fenton MJ, Golenbock DT (1998). LPS-binding proteins and receptors. *J Leukoc Biol* 64:25–32.

Ferrari D, Stroh C, Schulze-Osthoff K (1999). P2X7/P2Z purinoreceptor-mediated activation of transcription factor NFAT in microglial cells. *J Biol Chem* 274:13205–13210.

Ferrari D, Wesselborg S, Bauer MKA, Schulze-Osthoff K (1997). Extracellular ATP activates transcription factor NF-kappaB through the P2Z purinoreceptor by selectively targeting NF-kappaB p65. *J Cell Biol* 139:1635–1643.

Ferrari D, Villalba M, Chiozzi P, Falzoni S, Ricciardi-Castagnoli P, Di Virgilio F (1996). Mouse microglial cells express a plasma membrane pore gated by extracellular ATP. *J Immunol* 156:1531–1539.

Gale LM, McColl SR (1999). Chemokines: extracellular messengers for all occasions? *Bioessays* 21:17–28.

Gehrmann J, Matsumoto Y, Kreutzberg GW (1995). Microglia: intrinsic immuneffector cell of the brain. *Brain Res Brain Res Rev* 20:269–287.

Goghari V, Franciosi S, Kim SU, Lee YB, McLarnon JG (2000). Acute application of interleukin-1beta induces Ca^{2+} responses in human microglia. *Neurosci Lett* 281:83–86.

Grand RJ, Turnell AS, Grabham PW (1996). Cellular consequences of thrombin-receptor activation. *Biochem J* 313:353–368.

Hahn J, Jung W, Kim N, Uhm DY, Chung S (2000). Characterization and regulation of rat microglial Ca^{2+} release-activated Ca^{2+} (CRAC) channel by protein kinases. *Glia* 31:118–124.

Harrison JK, Jiang Y, Chen S, Xia Y, Maciejewski D, McNamara RK, Streit WJ, Salafranca MN, Adhikari S, Thompson DA, Botti P, Bacon KB, Feng L (1998). Role for neuronally derived fractalkine in mediating interactions between neurons and CX3CR1-expressing microglia. *Proc Natl Acad Sci U S A* 95:10896–10901.

He J, Chen Y, Farzan M, Choe H, Ohagen A, Gartner S, Busciglio J, Yang X, Hofmann W, Newman W, Mackay CR, Sodroski J, Gabuzda D (1997). CCR3 and CCR5 are co-receptors for HIV-1 infection of microglia. *Nature* 385:645–649.

Hegg CC, Hu S, Peterson PK, Thayer SA (2000). Beta-chemokines and human immunodeficiency virus type-1 proteins evoke intracellular calcium increases in human microglia. *Neuroscience* 98:191–199.

Herms JW, Madlung A, Brown DR, Kretzschmar HA (1997). Increase of intracellular free Ca2+ in microglia activated by prion protein fragment. *Glia* 21:253–257.

Hide I, Tanaka M, Inoue A, Nakajima K, Kohsaka S, Inoue K, Nakata Y (2000). Extracellular ATP triggers tumor necrosis factor-alpha release from rat microglia. *J Neurochem* 75:965–972.

Imai Y, Ibata I, Ito D, Ohsawa K, Kohsaka S (1996). A novel gene iba1 in the major histocompatibility complex class III region encoding an EF hand protein expressed in a monocytic lineage. *Biochem Biophys Res Commun* 224:855–862.

Inoue K, Nakajima K, Morimoto T, Kikuchi Y, Koizumi S, Illes P, Kohsaka S (1998). ATP stimulation of Ca2+ -dependent plasminogen release from cultured microglia. *Br J Pharmacol* 123:1304–1310.

Ito D, Imai Y, Ohsawa K, Nakajima K, Fukuuchi Y, Kohsaka S (1998). Microglia-specific localisation of a novel calcium binding protein, Iba1. *Brain Res Mol Brain Res* 57:1–9.

Kong C, Gill BM, Rahimpour R, Xu L, Feldman RD, Xiao Q, McDonald TJ, Taupenot L, Mahata SK, Singh B, O'Connor DT, Kelvin DJ (1998). Secretoneurin and chemoattractant receptor interactions. *J Neuroimmunol* 88:91–98.

Korotzer AR, Whittemore ER, Cotman CW (1995). Differential regulation by beta-amyloid peptides of intracellular free Ca^{2+} concentration in cultured rat microglia. *Eur J Pharmacol* 288:125–130.

Kostyuk P, Verkhratsky A (1994). Calcium stores in neurons and glia. *Neuroscience* 63:381–404.

Kreutzberg GW (1996). Microglia: a sensor for pathological events in the CNS. Trends *Neurosci* 19:312–318.

Lacy M, Jones J, Whittemore SR, Haviland DL, Wetsel RA, Barnum SR (1995). Expression of the receptors for the C5a anaphylatoxin, interleukin-8 and FMLP by human astrocytes and microglia. *J Neuroimmunol* 61:71–78.

Lin H, Zhu YJ, Lal R (1999). Amyloid beta protein (1–40) forms calcium-permeable, Zn^{2+}-sensitive channel in reconstituted lipid vesicles. *Biochemistry* 38:11189–11196.

Lin MC, Mirzabekov T, Kagan BL (1997). Channel formation by a neurotoxic prion protein fragment. *J Biol Chem* 272:44–47.

Maciejewski-Lenoir D, Chen S, Feng L, Maki R, Bacon KB (1999). Characterization of fractalkine in rat brain cells: migratory and activation signals for CX_3CR-1-expressing microglia. *J Immunol* 163:1628–1635.

Masaki T, Vane JR, Vanhoutte PM (1994). International Union of Pharmacology nomenclature of endothelin receptors. *Pharmacol Rev* 46:137–142.

McLarnon JG, Wang X, Bae JH, Kim SU (1999a). Endothelin-induced changes in intracellular calcium in human microglia. *Neurosci Lett* 263:9–12.

McLarnon JG, Zhang L, Goghari V, Lee YB, Walz W, Krieger C, Kim SU (1999b). Effects of ATP and elevated K^+ on K^+ currents and intracellular Ca^{2+} in human microglia. *Neuroscience* 91:343–352.

Minelli A, Lyons S, Nolte C, Verkhratsky A, Kettenmann H (2000). Ammonium triggers calcium elevation in cultured mouse microglial cells by initiating Ca^{2+} release from thapsigargin-sensitive intracellular stores. *Pflügers Arch* 439:370–377.

Möller T, Hanisch UK, Ransom BR (2000a). Thrombin-induced activation of cultured rodent microglia. *J Neurochem* 75:1539–1547.

Möller T, Kann O, Verkhratsky A, Kettenmann H (2000b). Activation of mouse microglial cells affects P2 receptor signaling. *Brain Res* 853:49–59.

Möller T, Nolte C, Burger R, Verkhratsky A, Kettenmann H (1997a). Mechanisms of C5a and C3a complement fragment-induced $[Ca^{2+}]_i$ signaling in mouse microglia. *J Neurosci* 17:615–624.

Möller T, Kann O, Prinz M, Kirchhoff F, Verkhratsky A, Kettenmann H (1997b). Endothelin-induced calcium signaling in cultured mouse microglial cells is mediated through ETB receptors. *Neuroreport* 8:2127–2131.

Morgan BP (2000). The complement system: an overview. *Methods Mol Biol* 150:1–13.

Morgan BP, Gasque P (1996). Expression of complement in the brain: role in health and disease. *Immunology Today* 17:461–466.

Morgan BP, Gasque P, Singhrao S, Piddlesden SJ (1997). The role of complement in disorders of the nervous system. *Immunopharmacology* 38:43–50.

Mori M, Aihara M, Kume K, Hamanoue M, Kohsaka S, Shimizu T (1996). Predominant expression of platelet-activating factor receptor in the rat brain microglia. *J Neurosci* 16:3590–3600.

Morigiwa K, Quan M, Murakami M, Yamashita M, Fukuda Y (2000). P2 Purinoceptor expression and functional changes of hypoxia-activated cultured rat retinal microglia. *Neurosci Lett* 282:153–156.

Mukherjee P, Pasinetti GM (2000). The role of complement anaphylatoxin C5a in neurodegeneration: implications in Alzheimer's disease. *J Neuroimmunol* 105:124–130.

Murdoch C, Finn A (2000). Chemokine receptors and their role in inflammation and infectious diseases. *Blood* 95:3032–3043.

Nakanishi S, Nakajima Y, Masu M, Ueda Y, Nakahara K, Watanabe D, Yamaguchi S, Kawabata S, Okada M (1998). Glutamate receptors: brain function and signal transduction. *Brain Res Rev* 26:230–235.

Noda M, Nakanishi H, Nabekura J, Akaike N (2000). AMPA-kainate subtypes of glutamate receptor in rat cerebral microglia. *J Neurosci* 20:251–258.

Nolte C, Möller T, Walter T, Kettenmann H (1996). Complement 5a controls motility of murine microglial cells in vitro via activation of an inhibitory G-protein and the rearrangement of the actin cytoskeleton. *Neuroscience* 73:1091–1107.

Nörenberg W, Cordes A, Blohbaum G, Frohlich R, Illes P (1997). Coexistence of purino- and pyrimidinoceptors on activated rat microglial cells. *Br J Pharmacol* 121:1087–1098.

Oberlin E, Amara A, Bachelerie F, Bessia C, Virelizier JL, Arenzana-Seisdedos F, Schwartz O, Heard JM, Clark-Lewis I, Legler DF, Loetscher M, Baggiolini M, Moser B (1996). The CXC chemokine SDF-1 is the ligand for LESTR/fusin and prevents infection by T-cell-line-adapted HIV-1 [published erratum appears in *Nature* 1996 Nov 21; 384(6606):288]. *Nature* 382:833–835.

Ohsawa K, Imai Y, Kanazawa H, Sasaki Y, Kohsaka S (2000). Involvement of iba1 in membrane ruffling and phagocytosis of macrophages/microglia. *J Cell Sci* 113:3073–3084.

Ozawa S, Kamiya H, Tsuzuki K (1998). Glutamate receptors in the mammalian central nervous system. *Prog Neurobiol* 54:581–618.

Parekh AB, Penner R (1997). Store depletion and calcium influx. *Physiol Rev* 77:901–930.

Paresce DM, Ghosh RN, Maxfield FR (1996). Microglial cells internalize aggregates of the Alzheimer's disease amyloid beta-protein via a scavenger receptor. *Neuron* 17:553–565.

Patel S, Joseph SK, Thomas AP (1999). Molecular properties of inositol 1,4,5-trisphosphate receptors. *Cell Calcium* 25:247–264.

Penner R, Fasolato C, Hoth M (1993). Calcium influx and its control by calcium release. *Curr Opin Neurobiol* 3:368–374.

Pettit EJ, Fay FS (1998). Cytosolic free calcium and the cytoskeleton in the control of leukocyte chemotaxis. *Physiol Rev* 78:949–967.

Prescott SM, Zimmerman GA, Stafforini DM, McIntyre TM (2000). Platelet-activating factor and related lipid mediators. *Annu Rev Biochem* 69:419–445.

Priller J, Haas CA, Reddington M, Kreutzberg GW (1995). Calcitonin gene-related peptide

and ATP induce immediate early gene expression in cultured rat microglial cells. *Glia* 15:447–457.
Putney JW, Jr. (1986). A model for receptor-regulated calcium entry. *Cell Calcium* 7:1–12.
Putney JW, Jr. (1990). Capacitative calcium entry revisited. *Cell Calcium* 11:611–624.
Putney JW, Jr., McKay RR (1999). Capacitative calcium entry channels. *Bioessays* 21:38–46.
Ralevic V, Burnstock G (1998). Receptors for purines and pyrimidines. *Pharmacol Rev* 50:413–492.
Rhee SK, Quist AP, Lal R (1998). Amyloid beta protein-(1-42) forms calcium-permeable, Zn^{2+}-sensitive channel. *J Biol Chem* 273:13379–13382.
Righi M, Letari O, Sacerdote P, Marangoni F, Miozzo A, Nicosia S (1995). *myc*-immortalized microglial cells express a functional platelet-activating factor receptor. *J Neurochem* 64:121–129.
Rossi D, Zlotnik A (2000). The biology of chemokines and their receptors. *Annu Rev Immunol* 18:217–242.
Rubanyi GM, Polokoff MA (1994). Endothelins: molecular biology, biochemistry, pharmacology, physiology, and pathophysiology. *Pharmacol Rev* 46:325–415.
Sanders VJ, Pittman CA, White MG, Wang G, Wiley CA, Achim CL (1998). Chemokines and receptors in HIV encephalitis. *Aids* 12:1021–1026.
Sanz JM, Di Virgilio F (2000). Kinetics and mechanism of ATP-dependent IL-1 beta release from microglial cells. *J Immunol* 164:4893–4898.
Sheng WS, Hu S, Hegg CC, Thayer SA, Peterson PK (2000). Activation of human microglial cells by HIV-1 gp41 and Tat proteins. *Clin Immunol* 96:243–251.
Silei V, Fabrizi C, Venturini G, Salmona M, Bugiani O, Tagliavini F, Lauro GM (1999). Activation of microglial cells by PrP and beta-amyloid fragments raises intracellular calcium through L-type voltage sensitive calcium channels. *Brain Res* 818:168–170.
Silei V, Fabrizi C, Venturini G, Tagliavini F, Salmona M, Bugiani O, Lauro GM (2000). Measurement of intracellular calcium levels by the fluorescent Ca^{2+} indicator Calcium-Green. *Brain Res Protoc* 5:132–134.
Sola C, Tusell JM, Serratosa J (1997). Calmodulin is expressed by reactive microglia in the hippocampus of kainic acid-treated mice. *Neuroscience* 81:699–705.
Tanabe S, Heesen M, Yoshizawa I, Berman MA, Luo Y, Bleul CC, Springer TA, Okuda K, Gerard N, Dorf ME (1997). Functional expression of the CXC-chemokine receptor-4/fusin on mouse microglial cells and astrocytes. *J Immunol* 159:905–911.
Taupenot L, Ciesielski-Treska J, Ulrich G, Chasserot-Golaz S, Aunis D, Bader MF (1996). Chromogranin A triggers a phenotypic transformation and the generation of nitric oxide in brain microglial cells. *Neuroscience* 72:377–389.
Thomas A, Gasque P, Vaudry D, Gonzalez B, Fontaine M (2000). Expression of a complete and functional complement system by human neuronal cells in vitro. *Int Immunol* 12:1015–1023.
Toescu EC, Möller T, Kettenmann H, Verkhratsky A (1998). Long-term activation of capacitative Ca^{2+} entry in mouse microglial cells. *Neuroscience* 86:925–935.
Van Beek J, Bernaudin M, Petit E, Gasque P, Nouvelot A, MacKenzie ET, Fontaine M (2000). Expression of receptors for complement anaphylatoxins C3a and C5a following permanent focal cerebral ischemia in the mouse. *Exp Neurol* 161:373–382.
Verkhratsky A, Orkand RK, Kettenmann H (1998). Glial calcium: homeostasis and signaling function. *Physiol Rev* 78:99–141.

Visentin S, Renzi M, Frank C, Greco A, Levi G (1999). Two different ionotropic receptors are activated by ATP in rat microglia. *J Physiol* (Lond) 519 Pt 3:723–736.

Walz W, Ilschner S, Ohlemeyer C, Banati R, Kettenmann H (1993). Extracellular ATP activates a cation conductance and a K+ conductance in cultured microglial cells from mouse brain. *J Neurosci* 13:4403–4411.

Wang X, Bae JH, Kim SU, McLarnon JG (1999). Platelet-activating factor induced Ca2+ signaling in human microglia. *Brain Res* 842:159–165.

Wang X, Kim SU, Van Breemen C, McLarnon JG (2000). Activation of purinergic P2X receptors inhibits P2Y-mediated Ca^{2+} influx in human microglia *Cell Calcium* 27:205–212.

Whittemore ER, Korotzer AR, Etebari A, Cotman CW (1993). Carbachol increases intracellular free calcium in cultured rat microglia. *Brain Res* 621:59–64.

Yamashita K, Niwa M, Kataoka Y, Shigematsu K, Himeno A, Tsutsumi K, Nakano-Nakashima M, Sakurai-Yamashita Y, Shibata S, Taniyama K (1994). Microglia with an endothelin ETB receptor aggregate in rat hippocampus CA1 subfields following transient forebrain ischemia. *J Neurochem* 63:1042–1051.

Yan SD, Chen X, Fu J, Chen M, Zhu H, Roher A, Slattery T, Zhao L, Nagashima M, Morser J, Migheli A, Nawroth P, Stern D, Schmidt AM (1996). RAGE and amyloid-beta peptide neurotoxicity in Alzheimer's disease. *Nature* 382:685–691.

Zhang L, McLarnon JG, Goghari V, Lee YB, Kim SU, Krieger C (1998). Cholinergic agonists increase intracellular Ca2+ in cultured human microglia. *Neurosci Lett* 255:33–36.

Zimmerman GA, Elstad MR, Lorant DE, McLntyre TM, Prescott SM, Topham MK, Weyrich AS, Whatley RE (1996). Platelet-activating factor (PAF): signalling and adhesion in cell-cell interactions. *Adv Exp Med Biol* 416:297–304.

Zlotnik A, Yoshie O (2000). Chemokines: a new classification system and their role in immunity. *Immunity* 12:121–127.

5

Microglia as a Source and Target of Cytokine Activities in the Brain

UWE-KARSTEN HANISCH

1. Introduction

Cytokines are regulatory proteins serving cellular communication. Released for auto- and paracrine short-distance signaling, membrane-associated for cell-cell interaction, or being distributed through body fluids to carry biological information to remote targets, these small proteins serve a variety of functions relating to the growth, survival, differentiation or the activities of virtually every cell type.

First reports describing cytokine actions can be traced to the 1950s. Since then, the number of structurally defined cytokines has exceeded one hundred. New entries are reported every year. Even though their molecular nature has remained enigmatic for a long time, cytokines, as well as their receptors and associated signaling components, are now irreversibly inserted into common concepts of immunology. Upon molecular characterization, it has also become apparent that the physiological roles of cytokines are not restricted to cells of the hematopoietic system and to mechanisms of innate and acquired immunity.

It has been a slow process to accept that certain cytokines and their receptors are present and functional in the nervous system. Developmental and constitutive functions of these immunoregulators in the physiology of the normal, immune previleged CNS are still not generally accepted. However, when CNS homeostasis is disturbed as a result of trauma, stroke, ischemia, infection or degenerative processes, certain cytokines can be detected in the affected region and in the cerebrospinal fluid. Increased CNS levels of cytokine proteins may result, in part, from blood-brain barrier disruption, or from local synthesis by invading immune cells. However, most, if not all, neuropathologies are associated with glial cell activation, and it is accepted that activated microglia and astrocytes can serve as endogenous sources of various cytokines. Activated microglia and astrocytes use cytokines to orchestrate cellular responses aimed at rapid re-establishment of tissue integrity and subsequent repair. Release products of activated glia are also thought to be crucial for initiating and guiding the infiltration of immune cells, and for coordinating their activities in the brain tissue. These glial responses may thus be considered beneficial mechanisms. However, when local production of cytokines exceeds the appropriate range, or escapes regulatory feedback, cytokine-mediated neurotoxicity may ensue.

Indeed, several cytokines have been shown to directly cause damage to neurons and oligodendrocytes *in vitro*. *In vivo,* increased CNS levels of cytokines would have several adverse consequences. First, they may directly induce cell and tissue damage. Second, they may induce the secretion of secondary factors with neurotoxic potential that impair the survival or the functions of neural cell populations. Third, some cytokines have apparently functional receptors on glial, endothelial, neuronal and neurosecretory cells that influence their physiological features, including neurotransmitter and neuropeptide release. Massively or chronically increased cytokine levels in the CNS could thus interfere with normal neurotransmission, or disturb the activity of neuroendocrine axes, which may result in physiological dysfunctions and detrimental consequences for the nervous tissue.

The dual involvement of microglia as a significant source and sensitive target within the cytokine network is the subject of this chapter. Emphasis will be placed on some general mechanisms and principles concerning microglial cytokines and responses to microglial cytokine receptor activation. The interested reader may consult cited literature for more detailed information. For additional surveys, a few reviews and books are recommended (refs. 1–3; Otero and Merrill (1994); Merrill and Benveniste (1996); Rothwell et al. (1996); Raivich et al. (1996, 1999a); Kreutzberg (1996); Stoll and Jander (1999); Streit et al. (1999); Streit (2000); Becher et al. (2000)).

2. Microglial cells as tissue macrophages

Microglial cells are commonly considered tissue macrophages of the CNS (Giulian (1995); Kreutzberg (1996); Stoll and Jander (1999)). Accordingly, microglial functions have been associated with innate defense mechanisms of the CNS and its ability to present antigen for specific immune reactions. The macrophage concept of microglia implies that these cells share many features with other tissue macrophages. Indeed, activated and fully reactive microglial cells exhibit typical macrophage-like characteristics, as seen in their extraneural counterparts. They can become highly motile, move along chemotactic gradients, synthesize and release an immense variety of soluble factors (including cytokines), reveal strong phagocytic activity and execute cytotoxic attacks. A series of cell surface antigens has been identified on both microglia and macrophages/monocytes, and only few markers may allow for a differential labeling. For developing macrophage-like characteristics, however, resting microglia need to be challenged to enter a stepwise transformation (Streit et al. (1999)). Thus, resident microglia in the normal brain represent cells with latent macrophagic potential. It is important to distinguish between resting and activated forms when considering cytokine activities.

2.1. Tissue-specific features affecting and controlling microglial properties

Morphologically, microglia of the normal brain differ from other macrophage populations by revealing a unique, ramified cell shape. This feature may stand for other, more subtle, differences in the cell biology of microglia and extraneural macrophages. Certain functional capacities may differ not only among the various peripheral macrophage populations (Wewers and Herzyk (1989)), but also especially between macrophages and microglia due to the special environmental conditions of the CNS (Smith et al. (1998); Ren et al. (1999)).

Microglia face a unique ensemble of neighboring cell types. Microglia are specifically adapted to this community and its intercellular signaling. Within this microenvironment, release and cellular activities of cytokines are likely subject to specific regulations. In addition, differences in the physiological features of microglial populations *per se* may exist. Microglial cells of gray *versus* white matter, or the various anatomical regions of the brain, may reveal certain specific features. These populational differences may extend to the capacity of producing and sensing cytokines (Ren et al. (1999)).

The CNS maintains its tissue homeostasis in relatively strict separation from the blood circulation. The formation of a blood-brain barrier throughout most of its vascular system guarantees a highly organized exchange of soluble molecules, such as nutrients and metabolites, between the CNS and extraneural compartments. Parenchymal microglia are embedded in this unique biochemical and ionic milieu. Microglia may normally not be confronted with factors in the serum, so that sudden exposure to certain serum components could initiate cellular reactions without additional signals. As a consequence, microglia might be mostly spared when cytokine levels in the serum rise. However, when confronted with those cytokines in their environment, or when challenged by factors normally absent in the CNS, microglia might respond in a somehow individual fashion, e.g. including their own cytokine release.

Mechanisms of CNS immune surveillance have attracted considerable interest. They are under investigation, but it is widely accepted that lymphocyte patrolling is a rare process in terms of actual cell numbers. Microglial cells of the normal brain may thus not have much contact with cellular components of the immune system and their release products.

The CNS has a very limited capacity for repair. Therefore, inflammatory processes and immune reactions are thought to be tightly controlled. The CNS cannot afford all the consequences of a defense battle, and extensive tissue swelling, as frequently associated with inflammation, is not tolerated well by the brain. The reluctance to generate inflammatory and immune reactions, requiring the collaboration of activated microglia, is probably an actively maintained process. Mechanisms of microglial activation may thus differ from those underlying activation of macrophages in non-CNS tissues.

2.2. Experimental limitations of microglial studies

Most of the experimental evidence for microglial cytokine production and cytokine effects on microglia are derived from studies in cell culture. Such studies, which often represent the most straightforward approach, have provided valuable insights into microglial secretory functions. However, studies on isolated cells may miss microglial features of cytokine production and recognition, since they may depend on the presence of other CNS cells, their soluble products, or other conditions provided only in the tissue context.

The relation between morphological appearance and functional features of microglial cells needs to be briefly described. The ramified shape of resting microglia in the normal brain along with a "downregulated immunophenotype," as indicated by low expression of several cell surface antigens, has been interpreted as a sign of a functionally quiescent status. However, morphological and functional properties may not strictly correlate. First, resting microglia do not necessarily represent a functionally inactive cell. Resting microglia are likely to sense not only activation signals derived from damaged cells, released upon tissue destruction or provided by foreign material, but may also constantly monitor the well-being of their environment through constitutive signaling. Little is also known about the actual contributions of resting microglia in terms of a constitutive release of factors with supportive activities, although the literature covers examples for microglial synthesis of trophic factors (e.g. Elkabes et al. (1996)). Second, a ramified (process bearing) phenotype can be induced experimentally in culture without consistent proof for a resting status, since other markers or cellular features can still indicate some activation. Third, the shape of microglia in homogenous cultures can differ considerably from that in the normal brain while the cells still do not produce detectable amounts of many cytokines until challenged by addition of an experimental stimulus. Finally, it has been shown in model stimulations of microglial cells *in vitro* that cytokine release can be selectively inhibited by manipulation of intracellular signaling pathways without affecting other activation-associated markers, such as changes in the electrophysiological properties (Prinz et al. (1999)). It thus remains unknown whether all possible functional changes of microglia are always triggered as a stereotypical response to any stimulation, or whether these cells may adapt their responses *in vivo* to the type and the context of a given signal. Certain functions might be induced independently of each other. Microglial activation will thus result in morphological changes, but cell shape may not strictly reflect the repertoire of ongoing functional reactions.

Studies on microglial cytokine production, release and activities have been based on cell cultures, tissue preparations and brain slice material. Primary cultures offer the advantage of easy experimental manipulation. Microglial features and responses can be studied over hours and days. Cytokine release can be monitored for periods before and after stimulation. Still, these cells cannot be regarded as a suitable model for microglia within their normal environment. They are deprived of cellular contacts and lack factors produced by the other CNS cell types. Moreover, any distinct features of microglia from different regions of the brain would be lost or "blended" when cultures are routinely made from larger anatomical structures. Cultures may thus represent a pool of alerted microglia, cells at a certain degree of activation. Some of the limitations of isolated microglial cultures are avoided by co-culturing the cells with other glia or neurons.

Cell lines offer access to larger numbers of cells with less effort than primary cultivation. Cell lines are especially useful for biochemical studies that may depend on a certain amount of protein for further analyses. However, cell lines run the risk of deviated physiological properties. For example, the microglia-derived BV-2 cell line fails to produce the full spectrum of factors that can be induced in microglia (Stohwasser et al. (2000)).

Microglial properties can also be studied in acute brain slice preparations (Brockhaus et al. (1993); Boucsein et al. (2000)). This approach, which is suitable for electrophysiological recordings or imaging-based studies, has its own limitations. Acute tissue slices suffer from unavoidable tissue damage during sectioning. Fast physiological reactions, such as rapid phosphorylation events, are probably initiated before the cells are examined, and thus acute slice preparations are not ideal for studies on the basal and stimulus-evoked release of cytokines.

In the past, several techniques have been developed to study cellular properties in organotypic cultures. In these models, brain tissue is kept *in vitro* for many days or weeks. It was shown that microglia in these tissue cultures first undergo a transient activation, as revealed by morphological changes and enhanced expression of activation-related surface markers, but after a few days the cells revert to a ramified phenotype as expected in the normal brain. Experimental manipulation is then more likely to start from an *in vivo*-like resting state. Organotypic cultures are also potentially useful for investigating cytokine induction and release. For example, cultures of rodent retina have revealed a mild and transient cytokine release soon after preparation, a return to baseline values during maturation over days in culture, and a strong induction upon experimental stimulation (Mertsch et al. (2001)). However, cytokine release measured in these tissue preparation reflects the total release capacity of all potential cellular sources. Cellular identification of the major contributors then requires intracellular detection of the respective message or protein.

Major evidence for the activation of microglia and its involvement and role in pathological processes has been delivered by histological examination of human and animal brain sections. Cytokine proteins and mRNAs have been successfully localized in fixed brain sections by immunocytochemistry and *in situ* hybridization. Extracts of brain tissue allowed for the subsequent detection of the proteins and their messages on blotting membranes or in respective gels. ^{125}I- or biotin-labeled cytokines revealed the presence of cytokine binding sites in the CNS, assisted and complemented by various studies using antibodies against cytokine receptor molecules. Measurements of cytokines are also made on samples taken from the cerebrospinal fluid, or collected in microdialysis experiments. These studies are much closer to the *in vivo* situation, but the regional and cellular resolution varies, or remains poor. They are technically more difficult than culture-based work. Ultimately, it is the combination of *in vitro* and *in vivo* preparations that together generate the closest view on the actual behavior of microglia as sensors and producers of cytokines.

2.3. Signals for microglial activation

Resting microglia may be in an actively maintained standby mode from which the cells can readily transform into "activated" or "reactive" microglia, and into full-blown brain macrophages (Raivich et al. (1999a); Streit (2000)). The initial stimulus for such a transformation could involve the triggering of receptor signaling, or perhaps the disappearance of a constitutive signal. In the former scenario, foreign material or molecules normally not occurring in the intercellular space or increases in the tissue level of such compounds could serve as an "on signal." In the latter case, it might be the loss of a constitutive inhibitory interaction that allows activation to proceed.

Viral envelope structures and bacterial cell wall and surface components of either gram-negative or gram-positive origin are known stimulators of macrophage and microglia activation in systemic and CNS infection. Bacterial cell wall-derived factors, such as lipopolysaccharide (LPS), represent the standard agents for *in vitro* mimicking of microglial activation in the brain.

CNS trauma may result in the production of still unknown molecules that appear in unusual concentrations in the interstitium when neural cells are damaged. Some tangible candidates in this scenario are neurotransmitters or cotransmitters. ATP, a commonly co-released molecule in neurotransmission, and purinoreceptors could play a significant role in microglial physiology because they would allow not only for the monitoring of increased and excessive neuronal activity, but also for the detection of changing ATP levels related to neuronal disintegration. In this context, ions exceeding their normal extracellular CNS concentrations would have similar meaning, potassium being a prime candidate. Signals sent from endangered or dying neurons and causing microglial responses can be subtle and affect glia in the vicinity of neuronal somata while the primary insult is set farther away, or even outside of the brain (Raivich et al. (1996); Kreutzberg (1996); Graeber et al. (1998); Streit et al. (1999)). The facial nerve transection model represents one of the most elegant and powerful approaches for studying the various qualitative and quantitative aspects of neuron-microglia interactions during neuronal regeneration and degeneration. Because the blood-brain barrier is not compromised, this model has helped to distinguish between microglial and macrophagic features.

Proteins with disease-related production, such as amyloid β (Aβ), can reportedly stimulate microglia and induce cytokine release. In conjunction with additional stimuli, Aβ may be one of the crucial factors that concentrate and irritate these cells at sites of their accumulation (plaques). In turn, the activated microglia may produce factors (cytokines, such as IL-1) that synergistically drive additional, potentially neurotoxic, cascades that, in turn, stimulate more microglia.

Serum-derived proteins, including proteases known mainly for their involvement in blood coagulation, have been neglected for a long time in terms of direct and indirect effects on CNS cells and extracellular matrix composition. However, it now appears that circulating proteases can affect molecular substrates and cells of the CNS once they penetrate the CNS parenchyma proper, and after they overrun protective antiproteolytic mechanisms (Gingrich and Traynelis (2000); Coughlin (2000)). Thrombin is one example of these serum-derived enzymes (Möller et al. (2000)). In addition, complement factors may bind to microglia and cause activation.

Although these factors increase in concentration following CNS injury, there are other molecules that could generate microglia-activating signals by their disappearance. Fractalkine, as a cell-attached chemokine, is produced by neurons and the fractalkine receptor is present on microglia (Harrison et al. (1998)). Although soluble versions of fractalkine exist as well, even having microglia as a source (Zujovic et al. (2000)), this chemokine-receptor pair, and the functional effects that fractalkine and anti-fractalkine antibodies show on microglial behavior, raise the hypothesis that the chemokine may play a role in controlling microglial activation. If so, this would be an example of factors that maintain the microglial resting status.

3. Cytokines as signaling molecules

An immense body of knowledge has accumulated regarding the molecular biology, biochemistry and biological function of cytokines. Beginning with their discovery as factors that regulate the growth, differentiation and activities of leukocytes, or interfere with viral replication, their major physiological importance is most often seen in relation to innate defense mechanisms and immune responses. However, cytokines also function as signaling molecules in the CNS. In particular, disturbances of CNS homeostasis and integrity will cause cells in the brain to synthesize certain cytokines. Microglia are such a cell type, probably representing the most important endogenous source of cytokines. Microglia will also react to the presence of cytokines in their environment, and thus these cells are pivotal elements in mechanisms of CNS tissue defense and restoration during emergencies.

3.1. Nomenclature and principles of cytokine action

During early cytokine research, individual "factors" were discovered in culture supernatants of lymphocytes that showed various growth-regulating and function-modulating activities in leukocytes. These factors were individually named according to their first known source or reported physiological effect, whereas the terms "lymphokines" and "monokines" were introduced to group factors obtained from activated lymphocytes and monocytes. However, subsequent purification and biochemical identification often revealed that several biological activities of those factors could be assigned to a single molecule. Pleiotropy, the principle that one factor can have various cell types as target to elicit several cellular effects, emerged as a general feature of cytokines (Vilcek (1998)). The original subclassification into lympho- and monokines lost its descriptive value, since it soon became clear that different cell types had the capacity for synthesizing the same factor. Twenty-five years ago, the now widely used term *cyto*kine was coined to better reflect this feature.

Starting with interleukin-1 (IL-1) and IL-2 in the 1970s, a more systematic nomenclature has been applied to unify the ever-expanding cytokine family. As the latest member, IL-18 was added to the list of interleukins. However, interferons (IFN), colony-stimulating factors (CSF) and several other growth factors still keep their original names. More recently, the term "chemokine" was introduced for the rapidly growing number of chemoattractive cytokines, but there are still inconsistencies in the nomenclature. For example, IL-8 is a chemokine, and chemokines may have effects that are not restricted to chemoattraction.

Often the biological activity of a given cytokine is duplicated by another cytokine that may not necessarily be related in sequence. This redundancy may reflect a natural backup for pivotal functions. Occasionally, the surprisingly mild effect of a cytokine knock-out has pointed to the existence of another cytokine with similar activity (Hanisch (2001)). Cytokines act in parallel with other regulatory compounds, including other cytokines. Cytokines may thus act in synergy, or antagonize each other. Synergistic actions are seen when two cytokines induce the same, or a similar, executive output through independent mechanisms, since they are involving different receptors, signaling pathways and transcription factors. Synergy is less likely and obvious when the intracellular signaling pathways of two cytokines join (Vilcek (1998)). Cytokines can promote each other also by induction of receptors (Vilcek (1998)). Frequently, a cytokine can induce or upregulate the expression of receptors, or receptor subunits, to make a cell (more) responsive to the action of another cytokine. In addition, cytokines can enhance their own production, or cause the induction of other cytokines. Antagonistic cross-regulations exist as well, and may result from interference in intracellular signaling (receptor-proximal, transcriptional and post-transcriptional events), from inhibition of cytokine synthesis, or from receptor transmodulation.

3.2. Structural characteristics of cytokines and their receptors

Cytokines are small proteins or glycoproteins comprising single polypeptides, with a size of usually less than 30 kD, occasionally up to 40 kD. IL-12 is an exception because it occurs as a heterodimer of 35 and 40 kD subunits, with homodimeric $p40_2$ existing as well. The p40 unit may also associate with another recently identified protein to form a novel IL-12-related cytokine (Oppmann et al. (2000)). Certain cytokines may form multimeric aggregates. For example, tumor necrosis factor α (TNFα) can form a trimer that may be required for bioactivity (Zhang M. and Tracey (1998)). Cytokines are mostly secreted and travel to cells in the vicinity of the producer. Some versions of cytokines remain membrane-associated. The extracellular matrix may provide a reservoir for certain cytokines from which they can be released upon tissue injury.

Structurally, cytokines can be grouped on the basis of sequence homology or similarities in their folding topology (Vilcek (1998)). The family of IL-1 cytokines consists of the cytokines, IL-1α and IL-1β, the IL-1 receptor antagonist (IL-1ra) and IL-18 (which was temporarily named IL-1γ). All resemble each other and share certain characteristic sequence motifs. Recently, five new IL-1-like molecules were discovered, expanding the known structural diversity of the IL-1 (super)family. Certain cytokines, such as IL-2, IL-4, IL-5 and GM-CSF, show structural similarities in their folding and domain arrangements. Other families of cytokines include the interferons IFNα, IFNβ, IFNω and IFNτ, or the TNF-like cytokines TNFα, TNFβ, lymphotoxin(β), Fas ligand, CD40 ligand or TNF-related apoptosis-inducing ligand (TRAIL) (Vilcek (1998)). Chemokines, the rapidly expanding group of structurally related small (up to 10 kD) chemoattractive proteins, are further subclassified in the CXC (α), CC (β), C (γ) and CXXXC (δ) families, based on characteristic Cys-containing sequence motifs (Asensio and Campbell (1999)).

Groups of cytokines can be defined by structural similarities of their receptors, such as sequence similarities, especially within characteristic motifs and domains in extra- or intracellular portions, or the shared use of receptor subunits, in particular, signal-transmitting chains as part of cytokine receptor complexes. These complexes can consist of heteromeric assemblies of receptor polypeptides, some subunits strictly depending on induction, whereas others are found constitutively on cells (see the IL-2Rα, IL-2Rβ and IL-2Rγ).

Major receptor families are currently grouped as the class I cytokine (hematopoietin group) receptors, class II (interferon group) receptors, the IL-1-related receptors, TNF-related receptors, transforming growth factor (TGF)-related receptors and the seven-transmembrane domain, G-protein-coupled receptors for chemokines (Vilcek (1998)). Subgrouping is carried out in the bigger families of class I receptors and chemokine receptors. For example, receptors for IL-2, IL-4, IL-7, IL-9 and IL-15 share the IL-2Rγ chain (common γ, γ_c), which is part of specific receptor heterodimers or heterotrimers. Receptor complexes for IL-2 and IL-15 even share two subunits, IL-2Rβ and IL-2Rγ. Separate α units thereby represent the receptor element which confers ligand specificity. However, certain cells may express a distinct IL-15R (Hanisch (2001)).

Chemokines currently number more than 50 molecules that exert their cellular activities by binding to a large number of chemokine receptors. Unlike most cytokines, chemokines are less restricted to a specific receptor. Several chemokines are known to bind to different receptors, and several chemokine receptors are promiscuous, since they accept various chemokines (Asensio and Campbell (1999)).

3.3. Principles of cytokine induction

Signals that can trigger the production of cytokines are as diverse as antigens, inflammatory mediators, infectious agents, mechanical tissue damage, radiation or other physical stressors, chemical toxins, and cytokines themselves. Infectious diseases may cause induction of cytokines in cells that are able to recognize surface structures as non-self. Several bacterial cell wall components, most notably the lipopolysaccharide (LPS) of gram-negative strains, have been employed to stimulate cytokine production. Structural determinants on gram-positive bacteria are also powerful cytokine inducers, although their molecular nature is more complex (Draheim et al. (1999); Prinz et al. (1999)). Teichoic and lipoteichoic acid, as well as proteoglycans, are among those molecules. Viral envelope proteins are known to provide the signal for cytokine production. Some viruses can confuse the cytokine network of the host by encoding a cyto/-chemokine-like molecule, a cyto/chemokine receptor homologue, or a cytokine binding protein. This type of mimicry will allow a virus to dock on mammalian cells, or to escape from an immune attack. Inflammatory mediators, such as platelet-activating factor (PAF), lipids or complement factors are also strong cytokine-inducing compounds. There are many other classes of soluble factors that influence cytokine production and release, including prostaglandins, neurotransmitters, neuropeptides, classical hormones and neuroendocrine releasing factors. Most importantly, glucocorticoids are powerful regulators of the immune system, and a significant portion of their anti-inflammatory and immunoregulatory capacity relates to the inhibition of cytokine release (Besedovsky and Del Rey A. (1996)).

For the most part, cytokines are released immediately following synthesis. However, preformed and stored pools of cytokines have been described, the reservoir being related to either cytoplasmic granulae, the plasma membrane, extracellular matrix proteins, or complexes between the cytokine and specific binding proteins. Endogenous binding proteins for cytokines can be derived from the extracellular domains of cytokine receptors. Soluble cytokine receptors are known, for example, for IL-2 (sIL-2Rα) or TNFα (sTNF-R I and II). These soluble receptors and binding proteins as well as cytokine antagonists (IL-1ra) likely serve in the buffering and inhibition of cytokine activities. They may temporarily bind cytokines, and slow release from these complexes may prolong their actions. Cytokines are thought to function mainly in an autocrine and paracrine fashion, limiting their action to short distances. However, certain cytokines may also travel through the circulation to reach remote targets (Cartmell et al. (2000)).

3.4. Cytokine signaling

Receptor subunit sharing is one of the molecular principles by which several cytokines can develop overlapping cellular effects. Using the same signal-transducing receptor chain implies that several cytokines may use similar signaling pathways. Nonetheless, the intracellular mechanisms triggered by cytokine binding to the cell surface have remained enigmatic for two reasons. First, "classical" second messenger systems could not be shown to connect receptor activation to all the cytosolic and transcriptional consequences and cellular responses. Second, many cytokine receptors do not contain functional kinase domains in their cytoplasmic extension, although protein phosphorylations are a common cytosolic response to cytokine exposure.

The discovery of two new protein families finally provided a breakthrough. The Janus kinases (JAK) represent soluble (nonreceptor) proteins that are recruited to the cytoplasmic tail of many cytokine receptors. Their association with the ligand-stimulated receptors causes their own activation associated with (cross)phosphorylations of the kinases and the receptor. Subsequently, substrates of the JAK, signal transducer and activator of transcription (STAT) proteins can bind to specific sites in the receptor protein, become phosphorylated, dissolve from the complex and travel as homo- and heterodimers to the nucleus to act as transcription factors. The JAK-STAT principle has offered an explanation of how many cytokine receptors cause gene induction, and how cytokine specificity and cross-talk are organized at intracellular levels. IFN receptors are prototypic examples for the JAK-STAT signaling, but many other cytokines base their cytosolic consequences on the same system. Molecules of the JAK-STAT cascades and their functional induction have meanwhile also been shown in cells and tissues of the nervous system, including microglia.

Nevertheless, cytokine receptor signaling shows great variety, and is by no means restricted to the JAK-STAT principle. Receptors of the cytokines of the IL-1 family (IL-1α, IL-1β and IL-18) use similar, but distinct cascades of kinases, which have been intensely studied. Binding of IL-1 causes complex formation between the receptor subunits, adapters and kinases, such as MyD88, IL-1R-activating kinase (IRAK) and Toll-interacting protein (Tollip) (Dinarello (1997a); Adachi et al. (1998); Dinarello (1998a); Burns et al. (2000)). Subsequently, IRAK associates with TNF receptor-associated factor 6 (TRAF6), which phosphorylates NFκB-inducing kinase (NIK) and, in turn, results in activation of IκB kinase (IKK) that phosphorylates IκB. Degradation of IκB by an ubiquitin pathway releases NFκB to translocate to the nucleus, where it binds to κB binding sites. Alternatively, IL-1 signaling relies on additional protein kinase pathways that recruit additional transcription factors (Freshney et al. (1994)). IL-1 and IL-18 share some recruitment of cytosolic adapters and kinases, which translates into overlapping activation of downstream elements and transcription factors (Tsuji-Takayama et al. (1997); Matsumoto et al. (1997); Gillespie and Horwood (1998); Thomassen et al. (1998); Kojima et al. (1998); Adachi et al. (1998); Tsuji-Takayama et al. (1999); Kanakaraj et al. (1999); Dinarello (1999a); Dinarello (1999b)), but IL-18 also shows distinct biological activities and specificity for gene activation as a consequence of its specific receptor engagement (Dinarello (1997a)).

As another example, TNF-receptor signaling reveals a similarly complex pattern that relies on sequential activation of various cytosolic proteins (Zhang M. and Tracey (1998)). Through different domains of its receptors, TNFα activates different pathways with divergent consequences, some involving characteristic kinases and domain-interacting proteins, some turning into pathways as linked to other cytokine and non-cytokine receptors. Finally, chemokines markedly differ in features of their cytosolic effectors from other cytokines as their receptors belong to the G-protein-coupled, seven-transmembrane-domain type. Accordingly, changes in the intracellular calcium concentration ($[Ca^{2+}]_i$) are a major effector system for chemokine receptor activation. In contrast, cytokines rarely show signaling through fast transient increases in $[Ca^{2+}]_i$.

4. Microglia as a source and target of cytokine activities

Several cytokines and their receptors are found in the nervous system (Rothwell and Hopkins (1995); Hopkins and Rothwell (1995); Rothwell et al. (1996); Rothwell et al. (1997)). Among them are TNFα, the interferons, IL-1(α and β), IL-2, IL-6 and TGFβ (Hanisch and Quirion (1995); Hanisch (2001)). Other reports provide evidence for presence of IL-3, IL-4, IL-10, IL-12, IL-15 and IL-18, or growth factors, such as platelet-derived growth factor (PDGF), epidermal-growth factor (EGF), fibroblast-growth factor (FGF), insulin-like growth factors (IGF), nerve-growth factor (NGF), and brain-derived neurotrophic factor (BDNF) (Shadiack et al. (1993); Lapchak et al. (1993); Hopkins and Rothwell (1995); De Pablo and De la Rosa (1995); Shadiack et al. (1998); Dore et al. (2000)).

Increased cytokine expression and responsiveness occurs in the pathological CNS (Raivich et al. (1999a); Stoll et al. (2000); Streit et al. (1999); Hanisch (2001)). Regarding capacity for cytokine synthesis, astrocytes and microglia represent the most prominent cellular sources. They also represent major endogenous cytokine-responsive elements, which makes them important cellular mediators of indirect cytokine consequences that affect, disturb, or even damage the brain.

4.1. Constitutive versus inducible cytokine production

The idea that microglia represent brain-endogenous factories of cytokines is supported by the potentially macrophagic nature of these cells (Cavaillon (1994)). Monocytes/macrophages are known for their wide spectrum of release products, an immense cytokine production capacity, and rapid kinetics of synthesis onset and effective release.

Concerning the microglial potential for cytokine synthesis and release, the majority of *in vitro* studies shows more or less similar release activities in response to various stimuli. However, to draw conclusions about microglial behavior using findings obtained with peritoneal or alveolar macrophages is not recommended. Even cultures of microglia from the same species can display varying features depending on the ontogenetic stage of the tissue used for culture preparation. Certain properties are acquired during cellular maturation within the tissue and this developmental process may not continue after isolating the cells. *In vitro* experiments can easily miss the critical influence of the cellular environment found within the tissue. In this regard, the histological demonstration of cytokine mRNA and protein in identified microglia of brain-tissue specimens probably provides the most relevant proof.

Histological studies using immunocytochemistry and *in situ* hybridization suggest that the constitutive production of cytokines is very low. Even *in vitro*, cytokine protein synthesis and release depend on addition of macrophage/microglia-activating stimuli, although increased levels of mRNA (such as for IL-1) may be found in response to the non-physiological conditions of a culture system. Exceptions seem to exist with regard of certain trophic factors that can be localized to microglia in tissue sections (Elkabes et al. (1996)). Nevertheless, approaches based on brain sections can not reveal all dynamic aspects of cytokine release, induction and modulation, leaving the question unanswered whether the process of microglial activation is a stereotypical or adapted cellular response.

4.2. Microglial cytokines and cytokines acting on microglia

Microglial cells have the ability to produce a large number of cytokines and chemokines, as well as cytokine and chemokine receptors (Raivich et al.

TABLE 1: Cytokines and chemokines produced by microglial cells.

Abbreviation	Full name
IL-1α/IL-1β	interleukin-1α/-1β
IL-3	interleukin-3
IL-6	interleukin-6
IL-8	interleukin-8
IL-10	interleukin-10
IL-12	interleukin-12
IL-15	interleukin-15
IL-18	interleukin-18, also interferon-γ inducing factor (IGIF)
M-CSF	macrophage-colony stimulating factor
TGFβ	transforming growth factor
TNFα	tumor necrosis factor α
GROα	growth-regulated oncogene
MCP-1	monocyte chemoattractant protein-1
MIP-1α/MIP-1β	macrophage inflammatory protein-1α/-1β
MIP-2	macrophage inflammatory protein-2
RANTES	regulated on activation, normal T cell expressed and secreted

(1999a)) (Table 1). This makes them key regulators of cellular responses during challenges of CNS homeostasis. The rapid onset of cytokine synthesis and the effective release, as well as the amounts released, are impressive. Upon stimulation in cultures, increased cytokine protein (e.g. TNFα, IL-6) can be measured within 3 to 4 hours in the supernatants. This is preceded by induction or increased stability of respective mRNAs. Using prolonged stimulation with bacterial agents, cytokine release can be sustained over many hours, up to days. However, continuous presence of the stimulus is not always required for evoking a lasting cytokine output. In some cases, stimulation with cell wall components of gram-positive *Streptococcus pneumoniae* in the range of minutes to hours results in considerable amounts of TNFα, IL-6 or IL-12 as measured over 24 hours. This contrasts with LPS stimulation, which is less effective on transient exposure (unpublished). Most *in vitro* studies consider time periods of microglial cytokine release in the range of hours to days. It would be of considerable interest to determine the actual time periods for which microglia can tolerate a chronic challenge and sustain cytokine production without exhaustion or damage. Moreover, it would be interesting to determine how auto- and paracrine exposure would feedback on the pattern and quantity of release since self-regulatory mechanisms and influences from other cells may change and eventually terminate the release *in vivo*. Chronic stimulation of microglia may cause exhaustion of the response, and may become detrimental. Chronically elevated $[Ca^{2+}]_i$ levels are one conceivable mediator for altered enzyme activities, cytoskeletal arrangements and ion channel functions, and may drive the cells into apoptosis. Vicious cycles may exist in which microglia, astrocytes and peripheral immune cells fuel microglial cytokine release.

Several factors are released in large quantities, in terms of both effective concentrations that can trigger cellular responses, and the microglial production rate based on total microglial cell protein. Rates of ng per µg of total cell protein over 24 hours have been measured for IL-12, KC (GROα), MCP-1, MIP-1α, MIP-2 and RANTES (Table 1). These quantitative measurements show the great potential of activated microglia to send out cytokine and chemokine signals within CNS tissues. In the following sections, several specific cytokines are discussed as products and modulators of microglia.

Cytokines of the IL-1/IL-18 family

IL-1 has been frequently described as a crucial microglial effector cytokine. As an immunostimulatory/pro-inflammatory signal IL-1 has a strategic position in innate defense and immune responses (Dinarello (1997a); Dinarello (1998a); Dinarello (1998b)). Cells of lymphoid and myeloid lineage are main sources, IL-1α being the mostly cell-associated molecular variant, and IL-1β being the major soluble form. The numerous targets include T and B cells, monocytes and macrophages (Dinarello (1997a)). The function of natural IL-1 receptor antagonist (IL-1ra), the third family member, is unique among the cytokines and underscores the importance of controlling IL-1 activity.

IL-1 appears to participate in most pathological scenarios, including infection, ischemia, stroke, excitotoxicity, mechanic, injury, or scar formation (Rothwell (1997); Rothwell et al. (1997); Hays (1998); Stoll et al. (1998); Vitkovic et al. (2000); Rothwell and Luheshi (2000); Fassbender et al. (2000)). Even though the etiology of CNS disorders varies, IL-1 surfaces as a common link in several processes that lead to neuronal death (Rothwell et al. (1997); Loddick et al. (1998); Rothwell and Luheshi (2000)). Indeed, suppression of endogenous IL-1 has neuroprotective outcomes (Rothwell et al. (1997); Loddick et al. (1998)). Activated glial cells, namely microglia, and invading immune cells serve as major sources for IL-1 synthesis under pathological conditions (Woodroofe et al. (1991); Yao et al. (1992); Otero and Merrill (1994); Kreutzberg (1996); Raivich et al. (1999a)). The appearance of IL-1β following neurotrauma has been shown to be extremely fast, faster than expected for *de novo* synthesis through gene induction, and hence suggesting enzymatic conversion of its pro-form (Fassbender et al. (2000)).

Several cytokines and cytokine receptors are markedly elevated in Alzheimer's disease brain, but IL-1 in particular is increasingly being implicated as a pathogenetic factor (Akiyama et al. (2000); Cooper et al. (2000a); Cooper et al. (2000b)). This idea is based on numerous reports providing evidence of chronic inflammation and cytokine presence in AD brains, as well as on the prevalence of AD being markedly reduced in patients treated with anti-inflammatory drugs (McGeer and McGeer (1999); McGeer et al. (2000); McGeer and McGeer (2000)). Interestingly, IL-1 and IL-6 have been suggested as critical factors in AD etiology for quite some time (Vandenabeele and Fiers (1991)). Recent studies on IL-1 as a risk factor for AD suggest that IL-1-driven inflammatory cascades and neurotoxic loops could even precede AD pathology (Mrak and Griffin (2000); Nicoll et al. (2000); Schubert et al. (2000)).

Various IL-1-like contributions to the progression or sequelae of AD have been postulated (Blume and Vitek (1989); Cacabelos et al. (1994); Sheng et al. (1995); Das and Potter (1995); Grilli et al. (1996); Sheng et al. (1996); Rothwell et al. (1997); Hays (1998); Sheng et al. (1998)), but despite the clinical importance and potential therapeutic value of anti-inflammatory treatments, mechanisms and causal relations between inflammatory cytokines, glial activation, amyloid deposition, disturbed neurotransmission and neuronal death remain largely unresolved (Rogers et al. (1996)). Interestingly, Aβ would not only induce microglial activation directly, it may also act synergistically with proinflammatory cytokines to induce further microglial release of proinflammatory agents, including IL-1 itself. Aβ preparations were shown to markedly enhance IL-1β, IL-6, TNFα, MIP-1α, MIP-1β and IL-8 production when microglial and THP-1 cells were stimulated by LPS (Yates et al. (2000); Cooper et al. (2000b)). Vicious cycles could be generated on such a basis. Factors released from dying neurons, such as ATP, could further fuel cytokine release from already activated microglia (Ferrari et al. (1997); Sanz and Di Virgilio (2000); Hide et al. (2000)) while affecting the consequences of IL-1 and TNFα on other neural cells (Liu et al. (2000)).

IL-1 also serves some constitutive functions in the brain, apart from inflammation. IL-1 affects cell proliferation and differentiation during development (Giulian et al. (1988a)). It influences neuro-endocrine-immune circuits by enhancing the firing and release activity of hypothalamic neurons and pituitary cells (Berkenbosch et al. (1987); Berkenbosch et al. (1989); Besedovsky and del Rey A. (1996); McCann et al. (1997a); Tringali et al. (1998)). Fever induction, soporific effects, suppression of food intake and behavioral changes are thought to be part of an IL-1-regulated "sickness behavior" involving central IL-1 action and afferent pathways stimulated by peripheral IL-1. Circulating IL-1 may penetrate the blood-brain barrier at certain regions, assisted by a proposed transporter (Banks et al. (1991); Banks and Kastin (1997)). Some of these effects may be "normal" control mechanisms (Alheim and Bartfai (1998); Maier et al. (1998); Luheshi et al. (1999); Cartmell et al. (1999); Takahashi et al. (1999)). Indirect influences on regulatory nuclei are achieved by virtue of IL-1-inducible mediators, including IL-6 and releasing factors. IL-1 modulates neurotransmission and synaptic efficacy in neuronal populations (Dunn et al. (1999); Vitkovic et al. (2000)), especially of the hippocampus (Katsuki et al. (1990); Schneider et al. (1998); O'Connor and Coogan (1999); Luk et al. (1999)). Modulation of neuronal and glial properties might be the brain-intrinsic part of its pleiotropic activity spectrum. Those functions are still poorly understood, and results from studies on the IL-1/IL-1R system have been confusing (Hopkins and Rothwell (1995)), but their dysregulation would contribute to homeostatic disturbances involving pathologically increased IL-1 levels. "IL-1R accessory protein-like" molecule (IL1RAPL), another new structure relating to IL-1R related proteins (Carrie et al. (1999)), deserves special interest. Highly enriched in the hippocampal formation, mutations in its gene are found in X-linked mental retardation (Carrie et al. (1999)). This links a molecular defect in an IL-1R-like molecule directly to disturbances of higher CNS function.

IL-18, identified as IFNγ-inducing factor (IGIF), was recently shown to be IL-1-like by sequence, folding and certain activities on cells serving innate and adaptive immunity (Okamura et al. (1995); Bazan et al. (1996); Gillespie and Horwood (1998); Dinarello et al. (1998); Kohno and Kurimoto (1998); Dinarello (1999a); Dinarello (1999b); Akira (2000); McInnes et al. (2000)). Mainly produced by macrophages/monocytes, IL-18 affects T, B and NK cells. Synergistic IFNγ induction by IL-12 and IL-18 drives Th1 responses. IL-1 and IL-18 share recruitment of cytosolic adapters and kinases, which translates into overlapping activation of downstream elements and transcription factors (Tsuji-Takayama et al. (1997); Matsumoto et al. (1997); Gillespie and Horwood (1998); Thomassen et al. (1998); Kojima et al. (1998); Adachi et al. (1998); Tsuji-Takayama et al. (1999); Kanakaraj et al. (1999); Dinarello (1999a); Dinarello (1999b)). Our laboratory and others have demonstrated IL-18 and IL-18R signaling in rodent CNS cells and tissues (Conti et al. (1997); Culhane et al. (1998); Prinz and Hanisch (1999); Conti et al. (1999); Wheeler et al. (2000)). Astrocytes and microglia appear to be sources whereas localization to the hypothalamus, pituitary and adrenal cortex suggests neuroendocrine involvement. IL-18 transcripts in the cortex, striatum, hippocampus and cerebellum suggest additional physiological roles (Culhane et al. (1998)). The developmental regulation of IL-18 points to a participation in brain ontogeny (Prinz and Hanisch (1999)). Effects on immune cells and the strong induction of IFNγ and other factors (TNFα) suggest that IL-18 is critical for driving neuroimmune/inflammatory circuits. Supportive data relate to infection and multiple sclerosis (Wildbaum et al. (1998); Fassbender et al. (1999)). When overproduced in the CNS, IL-18 may also lead to cell death without primarily involving immune cells (Rothwell et al. (1997); Loddick et al. (1998)).

It has become evident that both IL-1 and IL-18 belong to an ancient superfamily with unmatched diversity. Emerging concepts suggest that IL-1-like signaling participates in cellular communication and innate defense mechanisms throughout the phylogenetic tree, with structural conservation among vertebrates, insects and even plants (Gay and Keith (1991); Yang et al. (1998); O'Neill and Greene (1998); Hoffmann et al. (1999); Yoshimura et al. (1999); O'Neill and Dinarello (2000)). The family of related IL-1/IL-18 structures has been expanded by five "family of IL-1" (FIL1) or "IL-1 homologue" (IL-1H) molecules, FIL1ε, FIL1δ/IL-1H3, FIL1ξ/IL-1H4, FIL1η/IL-1H2 and IL-1H1 (Smith et al. (2000); Kumar et al. (2000)). Interestingly, FIL1δ/FIL1ε mRNAs are found in the human brain.

IL-2/IL-15

Receptors for IL-2 and IL-15 are widely expressed throughout the brain. Messages for IL-2Rα, IL-2Rβ, IL-2Rγ and IL-15Rα subunits are found in cortical regions, the hippocampus, as well as the cerebellum at various ontogenetic time points (Hanisch et al. (1997a)). Indeed, multiple CNS effects of IL-2 are reported with various lines of biochemical support for the presence of the receptor complexes, as further illustrated below (Hanisch and Quirion (1995); Hanisch (2001)). Whether certain CNS effects *in vivo* are preferentially mediated by IL-2 or IL-15 is still unknown because little comparative information exists on IL-15. Microglial cells seem to express the four receptor molecules as well, and they respond to IL-2 and IL-15 with cytosolic events and increased survival in culture. In combination with LPS or IFNγ, IL-2 may enhance microglial NO production whereas IL-15 attenuates LPS-evoked NO release (Hanisch and Quirion 1995; Hanisch et al. 1997a). This illustrates that the two cytokines, even though they share IL-2Rβ and IL-2Rγ subunits, can have different effects on the same cell type. IL-15 is produced by microglia, as shown by mRNA and protein detection (Hanisch et al. (1997a); Prinz et al. (1998)). However, relatively little is known about the microglial IL-2/IL-15/receptor system *per se* as it participates in CNS pathology. The property of IL-2 to act as a T-cell growth factor, to be toxic at elevated CNS tissue levels and to have a variety of neuroregulatory and neuroendocrine consequences suggests that microglia as a potential producer of these cytokines could have multiple influences on the CNS (Hanisch et al. (1993); Hanisch et al. (1994); Hanisch et al. (1996a); Hanisch et al. (1996b); Hanisch et al. (1997b); Hanisch (2001)).

IL-4/IL-10/IL-13/TGFβ

The anti-inflammatory and neuroprotective effects of IL-4, IL-10, IL-13 and TGF-β are thought to be largely mediated by modulation of microglial IL-1 and TNFα production, and the attenuation of IL-1- and TNFα-inducible secondary release effects (Chao et al. (1993); Suzumura et al. (1994); Wei and Jonakait (1999); Sawada et al. (1999); Raivich et al. (1999a); Szczepanik et al. (2001)). IL-4 and IL-13 also interfere with IL-1 bioactivity by enhancing IL-1ra synthesis (Dinarello (1997c); Dinarello (1997d)).

IL-6

IL-6 can be produced by various CNS cell types, including microglia (Lee et al. (1993)). Microglial cells also express IL-6R. Like IL-1 and TNFα, IL-6 is considered a proinflammatory cytokine acting in the initiation and coordination of inflammatory responses and helping to limit the spread of infectious agents. IL-6 particularly helps to initiate and regulate acute phase responses, a complex of adjustments in metabolic and executive organ functions (liver, immune cells) and circulating serum components that together serve in host defense. Centrally mediated effects are fever induction, sleep, increased pain perception, reduced food intake and neuroendocrine activation of energy stores. IL-1 is also a signal for triggering these responses, but IL-6 may mediate many of the physiological IL-1 effects because it is induced by IL-1. IL-6 can have both pro- and anti-inflammatory outcomes (Campbell (1998); Raivich et al. (1999a)). The net effect is likely determined by the presence of other cytokines, but IL-6 production by microglia may account for increased IL-6 activity in various types of CNS insults, especially during early phases. Subsequently, IL-6 may act on astrocytes to involve these cells in the orchestration of tissue responses, including attempts at repair (Raivich et al. (1999a)). IL-6, together with M-CSF, has been discussed as neuronally derived mediators of microglial activation (Raivich et al. (1996); Streit et al. (2000)).

Interferons

IFNα, IFNβ, IFNω and IFNτ are grouped together as type I interferons because of similarities in structure and receptor binding, whereas IFNγ (type II interferon) is structurally distinct and has its own receptor (Hanisch (2001)). Activities of the IFNα/β family relate to the resistance of mammalian cells against viral infection, but additional functions for immune cells become apparent (Akbar et al. (2000)). IFNγ, the "immune interferon," is mostly triggered by antigenic stimulation and T-cell activation, $CD4^+$ and $CD8^+$ T-cells and natural killer (NK) cells being the main cellular sources. However, macrophages have been found to also synthesize IFNγ upon costimulation with IL-12 and IL-18. IFNγ is an inducer of major histocompatibility (MHC) class I and II antigens, cell adhesion molecules and cytokines, and thus supports antigen presentation and induction of humoral and cell-mediated immune responses, as well as interaction of lymphocytes with the vascular endothelium.

With regard to microglia, IFNγ causes induction and upregulation of cell surface molecules (MHC class I and II, intercellular adhesion molecule I, immune-accessory molecules B7, leukocyte-function-associated molecule 1, CD14, Fc receptors, complement receptors), immunoproteasomes, release of cytokines (TNFα, IL-1, IL-6), complement (C1q, C2, C3, C4), and NO and modulates the cytokine-inducing effects of various microglia-stimulating agents (Stohwasser et al. (2000); Hanisch (2001)). We recently studied IFNγ-mediated modulation of glial cytokine and chemokine release and found that the release of several chemokines, most notably KC (GROα) and MIP-2, was significantly reduced in microglial cells. In other cases, augmentation was observed. The effects of IFNγ in either direction had a clear dependence on the bacterial stimulus, pointing to different net outcomes for gram-negative *versus* gram-positive infections. This indicates that the cyto/chemokine production by microglia is under a much more complex control of IFNγ than previously thought, and, in fact, it appears to be as complex as similarly revealed by the IFNγ shaping of CNS immune invasion in experimental autoimmune encephalomyelitis (Tran et al. (2000)). In addition to its microglia-activating effects, IFNγ may also induce apoptosis through simultaneous upregulation of Fas and FasL (Badie et al. (2000)). However, in general, the effects of IFNγ on microglia resemble those for other macrophages and relate primarily to the development of cytotoxic, phagocytic and antigen-presenting features (Raivich et al. (1999a); Hanisch (2001)).

TNF and relatives

TNFα production in the CNS has been attributed to neurons, astrocytes and microglia (Cheng et al. (1994); Zhang and Tracey (1998)). Cultures of astrocytes and microglia can synthesize and release TNFα upon stimulation (Lee et al. (1993); Prinz et al. (1999)). In severe brain injury, activated microglia appear to be an early and prominent source of TNFα (Gabay et al. (1997)). Interestingly, microglial cells also produce soluble (s)TNF-R II. Depending on the type of stimulus, the induction of microglial TNFα release can be very effective. In models of microglial confrontation with gram-positive bacteria, short exposure periods can result in a significant and lasting release response. Moreover, the ratios of microglial TNFα and sTNF-R production can also vary with the type of stimulation. Following CNS injury, the rapid and early induction of microglial TNFα production may thus be critical for influencing subsequent events. Concomitant microglial release of sTNF-R would provide added control over TNFα-mediated consequences.

Increased levels of TNFα in the brain have been observed especially after traumatic brain injury and/or ischemia, in bacterial and viral infections, as well as in multiple sclerosis and AD. Potentially harmful outcomes of increased TNFα could relate to its ability to promote inflammation and edema, although TNFα administration in animals has been reported to cause only minor inflammatory changes unless paired with a bacterial stimulus (Angstwurm et al. (1998)). TNFα has also direct toxic effects on neural cells and effects on neuronal structures and myelin have been reported. High levels in the cerebrospinal fluid indicate a poor prognosis in meningitis. However, TNFα has also been shown to promote neural cell survival, to serve as a proliferation factor, and to be neuroprotective. Animals deficient in TNF-R molecules are much more susceptible to conditions causing neuronal damage. Thus, TNFα is a good example of how cytokines can be a double-edged sword

(Carlson et al. (1999)). Apparently, low levels of TNFα have neuroprotective and repair-supporting consequences, whereas high levels may cause harm.

TNFα and TNF-R comprise families of ligands and related receptors, including TNFα, lymphotoxins (LTs), Fas ligand (FasL), CD40 ligand (CD40L) and TRAIL as ligands and TNF-R I and II, LTR, Fas (CD95), CD40, TRAIL-R1 and TRAIL-R2 as well as p75NTR (NGF, neurotrophin receptors) as receptors (Ware et al. (1998); Ashkenazi and Dixit (1998); Ashkenazi and Dixit (1999)). Activation of these receptors by their respective ligands causes multiple and heterogeneous effects, including apoptosis, survival signals, and proliferation depending on the cell and the circumstances. These processes may affect microglia-mediated cell communication and toxicity in the diseased adult brain. Special attention has been paid to the CD40L-CD40 system in microglial activation as a result of Aβ stimulation (Tan et al. (1999)).

Chemokines

Chemokines are chemotactic cytokines acting through G-protein-coupled receptors, and this field is currently undergoing enormous expansion in terms of identified structures and functions (Luster (1998); Locati and Murphy (1999)). The expression of chemokines and chemokine receptors in CNS cells and tissues has been shown in relation to various CNS diseases and pathologies (Xia et al. (1998); Glabinski and Ransohoff (1999); Asensio and Campbell (1999); Xia and Hyman (1999); Zhang et al. (2000)). IL-8R, CXCR4, CCR3, CCR5 and CX_3CR1 are among those receptors reported to be relevant for microglia *in vitro* and *in vivo* (Glabinski and Ransohoff (1999)).

CCR and CXCR receptors also occur on other neural cell populations, including neurons (Glabinski and Ransohoff (1999); Asensio and Campbell (1999)). Similarly, chemokines of both the α and β subfamilies are located in neural cells and brain tissue under normal *in vitro* conditions and in sections of normal brain. Neurons, in particular, express an array of CCR and CXCR (Asensio and Campbell (1999)). Regional differences suggest specific involvement in the functions of individual neuronal populations, although the cytoarchitectural resolution and correlation with transmitter phenotypes and neuronal connectivity awaits investigation. Although these neuronal expression patterns suggest neuromodulatory chemokine effects, chemokine receptors on glial cells could be involved in migratory events upon pathological changes. CCR3, CCR5 and CX_3CR1 are observed on microglia of the normal CNS.

Microglia in culture can produce GROα (KC), MIP-1α, MIP-1β, MIP-2, MCP-1, RANTES, IP-10 and IL-8 in response to experimental stimulation with bacterial agents, Aβ peptides, or with cytokines, such as TNFα and IL-1. Our own findings indicate that certain cytokines can dramatically influence the pattern of evoked microglial chemokine release (unpublished). For some chemokines, little change was observed following addition of IFNγ, but in other cases chemokine release was either increased or suppressed.

The fractalkine-CX_3CR chemokine-chemokine receptor pair may have special implication for microglia. Fractalkine, which occurs in both soluble and membrane-bound forms, is predominantly found in neuronal cells of various brain regions, namely the cortex, hippocampus, caudate putamen or the thalamus, whereas its receptor, CX_3CR1 localizes mainly on microglia throughout the brain (Harrison et al. (1998); Nishiyori et al. (1998); Maciejewski-Lenoir et al. (1999)). Neuronal damage is reported to decrease neuronal fractalkine message, to change the molecular form, and to result in the accumulation of CX_3CR1-expressing microglial cells (Harrison et al. (1998)). TNFα and IL-1 can upregulate fractalkine in astrocytes, and fractalkine can induce microglial migration (Maciejewski-Lenoir et al. (1999)).

Astrocytes and hippocampal neurons can also express CX_3CR1, and receptor activation in neurons stimulates Akt-mediated survival signaling (Meucci et al. (2000)). On the other hand, microglia may themselves produce fractalkine (Zujovic et al. (2000)), allowing for autocrine and reciprocal signaling through this system. Fractalkine may support microglial survival (Boehme et al. (2000)) and is thought to control microglia tonically (Zujovic et al. (2000)). When added during microglial stimulation, fractalkine attenuated TNFα release, and its neutralization drastically augmented TNFα output. Thus, neuronal and microglial fractalkine forms may organize a constitutive "calming" effect on microglia. Disturbance of this signal may be sufficient to trigger or enhance microglial activation. It remains to be unraveled how soluble and membrane forms of neuronal and microglial fractalkine organize the survival, silencing and migratory effects on microglia *in vivo* and how recruitment of astrocytes by TNFα and IL-1 and their contribution to the pool of fractalkine would influence the picture. However, this chemokine may contribute to the understanding of conditions of microglial activation due to altered neuron-glia signaling.

Growth factors

Neurons may constitutively signal to microglial cells to inform them about their well-being. Endangered or injured neurons may require microglia for trophic, or other survival-supporting, activity. It has been shown that microglia produce neurotrophic factors of the nerve growth factor (NGF) family, such as NGF, BDNF, neurothrophin-3 (NT-3) and NT-4, both *in vitro* and *in vivo,* and microglial expression of neurotrophins in the normal brain reveals some regional specificity as well as intraregional heterogeneity (Elkabes et al. (1996)). Constitutive or inducible synthesis of NGF, BDNF, PDGF, EGF, FGF, IGF, NT-3 or NT-4 by microglia could offer support for various neuronal cell populations. The neurosupportive activity of microglia may be switched to neurotoxic activity when signals from neurons indicate irreversible cell damage. This idea is discussed in more detail elsewhere (Streit et al. (1999)).

4.3. Microglial cytokine effects on glial populations and the endothelium

Microglial cytokines can have profound effects on astrocytes and oligodendrocytes. Astroglial hypertrophy with enhanced expression of glial fibrillary acidic protein (GFAP) follows traumatic and other injuries of the CNS (Norton et al. (1992); Mucke and Eddleston (1993); Eddleston and Mucke (1993)). Astroglial scar formation is an important tissue reaction for wound closure and repair, to isolate necrotic areas, and to stabilize regions by filling space created through neural cell loss (Norton et al. (1992); Eddleston and Mucke (1993)). Astrocytes also contribute to protective mechanisms by secreting neurotrophic factors and by removing neurotoxins and excitatory neurotransmitters.

Various cytokines, including IFNγ, IL-1, IL-2, IL-6, TNFα, and M-CSF have been shown to cause, or to be associated with, astrogliosis (Giulian et al. (1988b); Selmaj et al. (1990); Selmaj et al. (1991a); Balasingam et al. (1994); Rostworowski et al. (1997)). Activated microglia/macrophages seem to be required for astrogliosis to occur (Balasingam et al. (1996)). It has been suggested that soluble factors produced by activated microglia are needed for the evolution of the astrocytic reaction. Cytokines, especially IL-1, are likely candidates and IL-1ra is, indeed, sufficient to prevent astroglial proliferation (Giulian et al. (1988b); Giulian et al. (1988c); Giulian et al. (1994a); Giulian et al. (1994b)). On the other hand, IL-10 can also effectively attenuate astrogliosis (Balasingam and Yong (1996)). This effect seems to be indirect and mediated by a downregulation of miroglial cytokines.

Oligodendrocytes are a sensitive target of cytokine activities, and this may play a significant role in demyelinating diseases, such as multiple sclerosis (Merrill and Benveniste (1996); Antel et al. (1996); Merrill and Scolding (1999); Martino et al. (2000)). TNFα can be toxic to oligodendrocytes and damage myelinated structures, or prevent remyelination (Selmaj et al. (1991b); Hartung et al. (1992); D'Souza et al. (1995); Renno et al. (1995); Madigan et al. (1996); Raine et al. (1998); Cammer and Zhang (1999); Gu et al. (1999); Kita et al. (2000)). Oligodendrocytes respond variably to IL-1, IL-2, IL-6 and IFNγ and may themselves express some cytokines under certain conditions (Hartung et al. (1992); Torres et al. (1995); Otero and Merrill (1997); Brogi et al. (1997); Blasi et al. (1999); Hanisch (2001)). Microglial cells thus affect the survival and the functional activities of other glial cell types in a rather complex fashion.

Microglial cytokines will also affect endothelial cells. Cytokines and chemokines may induce the release of endothelial mediators and stimulate the expression of adhesion molecules required for the extravasation of leukocytes. Blood-brain barrier

TABLE 2: Possible effects of microglia-derived cytokines on CNS cells and functions.

Domain of action	Effect
Survival, growth, differentiation of neural cells	Increase in proliferation and support of survival of neuronal and glial cells
Promotion of axonal outgrowth and stimulation of glial maturation
Induction of astrogliosis
Reduction in the number of reactive astrocytes and astrocytic hypertrophy
Suppression of neuronal and glial cell growth, toxicity and induction of morphological abnormalities
Neurotransmitter, neuromodulator and neuropeptide release	Suppression and enhancement of neurotransmitter release (acetylcholine, noradrenaline, serotonin, dopamine, GABA), NO, and neuropeptide release
Increase in the expression and activity of neurotransmitter-synthesizing enzymes and NO synthase
Induction and modulation of neuropeptide and releasing factor release (CRF, AVP, luteinizing hormone releasing hormone, growth hormone releasing hormone, somatostatin, oxytocin, Met-enkephalin, β-endorphin)
Modulation (enhancement or reduction) of the release of pituitary hormones (ACTH, prolactin, thyrotrophic hormone, follicle stimulating hormone, luteinizing hormone, gowth hormone
Bioelectric activity, body homeostasis and behavior	Induction, upregulation and downmodulation of cell surface antigens and ion currents in neural cells
Modulation of ion channel activities
Increase and decrease in neuronal discharge frequencies
Reduction of short-term potentiation, inhibition of and interference with LTP
Suppression of afferent sensory transmission
Changes in EEG
Induction of fever and regulation of body temperature
Effects on locomotor and exploratory activity
Sleep induction
Decrease in food intake
Impairment of memory functions and performance in spatial learning

functions are likely affected as well when microglia become focally activated. Several cytokines and chemokines with angiogenic and angiostatic properties are known to be produced by activated microglia.

4.4. Cytokines as neuroregulatory factors

Since cytokines can affect the brain in many ways, local microglial release of cytokines could have many consequences (see Table 2). Stereotactic cytokine injections into different brain structures clearly cause neurophysiological reactions and even behavioral changes (Vitkovic et al. (2000); Hanisch (2001)). However, cytokine-mediated regulation of rapid events, such as neurotransmission, appears unlikely, considering that cytokines exert their actions through gene induction (Hopkins and Rothwell (1995)). Nevertheless, there is an increasing list of reports that shows modulation of neurotransmitter release and neuronal activity via constitutive cytokine receptors and within minutes. The hippocampus and the hypothalamus are the major structures for those reported cytokine actions. Physiological and pharmacological observations are supported by data on mRNA, protein and binding-site localization for cytokines and their receptors.

Release of acetylcholine, noradrenaline, dopamine, serotonin, and GABA has been shown to change following cytokine administration to cells and to *ex vivo* tissue preparations. While basal neurotransmitter release usually remains stable, cytokines modulate transmitter release evoked by extracellular potassium, or by addition of veratridine. The effect of a given cytokine can depend on the administered dose. For example, IL-2 was shown to suppress the release of acetylcholine in hippocampal slices at high doses (nM), whereas low-dose treatment (<pM) enhanced the release (Hanisch et al. (1993); Seto et al. (1997)). IL-2's effect on acetylcholine release can be region-specific, since the frontal cortex was found to be sensitive for IL-2-mediated release inhibition, but parietal cortex and striatal preparations were not (Hanisch et al. (1993)). The distribution of IL-2-like and various IL-2R-related molecules matches the regional effect, although a more widespread localization of these molecular structures indicates other potential cytokine roles (Lapchak et al. (1991)). Chronically increased IL-2 levels in cytokine-infused rats showed not only some selective changes in muscarinic acetylcholine receptors, but also functional impairment in a spatial learning and memory task (Hanisch et al. (1997b)). IL-2-mediated influences or disturbances in hippocampal cholinergic transmission may thus surface as impaired performance in behavioral tasks, and IL-2 gene deletion has been shown to disrupt normal learning and memory (Petitto et al. 1999).

Other cytokines known to modulate neurotransmitter release include IL-1, TNFα, and the interferons, cytokines characteristically synthesized and/or sensed by microglia (Rothwell and Hopkins (1995); Hanisch (2001)). Tissue levels of neurotransmitters have also been affected following *in vivo* administration of these cytokines (Hanisch and Quirion (1995); Hanisch (2001)), and their release-modulating activities include neuropeptides and neuroendocrine releasing factors (see below), as well as NO (Rothwell and Hopkins (1995); McCann et al. (1997b); McCann et al. (1998)).

4.5. Cytokines as mediators of neural-immune-endocrine communication

Cytokines may serve as mediators in the communication between the immune, endocrine and nervous systems (McCann et al. (1994a); McCann et al. (1994b); Besedovsky and Del Rey A. (1996); Sternberg (1997); Licinio (1997); Besedovsky and Del Rey A. (2001)). Many of them are produced by activated microglia in increasing amounts following CNS injury.

Trauma or infection may induce increased production of cytokines in the periphery and in the CNS. IL-1, IL-2, IL-6, IFNs and TNFα have been reported to affect neuronal firing and neurosecretory activity in body homeostasis-regulating structures of the brain that express their receptors, including various hypothalamic nuclei, where fever induction, somnolence or reduced food intake can be triggered (Hopkins and Rothwell (1995); Luheshi and Rothwell (1996); Zhang and Tracey (1998)).

In turn, the CNS participates in immune regulations that also cover the production of cytokines in peripheral cells and tissues. One control mechanism by which the CNS can regulate cellular and humoral immune functions is based on neuroendocrine cascades (McCann et al. (1994a); McCann et al. (1994b); Besedovsky and Del Rey A. (1996)). Activation of the HPA axis will cause the release of glucocorticoids from the adrenal gland. These hormones are known for multiple actions on the immune system, and for their anti-inflammatory effects. Massive activation of the HPA axis, such as in stress, can rapidly and massively increase the circulating levels of glucocorticoids, and therefore also affect the functioning of the immune system.

Neuroendocrine activities are subject to many feedback mechanisms. Again, cytokines are among the soluble factors shown to provide direct and indirect influences (Besedovsky and Del Rey A. (1996)). Several have been shown experimentally to cause activation of the HPA axis at its various hierarchical levels (Hanisch and Quirion (1995); Besedovsky and Del Rey A. (1996); Raber et al. (1998); Smith et al. (1999); Hanisch (2001)). Candidates are IL-1, IL-2, IL-6, IL-10 and TNFα. Moreover, therapeutic use of certain cytokines in cancer patients has been frequently reported to have not only neurological and neuropsychiatric side effects, but also to result in massive neuroendocrine disturbances (Hanisch and Quirion (1995); Hanisch (2001)). Cytokines can directly act on the adrenal cortex. Cytokines have also powerful effects on the pituitary gland, stimulating the release of adrenocorticotrophic hormone (ACTH) which, in turn, stimulates the adrenal release of glucocorticoids. In addition, cytokines can indirectly trigger the release of ACTH by stimulating the release of ACTH-releasing factors, such as corticotropin releasing factor (CRF) and arginine vasopressin (AVP) from the hypothalamus and the amygdala. For some cytokines, the molecular and cellular mechanisms of how they stimulate neuroendocrine function have been partly unraveled with detail and the experimental findings have been integrated in attractive models (McCann et al. (1994b); McCann et al. (1997b)). Furthermore, functional implications of cytokine (IL-1, -2, -6, -8)-induced CRF can expand to other central mechanisms, including thermogenesis.

Together, these actions of cytokines may help to integrate neuroendocrine functions to produce metabolic conditions for an appropriate defense against infections or malignant challenge. Neuroendocrine feedbacks may also serve in the control of immune-cell proliferation and the termination of an immune response. Being mostly anti-inflammatory and immunosuppressive, glucocorticoids can spare activated T-cells and even stimulate activated B-cells. This discriminating behavior with inhibitory and permissive effects, the pronounced circadian rhythm in plasma levels, and a timed induction following immune system activation, could participate in the clonal selection of activated immune cells, prevention of immune hyperreactivity, the principle of sequential antigenic competition and the preferential support of Th1 *versus* Th2 types of immune reactions. Indeed, the circadian profile of hormonal secretion could underlie the experimental and clinical observation that the action of immunomodulatory agents is highly susceptible to the day time of administration. Similar to the HPA axis, cytokines seem to modulate (mostly suppress) the hypothalamic-pituitary-gonadal axis.

4.6. Relations between peripheral and central production, and CNS effects of cytokines

Whether peripherally produced cytokines act directly on CNS cells and central regulatory structures of the neuroendocrine axis remains to be verified. A major issue is how proteins of this size cross the blood-brain barrier (Hopkins and Rothwell (1995)). At certain sites, the blood-brain barrier is less tight, and those regions may represent "gates" of cytokine entry (Blatteis (2000)). Pathological conditions may cause a disruption of the blood-brain barrier integrity, permitting larger molecules to penetrate the CNS. Cytokines could enter the brain via cellular trafficking. When leukocytes extravasate, they produce their cytokines locally in the CNS. Beside these routes, certain cytokines apparently enter the brain parenchyma by a kind of transport mechanism (Banks and Kastin (1992); Banks et al. (1993); Gutierrez et al. (1993); Waguespack et al. (1994); Gutierrez et al. (1994); Banks et al. (1994); Banks et al. (1995); Pan et al. (1996); Blatteis (2000)).

Alternatively, some cytokines may circulate in the periphery while they become produced independently in the brain as well. During systemic infections, for example, bacteria-derived material with cytokine-inducing capacity may also reach central structures. A peripherally induced cytokine may induce its brain-endogenous production without itself being physically transported (Hopkins and Rothwell (1995). Other mediators may carry the respective signal to the brain, or may be generated at the blood-brain border (Blatteis (2000)). In addition, afferent fiber pathways may convey cytokine-inducing signals to (hypothalamic) brain nuclei.

4.7. Special considerations for the study of cytokines in cells and tissues of the CNS

A few critical aspects should be stressed that concern potential technical pitfalls and specific features of molecular detections and studies on the activities of cytokines in the CNS. Because very modest amounts of cytokines and their receptors can stimulate cellular responses, physical detection of these molecules is generally difficult. Cytokine effects on hippocampal acetylcholine and pituitary hormone release were observed at subpicomolar concentrations, making IL-2, for example, a most potent modulator of neurotransmission (Karanth and McCann (1991); Hanisch et al. (1993)).

Functional cytokine receptors, including their coupling to intracellular signaling cascades, can be demonstrated indirectly by measuring the effects of recombinant or purified cytokine proteins on specific cellular, neurophysiological or neuroendocrine responses. Specificity of the respective cytokine action can be verified by using (heat-) inactivated cytokine as a control, appropriate blocking antibodies against the cytokine or its receptor or, in some cases, by using cytokine antagonists. Demonstration of specific cytokine or antibody binding may serve as additional evidence for the presence of receptors in neural cells and tissues, but receptor protein densities are often too low to obtain a clear signal together with topographic resolution.

An unequivocal demonstration of endogenous cytokines in the CNS can be very difficult because of extremely low concentrations. Bioassays for cytokines are mostly sensitive enough to detect low level, but they leave the identity of the active molecule unconfirmed. Direct detection of cytokines from brain tissues can sometimes be achieved by immunoblot analyses of electrophoretically separated proteins, with or without prior immunoprecipitation. In addition, enzyme-linked immunosorbent or radioimmunoassays (ELISA, RIA) have been employed. However, these sensitive and specific protein detection methods can be compromised by the high lipid content of brain tissue. Biochemical protein detections may also raise the question of whether the source of production is in the CNS, or due to some penetration by peripheral cells or to vascular leakage. The latter conditions are particularly relevant for traumatic CNS injuries, which are accompanied by blood-brain barrier disruption and leukocyte infiltration.

Detection of mRNA offers high sensitivity, especially when using RT-PCR. However, amplification of mRNA from brain tissue runs the risk of being "contaminated" with RNA from blood-derived cells. This risk can be minimized by perfusing animals before resecting tissues.

Immunocytochemistry and *in situ* hybridization carried out on CNS tissue slices allow for anatomical localization, provided that they are sufficiently sensitive. Labeling at cellular resolution, ideally in combination with cell-specific markers, can then reveal the cellular identity of the intrinsic source. Under pathological conditions, especially when they are associated with an infiltration of the brain by leukocytes, a clear correlation of local cytokine synthesis with a cellular source can still be difficult. Double immunocytochemistry may offer a solution, but often remains ambiguous as far as differentiating between activated microglia and invading macrophages. Detecting cytokine-producing cells is also limited because synthesis of the protein does not produce cytosolic accumulation, as shown by studies on microglial cytokine release *in vitro*.

Cell cultures, by virtue of their reduced complexity, represent an easier approach than tissue analysis. *In vitro* studies can confirm the capacity of a cell type to produce a cytokine in response to a challenge, but they have limitations, as mentioned. Cytokine release studies *in vitro* have to consider whether the cultivation method employed allows for a proper release and whether the detection assay properly reveals the secreted cytokine. IL-1β and IL-18, for example, depend on post-translational proteolytic processing by interleukin-1 converting enzyme (ICE, also known as caspase-1).

Lastly, it should be emphasized that CNS-specific molecular forms of cytokines and their receptors may exist. Although in some cases molecular analyses of CNS samples have revealed virtual identity to extraneural counterparts (Petitto et al. (1998)), in other cases unique and brain-specific forms have been described. Thus, oligonucleotide and antibody probes designed for detection of "peripheral" cytokines and their receptors may not necessarily work in the CNS.

5. Microglial cytokines in CNS pathology

The integration of multiple extracellular signals for regulated responses may place microglia in a strategic position whereby they carry out protective, defensive and restoring mechanisms as part of the program designed for coping with CNS insults and malfunctioning. Microglial cytokines may help to initiate as well as shape an immune reaction within the CNS. Microglial chemokines may crucially determine the composition of leukocyte infiltrates. Overshooting responses of microglial cells, in terms of their cytokine release, could have destructive outcomes by driving the progression of a potentially harmful process. Poorly regulated cytokine production in microglia may represent an initial factor in the development of degenerative processes.

5.1. Cytokines in trauma, stroke, ischemia, infection and neurodegenerative processes

This section describes three aspects of what we know about microglial cytokines. First, there is microglial involvement in virtually all neuropathologies and, thus far, all reported conditions leading to microglial activation have resulted in a more or less well documented cytokine/chemokine release response. Second, information about microglial cytokine production in the diseased brain is still incomplete because it does not reveal all topographic or temporal patterns, or inventories of the induced spectrum. Third, while immunologists and cell biologists have unraveled countless activities of cytokines in peripheral cells and tissues, understanding of tissue-specific actions in the CNS is still fragmentary. Specifically, the integration of cytokines in central homeostatic functions, and their "daily duties" in the CNS, deserve investigation.

The numerous reports on microglial cytokines in pathology may lead to a unifying theme revealing that a few microglial products are invariant factors of CNS pathology. These factors may, indeed, function as crucial mediators for development of a final spectrum of CNS damage. Much progress has been made regarding some, such as IL-1 and TNFα (Rothwell (1997); Rothwell et al. (1997); Dinarello (1997b)). Of course, this does not mean that these microglial "key factors" are the only, or even the most relevant, effector molecules, nor does it imply that these factors are provided only by microglia. As stated, the microglial response to a cocktail of stimulating and modulating signals can be expected to shape a similarly complex, timely organized and adapted profile of cyto/chemokine release.

Rapid microglial activation in response to a challenge is certainly a beneficial event. Rapid production of cytokines should be seen as a phylogenetically confirmed mechanism. Only excessive or chronic release may have harmful outcomes.

Some of the consequences of uncontrolled microglial cytokine production could be instantaneous, some may develop with progression of pathological events, such as in AD, ischemia (Stoll et al. (1998); Dirnagl et al. (1999)), or HIV encephalopathy (Dickson et al. (1993); Gendelman and Tardieu (1994); Seilhean et al. (1997)). Others may turn on the CNS via long-loop feedbacks that may not be obvious (Besedovsky and Del Rey A. (2001)). Since the reciprocal communication between the nervous, immune and endocrine systems contributes to the maintenance of body homeostasis, any uncoupling or poor regulation of these interactions could be pathogenic. Chronically increased central IL-1, IL-2, IL-6 or TNFα could affect regulatory nuclei of the hypothalamus leading to major hormonal imbalances. A chronically activated HPA axis would not only drive the metabolism to a catabolic state, but could also mask signs of inflammation and lead to an impairment of immune functions.

5.2. Microglia as a target of pharmacological intervention

Considering microglial activation as a common feature in many neuropathologies, and keeping in mind that overshooting microglial activation can have neurotoxic outcomes, it is possible that manipulation of microglial functions could serve future clinical approaches. Although treatment of the primary events in neurodegenerative diseases would still be the preferred intervention, this may not always be possible. Brain lesions or spinal cord injury are sudden events that are apparently followed by secondary cascades of destruction. Invading macrophages and activation of intrinsic microglia may carry a significant portion of these cascades. Administration of anti-inflammatory drugs may offer help, and more specific tools could prove successful in the future. In CNS infections, the microglial activation is a protective response of the tissue, but can have devastating consequences if it becomes excessive or chronic. Blocking the microglial activation process at the cell surface receptor level could prevent recruitment of additional microglia populations, and perhaps prevent entry of pathogens into microglia. Neutralizing potentially harmful release products of microglia could limit undesired consequences for CNS cells and tissues. An alternative approach may focus on intracellular signaling cascades that are specific for activated microglia. A temporary inhibition of an enzyme or protein-protein interaction would probably allow for a rather selective effect on activated microglia, while respective functions in other cell types are less affected. Understanding the intracellular key pathways that drive microglial overactivation could open new routes for drug development. As an example, "microglia-silencing" compounds could be helpful in certain antibiotic treatments of CNS infection in order to prevent temporarily massive and potentially harmful microglia activation as a consequence of bacterial lysis.

Moreover, the constitutive microglial expression of certain cytokine/chemokine receptors may render these cells sensitive to certain immunotherapies that primarily aim at other cellular targets. Cytokines have already been introduced in clinical practice. For example, there are many reports regarding the use of cytokines in tumor therapies. The idea behind these applications is the assumption that immune-cell-mediated reactions to malignancies can be supported by introducing cytokines that drive lymphocyte proliferation and enhance their reactivity. Immunotherapeutic strategies may also aim at stem cell growth, or at immunosuppressive interference.

6. Conclusion

Microglia in the adult healthy brain represent cells with the potential for becoming macrophages. Even though the cell is normally not inactive, there is no evidence for the constitutive production of cytokine proteins in easily detectable quantities. Exceptions may exist, especially with regard to some growth factors. The capacity to respond to a stimulation with a rapid-onset release of numerous immuno- and neuromodulatory cytokines and chemokines in large quantities is then one of the macrophagic features of the activated microglia. *In vivo,* the activation of the microglial cytokine release program may follow a sequence and represent a graded response, both features probably being overlooked in studies on isolated cells. Microglial cytokines and chemokines will organize a variety of direct and indirect activities in neuronal, neurosecretory, glial and endothelial cells aiming at appropriate responses to tissue injury or infection, and establishing protective and restorative processes. In inflammatory processes, the communication between activated glial and infiltrating cells will largely be based on the reciprocal exchange of cytokines, since many cytokine receptors are expressed or upregulated in neural and invading cells. Much remains to be learned about the mechanisms that induce, maintain and eventually terminate the microglia-activation-associated cytokine release. The correlation of excessively or chronically activated microglia with the course and regional pattern of CNS cell and tissue damage in neuropathologies of varying cause suggests that microglial cells have a remarkable pathogenetic potential. It may often be difficult to distinguish disease-driving contributions of activated microglia from microglial activation as a consequence of more proximal and preceding events. Nevertheless, it transpires that various secondary outcomes of tissue trauma, infection or neurodegeneration are carried by a microglial component. In the worst-case scenario, microglia could aggravate cascades and cycles of destructive events by fueling effector functions that may be mediated by their cytokines. Therapeutic notions relying on microglial manipulation will depend on a refined understanding of the signals for microglial activation and the cellular mechanisms that organize the various responses of these cells. Selective interference with potentially harmful actions, while sparing the more beneficial functions, represents an ambitious yet promising goal for developing novel neuroprotective strategies based on a better understanding of microglial biology.

References

(1995) Immune responses in the nervous system. Oxford: BIOS Scientific Publishers.
(1996) Cytokines and the CNS. Boca Raton: CRC Press.
(2001) Psychoneuroimmunology. San Diego: Academic Press.
Adachi O, Kawai T, Takeda K, Matsumoto M , Tsutsui H, Sakagami M, Nakanishi K, Akira S (1998) Targeted disruption of the MyD88 gene results in loss of IL-1- and IL-18-mediated function. Immunity 9: 143–150.
Akbar AN, Lord JM, Salmon M (2000) IFN-alpha and IFN-beta: a link between immune memory and chronic inflammation. Immunol Today 21: 337–342.
Akira S (2000) The role of IL-18 in innate immunity. Curr Opin Immunol 12: 59–63.
Akiyama H, Barger S, Barnum S, Bradt B, Bauer J, Cole GM, Cooper NR, Eikelenboom P, Emmerling M, Fiebich BL, Finch CE, Frautschy S, Griffin WS, Hampel H, Hull M, Landreth G, Lue L, Mrak R, Mackenzie IR, McGeer PL, O'Banion MK, Pachter J, Pasinetti G, Plata-Salaman C, Rogers J, Rydel R, Shen Y, Streit W, Strohmeyer R, Tooyoma I, Van Muiswinkel FL, Veerhuis R, Walker D, Webster S, Wegrzyniak B, Wenk G, Wyss-Coray T (2000) Inflammation and Alzheimer's disease. Neurobiol Aging 21: 383–421.
Alheim K, Bartfai T (1998) The interleukin-1 system: receptors, ligands, and ICE in the brain and their involvement in the fever response. Ann N Y Acad Sci 840: 51–58.

Aloisi F, Penna G, Cerase J, Menendez IB , Adorini L (1997) IL-12 production by central nervous system microglia is inhibited by astrocytes. J Immunol 159: 1604–1612.

Aloisi F, Serafini B, Adorini L (2000) Glia-T cell dialogue. J Neuroimmunol 107: 111–117.

Angstwurm K, Freyer D, Dirnagl U, Hanisch UK, Schumann RR, Einhaupl KM, Weber JR (1998) Tumour necrosis factor alpha induces only minor inflammatory changes in the central nervous system, but augments experimental meningitis. Neuroscience 86: 627–634.

Antel JP, Becher B, Owens T (1996) Immunotherapy for multiple sclerosis: from theory to practice. Nat Med 2: 1074–1075.

Araujo DM, Cotman CW (1993) Trophic effects of interleukin-4, -7 and -8 on hippocampal neuronal cultures: potential involvement of glial-derived factors. Brain Res 600: 49–55.

Araujo DM, Cotman CW (1995) Differential effects of interleukin-1 beta and interleukin-2 on glia and hippocampal neurons in culture. Int J Dev Neurosci 13: 201–212.

Asensio VC, Campbell IL (1999) Chemokines in the CNS: plurifunctional mediators in diverse states. Trends Neurosci 22: 504–512.

Ashkenazi A, Dixit VM (1998) Death receptors: signaling and modulation. Science 281: 1305–1308.

Ashkenazi A, Dixit VM (1999) Apoptosis control by death and decoy receptors. Curr Opin Cell Biol 11: 255–260.

Badie B, Schartner J, Vorpahl J, Preston K (2000) Interferon-gamma induces apoptosis and augments the expression of Fas and Fas ligand by microglia in vitro. Exp Neurol 162: 290–296.

Balasingam V, Dickson K, Brade A, Yong VW (1996) Astrocyte reactivity in neonatal mice: apparent dependence on the presence of reactive microglia/macrophages. Glia 18: 11–26.

Balasingam V, Tejada-Berges T, Wright E, Bouckova R, Yong VW (1994) Reactive astrogliosis in the neonatal mouse brain and its modulation by cytokines. J Neurosci 14: 846–856.

Balasingam V, Yong VW (1996) Attenuation of astroglial reactivity by interleukin-10. J Neurosci 16: 2945–2955.

Banks WA, Kastin AJ (1992) The interleukins-1 alpha, -1 beta, and -2 do not acutely disrupt the murine blood-brain barrier. Int J Immunopharmacol 14: 629–636.

Banks WA, Kastin AJ, Gutierrez EG (1993) Interleukin-1 alpha in blood has direct access to cortical brain cells. Neurosci Lett 163: 41–44.

Banks WA, Kastin AJ, Gutierrez EG (1994) Penetration of interleukin-6 across the murine blood-brain barrier. Neurosci Lett 179: 53–56.

Banks WA, Plotkin SR, Kastin AJ (1995) Permeability of the blood-brain barrier to soluble cytokine receptors. Neuroimmunomodulation 2: 161–165.

Banks WA, Kastin AJ (1997) Relative contributions of peripheral and central sources to levels of IL-1 alpha in the cerebral cortex of mice: assessment with species-specific enzyme immunoassays. J Neuroimmunol 79: 22–28.

Banks WA, Ortiz L, Plotkin SR, Kastin AJ (1991) Human interleukin (IL) 1 alpha, murine IL-1 alpha and murine IL-1 beta are transported from blood to brain in the mouse by a shared saturable mechanism. J Pharmacol Exp Ther 259: 988–996.

Bazan JF, Timans JC, Kastelein RA (1996) A newly defined interleukin-1? Nature 379: 591–591.

Becher B, Prat A, Antel JP (2000) Brain-immune connection: immuno-regulatory properties of CNS-resident cells. Glia 29: 293–304.

Berkenbosch F, de Goeij DE, del Rey AD, Besedovsky HO (1989) Neuroendocrine, sympathetic and metabolic responses induced by interleukin-1. Neuroendocrinology 50: 570–576.

Berkenbosch F, van Oers J, del Rey A, Tilders F, Besedovsky H (1987) Corticotropin-releasing factor-producing neurons in the rat activated by interleukin-1. Science 238: 524–526.

Besedovsky HO, del Rey A. (1996) Immune-neuro-endocrine interactions: facts and hypotheses. Endocr Rev 17: 64–102.

Besedovsky HO, del Rey A. (2001) Cytokines as mediators of central and peripheral immune-neuroendocrine interactions. In: Psychoneuroimmunology (Ader R, Felten DL, Cohen N, eds), pp 1–17. San Diego: Academic Press.

Blasi F, Riccio M, Brogi A, Strazza M, Taddei ML, Romagnoli S, Luddi A, D'Angelo R, Santi S, Costantino-Ceccarini E, Melli M (1999) Constitutive expression of interleukin-1beta (IL-1beta) in rat oligodendrocytes. Biol Chem 380: 259–264.

Blatteis CM (2000) The afferent signalling of fever. J Physiol 526 Pt 3: 470.

Blume AJ, Vitek MP (1989) Focusing on IL-1-promotion of beta-amyloid precursor protein synthesis as an early event in Alzheimer's disease. Neurobiol Aging 10: 412–414.

Boehme SA, Lio FM, Maciejewski-Lenoir D, Bacon KB, Conlon PJ (2000) The chemokine fractalkine inhibits Fas-mediated cell death of brain microglia. J Immunol 165: 397–403.

Boucsein C, Kettenmann H, Nolte C (2000) Electrophysiological properties of microglial cells in normal and pathologic rat brain slices. Eur J Neurosci 12: 2049–2058.

Brockhaus J, Ilschner S, Banati RB, Kettenmann H (1993) Membrane properties of ameboid microglial cells in the corpus callosum slice from early postnatal mice. J Neurosci 13: 4412–4421.

Brogi A, Strazza M, Melli M, Costantino-Ceccarini E (1997) Induction of intracellular ceramide by interleukin-1 beta in oligodendrocytes. J Cell Biochem 66: 532–541.

Burns K, Clatworthy J, Martin L, Martinon F, Plumpton C, Maschera B, Lewis A, Ray K, Tschopp J, Volpe F (2000) Tollip, a new component of the IL-1RI pathway, links IRAK to the IL-1 receptor. Nature Cell Biol 2: 346–351.

Cacabelos R, Alvarez XA, Fernandez-Novoa L, Franco A, Mangues R, Pellicer A, Nishimura T (1994) Brain interleukin-1 beta in Alzheimer's disease and vascular dementia. Methods Find Exp Clin Pharmacol 16: 141–151.

Cammer W, Zhang H (1999) Maturation of oligodendrocytes is more sensitive to TNF alpha than is survival of precursors and immature oligodendrocytes. J Neuroimmunol 97: 37–42.

Campbell IL (1998) Transgenic mice and cytokine actions in the brain: bridging the gap between structural and functional neuropathology. Brain Res Brain Res Rev 26: 327–336.

Campbell IL, Chiang CS (1995) Cytokine involvement in central nervous system disease. Implications from transgenic mice. Ann N Y Acad Sci 771: 301–312.

Campbell IL, Krucker T, Steffensen S, Akwa Y, Powell HC, Lane T, Carr DJ, Gold LH, Henriksen SJ, Siggins GR (1999) Structural and functional neuropathology in transgenic mice with CNS expression of IFN-alpha. Brain Res 835: 46–61.

Carlson NG, Wieggel WA, Chen J, Bacchi A , Rogers SW, Gahring LC (1999) Inflammatory cytokines IL-1 alpha, IL-1 beta, IL-6, and TNF-alpha impart neuroprotection to an excitotoxin through distinct pathways. J Immunol 163: 3963–3968.

Carr DJ, Campbell IL (1999) Transgenic expression of interleukin-6 in the central nervous system confers protection against acute herpes simplex virus type-1 infection. J Neurovirol 5: 449–457.

Carrie A, Jun L, Bienvenu T, Vinet MC, McDonell N, Couvert P, Zemni R, Cardona A, Van Buggenhout G, Frints S, Hamel B, Moraine C, Ropers HH, Strom.T., Howell GR, Whittaker A, Ross MT, Kahn A, Fryns JP, Beldjord C, Marynen P, Chelly J. (1999) A new member of the IL-1 receptor family highly expressed in hippocampus and involved in X-linked mental retardation. Nature Genetics 23: 25–31.

Cartmell T, Poole S, Turnbull AV, Rothwell NJ, Luheshi GN (2000) Circulating interleukin-6 mediates the febrile response to localised inflammation in rats. J Physiol 526 Pt 3: 653–661.

Cartmell T, Luheshi GN, Rothwell NJ (1999) Brain sites of action of endogenous interleukin-1 in the febrile response to localized inflammation in the rat. J Physiol Lond 518: 585–594.

Cavaillon JM (1994) Cytokines and macrophages. Biomed Pharmacother 48: 445–453.

Chao CC, Molitor TW, Hu S (1993) Neuroprotective role of IL-4 against activated microglia. J Immunol 151: 1473–1481.

Cheng B, Christakos S, Mattson MP (1994) Tumor necrosis factors protect neurons against metabolic-excitotoxic insults and promote maintenance of calcium homeostasis. NEURON 12: 139–153.

Conti B, Jahng JW, Tinti C, Son JH, Joh TH (1997) Induction of interferon-gamma inducing factor in the adrenal cortex. J Biol Chem 272: 2035–2037.

Conti B, Park LC, Calingasan NY, Kim Y, Kim H, Bae Y, Gibson GE, Joh TH (1999) Cultures of astrocytes and microglia express interleukin 18. Mol Brain Res 67: 46–52.

Cooper NR, Bradt BM, O'Barr S, Yu JX (2000b) Focal inflammation in the brain: role in Alzheimer's disease. Immunol Res 21: 159–165.

Cooper NR, Kalaria RN, McGeer PL, Rogers J (2000a) Key issues in Alzheimer's disease inflammation. Neurobiol Aging 21: 451–453.

Coughlin SR (2000) Thrombin signalling and protease-activated receptors. Nature 407: 258–264.

Culhane AC, Hall MD, Rothwell NJ, Luheshi GN (1998) Cloning of rat brain interleukin-18 cDNA. Mol Psychiatry 3: 362–366.

D'Souza S, Alinauskas K, McCrea E, Goodyer C, Antel JP (1995) Differential susceptibility of human CNS-derived cell populations to TNF-dependent and independent immune-mediated injury. J Neurosci 15: 7293–7300.

Das S, Potter H (1995) Expression of the Alzheimer amyloid-promoting factor antichymotrypsin is induced in human astrocytes by IL-1. Neuron 14: 447–456.

de Pablo F, de la Rosa EJ (1995) The developing CNS: a scenario for the action of proinsulin, insulin and insulin-like growth factors. Trends Neurosci 18: 143–150.

Dickson DW, Lee SC, Mattiace LA, Yen SH, Brosnan C (1993) Microglia and cytokines in neurological disease, with special reference to AIDS and Alzheimer's disease. Glia 7: 75–83.

Dinarello CA (1997c) Blocking interleukin-1 and tumor necrosis factor in disease. Eur Cytokine Netw 8: 294–296.
Dinarello CA (1997d) Induction of interleukin-1 and interleukin-1 receptor antagonist. Semin Oncol 24: S9-S9.
Dinarello CA (1997a) Interleukin-1. Cytokine Growth Factor Rev 8: 253–265.
Dinarello CA (1997b) Role of pro- and anti-inflammatory cytokines during inflammation: experimental and clinical findings. J Biol Regul Homeost Agents 11: 91–103.
Dinarello CA (1998b) Interleukin-1. In: The cytokine handbook (Thomson A, ed), pp 35–72. San Diego: Academic Press.
Dinarello CA (1998a) Interleukin-1, interleukin-1 receptors and interleukin-1 antagonist. Intern Rev Immunol 16: 457–499.
Dinarello CA (1999b) IL-18: A TH1-inducing, proinflammatory cytokine and a new member of the IL-1 family. J Allergy Clin Immunol 103: 11–24.
Dinarello CA (1999a) Interleukin-18. Methods 19: 121–132.
Dinarello CA, Novick D, Puren AJ, Fantuzzi G, Shapiro L, Muhl H, Yoon DY, Reznikov LL, Kim SH, Rubinstein M (1998) Overview of interleukin-18: more than an interferon-gamma inducing factor . J Leukoc Biol 63: 658–664.
Dirnagl U, Iadecola C, Moskowitz MA (1999) Pathobiology of ischaemic stroke: an integrated view. Trends Neurosci 22: 391–397.
Dore S, Kar S, Zheng WH, Quirion R (2000) Rediscovering good old friend IGF-I in the new millenium: possible usefulness in Alzheimer's disease and stroke. Pharm Acta Helv 74: 273–280.
Draheim HJ, Prinz M, Weber JR, Weiser T, Kettenmann H, Hanisch UK (1999) Induction of K^+ channels in mouse brain microglia: cells acquire responsiveness to pneumococcal cell wall components during late development. Neuroscience 89: 1379–1390.
Dunn AJ, Wang J, Ando T (1999) Effects of cytokines on cerebral neurotransmission. Comparison with the effects of stress. Adv Exp Med Biol 461: 117–127.
Durum S-K, Muegge K (1997) Contemporary immunology: cytokine knockouts. Totowa: Humunan Press .
Eddleston M, Mucke L (1993) Molecular profile of reactive astrocytes—implications for their role in neurologic disease. Neuroscience 54: 15–36.
Elkabes S, DiCicco BE, Black IB (1996) Brain microglia/macrophages express neurotrophins that selectively regulate microglial proliferation and function. J Neurosci 16: 2508–2521.
Fassbender K, Schneider S, Bertsch T, Schlueter D, Fatar M, Ragoschke A, Kuhl S, Kischka U, Hennerici M (2000) Temporal profile of release of interleukin-1beta in neurotrauma. Neurosci Lett 284: 135–138.
Fassbender K, Mielke O, Bertsch T, Muehlhauser F, Hennerici M, Kurimoto M, Rossol S (1999) Interferon-gamma-inducing factor (IL-18) and interferon-gamma in inflammatory CNS diseases. Neurology 53: 1104–1106.
Ferrari D, Chiozzi P, Falzoni S, Hanau S, Di Virgilio F (1997) Purinergic modulation of interleukin-1 beta release from microglial cells stimulated with bacterial endotoxin. J Exp Med 185: 579–582.
Freshney NW, Rawlinson L, Guesdon F, Jones E, Cowley S, Hsuan J, Saklatvala J (1994) Interleukin-1 activates a novel protein cascade that results in the phopshorylation of hsp27. Cell 78: 1039–1049.

Gabay C, Smith JMF, Eidlen D, Arend WP (1997) Interleukin 1 receptor antagonist (IL-1Ra) is an acute-phase protein. J Clin Invest 99: 2930–2940.

Gay NJ, Keith F (1991) Drosophila Toll and IL-1 receptor. Nature 351 : 355–356.

Gebicke-Haerter PJ, Appel K, Taylor GD, Schobert A, Rich IN, Northoff H, Berger M (1994) Rat microglial interleukin-3. J Neuroimmunol 50 : 203–214.

Gendelman HE, Tardieu M (1994) Macrophages/microglia and the pathophysiology of CNS injuries in AIDS. J Leukoc Biol 56: 387–388.

Gillespie MT, Horwood NJ (1998) Interleukin-18: Perspectives on the newest interleukin. Cytokine Growth Factor Rev 9: 109–116.

Gingrich MB, Traynelis SF (2000) Serine proteases and brain dam. Trends Neurosci 23: 399–407.

Giulian D (1993) Reactive glia as rivals in regulating neuronal survival. Glia 7: 102–110.

Giulian D (1995) Microglia and neuronal dysfunction. In: Neuroglia (Kettenmann H, Ransom BR, eds), pp 671–684. New York: Oxford University Press.

Giulian D, Li J, Leara B, Keenen C (1994b) Phagocytic microglia release cytokines and cytotoxins that regulate the survival of astrocytes and neurons in culture. Neurochem Int 25: 227–233.

Giulian D, Li J, Li X, George J, Rutecki PA (1994a) The impact of microglia-derived cytokines upon gliosis in the CNS. Dev Neurosci 16: 128–136.

Giulian D, Woodward J, Young DG, Krebs JF, Lachman LB (1988c) Interleukin-1 injected into mammalian brain stimulates astrogliosis and neovascularization. J Neurosci 8: 2485–2490.

Giulian D, Young DG, Woodward J, Brown DC, Lachman LB (1988b) Interleukin-1 is an astroglial growth factor in the developing brain. J Neurosci 8: 709–714.

Giulian D, Young DG, Woodward J, Brown DC, Lachman LB (1988a) Interleukin-1 is an astroglial growth factor in the developing brain. J Neurosci 8: 709–714.

Glabinski AR, Ransohoff RM (1999) Chemokines and chemokine receptors in CNS pathology. J Neurovirol 5: 3–12.

Gold LH, Heyser CJ, Roberts AJ, Henriksen SJ, Steffensen SC, Siggins GR , Bellinger FP, Chiang CS, Powell HC, Masliah E, Campbell IL (1996) Behavioral and neurophysiological effects of CNS expression of cytokines in transgenic mice. Adv Exp Med Biol 402: 199–205.

Graeber MB, Lopez-Redondo F, Ikoma E, Ishikawa M, Imai Y, Nakajima K, Kreutzberg GW, Kohsaka S (1998) The microglia/macrophage response in the neonatal rat facial nucleus following axotomy. Brain Res 813: 241–253.

Grilli M, Goffi F, Memo M, Spano P (1996) Interleukin-1beta and glutamate activate the NF-kappaB/Rel binding site from the regulatory region of the amyloid precursor protein gene in primary neuronal cultures. J Biol Chem 271: 15002–15007.

Gu C, Casaccia-Bonnefil P, Srinivasan A, Chao MV (1999) Oligodendrocyte apoptosis mediated by caspase activation. J Neurosci 19 : 3043–3049.

Gutierrez EG, Banks WA, Kastin AJ (1993) Murine tumor necrosis factor alpha is transported from blood to brain in the mouse. J Neuroimmunol 47: 169–176.

Gutierrez EG, Banks WA, Kastin AJ (1994) Blood-borne interleukin-1 receptor antagonist crosses the blood-brain barrier. J Neuroimmunol 55: 153–160.

Hanisch UK (2001) Effects of interleukin-2 and interferons on the nervous system. In: Psychoneuroimmunology (Ader R, Felten DL, Cohen N, eds), pp 585–631. San Diego: Academic Press.

Hanisch UK, Lyons SA, Prinz M, Nolte C, Weber JR, Kettenmann H, Kirchhoff F (1997a) Mouse brain microglia express interleukin-15 and its multimeric receptor complex functionally coupled to Janus kinase activity. J Biol Chem 272: 28853–28860.

Hanisch UK, Neuhaus J, Quirion R, Kettenmann H (1996b) Neurotoxicity induced by interleukin-2: involvement of infiltrating immune cells. Synapse 24: 104–114.

Hanisch UK, Neuhaus J, Rowe W, van Rossum D, Moller T, Kettenmann H, Quirion R (1997b) Neurotoxic consequences of central long-term administration of interleukin-2 in rats. Neuroscience 79: 799–818.

Hanisch UK, Quirion R (1995) Interleukin-2 as a neuroregulatory cytokine. Brain Res Brain Res Rev 21: 246–284.

Hanisch UK, Rowe W, Sharma S, Meaney MJ, Quirion R (1994) Hypothalamic-pituitary-adrenal activity during chronic central administration of interleukin-2. Endocrinology 135: 2465–2472.

Hanisch UK, Rowe W, van Rossum D, Meaney MJ, Quirion R (1996a) Phasic hyperactivity of the HPA axis resulting from chronic central IL-2 administration. Neuroreport 7: 2883–2888.

Hanisch UK, Seto D, Quirion R (1993) Modulation of hippocampal acetylcholine release: a potent central action of interleukin-2. J Neurosci 13: 3368–3374.

Harrison JK, Jiang Y, Chen S, Xia Y, Maciejewski D, McNamara RK, Streit WJ, Salafranca MN, Adhikari S, Thompson DA, Botti P, Bacon KB, Feng L (1998) Role for neuronally derived fractalkine in mediating interactions between neurons and CX3CR1-expressing microglia. Proc Natl Acad Sci U S A 95: 10896–10901.

Hartung HP, Jung S, Stoll G, Zielasek J, Schmidt B, Archelos JJ, Toyka KV (1992) Inflammatory mediators in demyelinating disorders of the CNS and PNS. J Neuroimmunol 40: 197–210.

Hays SJ (1998) Therapeutic approaches to the treatment of neuroinflammatory diseases. Curr Pharm Des 4: 335–348.

Hide I, Tanaka M, Inoue A, Nakajima K, Kohsaka S, Inoue K, Nakata Y (2000) Extracellular ATP triggers tumor necrosis factor-alpha release from rat microglia. J Neurochem 75: 965–972.

Hoffmann JA, Kafatos FC, Janeway CA, Ezekowitz RAB (1999) Phylogenetic perspectives in innate immunity. Science 284: 1313–1318.

Hopkins SJ, Rothwell NJ (1995) Cytokines and the nervous system. I: Expression and recognition. Trends Neurosci 18: 83–88.

Kanakaraj P, Ngo K, Wu Y, Angulo A, Ghazal P, Harris CA, Siekierka JJ, Peterson PA, Fung-Leung WP (1999) Defective interleukin (IL)-18-mediated natural killer and T helper cell type 1 responses in IL-1 receptor-associated kinase (IRAK)-deficient mice. J Exp Med 189: 1129–1138.

Karanth S, McCann SM (1991) Anterior pituitary hormone control by interleukin 2. Proc Natl Acad Sci U S A 88: 2961–2965.

Katsuki H, Nakai S, Hirai Y, Akaji K, Kiso Y, Satoh M (1990) Interleukin-1 beta inhibits long-term potentiation in the CA3 region of mouse hippocampal slices. Eur J Pharmacol 181: 323–326.

Kita T, Tanaka T, Tanaka N, Kinoshita Y (2000) The role of tumor necrosis factor-alpha in diffuse axonal injury following fluid-percussive brain injury in rats [In Process Citation]. Int J Legal Med 113: 221–228.

Kohno K, Kurimoto M (1998) Interleukin 18, a cytokine which resembles IL-1 structurally and IL-12 functionally but exerts its effect independently of both. Clin Immunol Immunopathol 86: 11–15.

Kojima H, Takeuchi M, Ohta T, Nishida Y, Arai N, Ikeda M, Ikegami H, Kurimoto M (1998) Interleukin-18 activates the IRAK-TRAF6 pathway in mouse EL-4 cells. Biochem Biophys Res Commun 244: 183–186.

Kreutzberg GW (1996) Microglia: a sensor for pathological events in the CNS. Trends Neurosci 19: 312–318.

Kumar S, McDonnell PC, Lehr R, Tierney L, Tzimas MN, Griswold DE, Capper EA, Tal-Singer R, Wells GI, Doyle ML, Young PR (2000) Identification and initial characterization of four novel members of the interleukin-1 familiy. J Biol Chem 275: 10308–10314.

Lapchak PA, Araujo DM, Beck KD, Finch CE, Johnson SA, Hefti F (1993) BDNF and trkB mRNA expression in the hippocampal formation of aging rats. Neurobiol Aging 14: 121–126.

Lapchak PA, Araujo DM, Quirion R, Beaudet A (1991) Immunoautoradiographic localization of interleukin 2-like immunoreactivity and interleukin 2 receptors (Tac antigen-like immunoreactivity) in the rat brain. Neuroscience 44: 173–184.

Lee SC, Liu W, Dickson DW, Brosnan CF, Berman JW (1993) Cytokine production by human fetal microglia and astrocytes. J Immunol 150: 2659–2667.

Licinio JWML (1997) Pathways and mechanism for cytokine signaling of the central nervous system. J Clin Invest 100: 2941–2947.

Liu JS, Amaral TD, Brosnan CF, Lee SC (1998) IFNs are critical regulators of IL-1 receptor antagonist and IL-1 expression in human microglia. J Immunol 161: 1989–1996.

Liu JS, John GR, Sikora A, Lee SC, Brosnan CF (2000) Modulation of interleukin-1beta and tumor necrosis factor alpha signaling by P2 purinergic receptors in human fetal astrocytes. J Neurosci 20: 5292–5299.

Locati M, Murphy PM (1999) Chemokines and chemokine receptors: biology and clinical relevance in inflammation and AIDS. Annu Rev Med 50: 425–440.

Loddick SA, Liu C, Takao T, Hashimoto K, De Souza EB (1998) Interleukin-1 receptors: Cloning studies and role in central nervous system disorders. Brain Res Rev 26: 306–319.

Luheshi G, Rothwell N (1996) Cytokines and fever. Int Arch Allergy Immunol 109: 301–307.

Luheshi GN, Gardner JD, Rushforth DA, Loudon AS, Rothwell NJ (1999) Leptin actions on food intake and body temperature are mediated by IL-1. Proc Natl Acad Sci U S A 96: 7047–7052.

Luk WP, Zhang Y, White TD, Lue FA, Wu C, Jiang CG, Zhang L, Moldofsky H (1999) Adenosine: a mediator of interleukin-1beta-induced hippocampal synaptic inhibition. J Neurosci 19: 4238–4244.

Luster AD (1998) Chemokines—chemotactic cytokines that mediate inflammation. N Engl J Med 338: 436–445.

Maciejewski-Lenoir D, Chen S, Feng L, Maki R, Bacon KB (1999) Characterization of fractalkine in rat brain cells: migratory and activation signals for CX3CR-1-expressing microglia. J Immunol 163: 1628–1635.

Madigan MC, Sadun AA, Rao NS, Dugel PU, Tenhula WN, Gill PS (1996) Tumor necrosis factor-alpha (TNF-alpha)-induced optic neuropathy in rabbits. Neurol Res 18: 176–184.

Maier SF, Goehler LE, Fleshner M, Watkins LR (1998) The role of the vagus nerve in cytokine-to-brain communication. Ann N Y Acad Sci 840: 289–300.

Martino G, Furlan R, Brambilla E, Bergami A, Ruffini F, Gironi M, Poliani PL, Grimaldi LM, Comi G (2000) Cytokines and immunity in multiple sclerosis: the dual signal hypothesis. J Neuroimmunol 109: 3–9.

Matsumoto S, Tsuji-Takayama K, Aizawa Y, Koide K, Takeuchi M, Ohta T, Kurimoto M (1997) Interleukin-18 activates NF-kappaB in murine T helper type 1 cells. Biochem Biophys Res Commun 234: 454–457.

McCann SM, Karanth S, Kamat A, Les DW, Lyson K, Gimeno M, Rettori V (1994b) Induction by cytokines of the pattern of pituitary hormone secretion in infection. Neuroimmunomodulation 1: 2–13.

McCann SM, Kimura M, Karanth S, Yu WH, Rettori V (1997b) Nitric oxide controls the hypothalamic-pituitary response to cytokines. Neuroimmunomodulation 4: 98–106.

McCann SM, Kimura M, Karanth S, Yu WH, Rettori V (1998) Role of nitric oxide in the neuroendocrine responses to cytokines. Ann N Y Acad Sci 840: 174–184.

McCann SM, Lyson K, Karanth S, Gimeno M, Belova N, Kamat A, Rettori V (1994a) Role of cytokines in the endocrine system. Ann N Y Acad Sci 741: 50–63.

McCann SM, Kimura M, Karanth S, Yu WH, Rettori V (1997a) Nitric oxide controls the hypothalamic-pituitary response to cytokines. Neuroimmunomodulation 4: 98–106.

McGeer EG, McGeer PL (1999) Brain inflammation in Alzheimer disease and the therapeutic implications. Curr Pharm Des 5: 821–836.

McGeer PL, McGeer EG (2000) Autotoxicity and Alzheimer disease. Arch Neurol 57: 789–790.

McGeer PL, McGeer EG, Yasojima K (2000) Alzheimer disease and neuroinflammation. J Neural Transm Suppl 59: 53–57.

McInnes IB, Gracie JA, Leung BP, Wei XQ , Liew FY (2000) Interleukin 18: a pleiotropic participant in chronic inflammation. Immunol Today 21: 312–315.

Mennicken F, Maki R, De Souza EB, Quirion R (1999) Chemokines and chemokine receptors in the CNS: a possible role in neuroinflammation and patterning. Trends Pharmacol Sci 20: 73–78.

Merrill JE, Benveniste EN (1996) Cytokines in inflammatory brain lesions: helpful and harmful. Trends Neurosci 19: 331–338.

Merrill JE, Scolding NJ (1999) Mechanisms of damage to myelin and oligodendrocytes and their relevance to disease. Neuropathol Appl Neurobiol 25: 435–458.

Mertsch K, Hanisch UK, Kettenmann H, Schnitzer J (2001) Characterization of microglial cells and their response to stimulation in an organotypic retinal culture system. J Comp Neurol 431: 217–227.

Meucci O, Fatatis A, Simen AA, Miller RJ (2000) Expression of CX3CR1 chemokine receptors on neurons and their role in neuronal survival. Proc Natl Acad Sci U S A 97: 8075–8080.

Möller T, Hanisch UK, Ransom BR (2000) Thrombin-induced activation of cultured rodent microglia. J Neurochem 75: 1539–1547.

Mrak RE, Griffin WS (2000) Interleukin-1 and the immunogenetics of Alzheimer disease. J Neuropathol Exp Neurol 59: 471–476.

Mucke L, Eddleston M (1993) Astrocytes in infectious and immune-mediated diseases of the central nervous system. FASEB J 7: 1226–1232.
Nicoll JA, Mrak RE, Graham DI, Stewart J, Wilcock G, MacGowan S, Esiri MM, Murray LS, Dewar D, Love S, Moss T, Griffin WS (2000) Association of interleukin-1 gene polymorphisms with Alzheimer's disease. Ann Neurol 47: 365–368.
Nishiyori A, Minami M, Ohtani Y, Takami S, Yamamoto J, Kawaguchi N, Kume T, Akaike A, Satoh M (1998) Localization of fractalkine and CX3CR1 mRNAs in rat brain: does fractalkine play a role in signaling from neuron to microglia? FEBS Lett 429: 167–172.
Norton WT, Aquino DA, Hozumi I, Chiu FC , Brosnan CF (1992) Quantitative aspects of reactive gliosis: a review. Neurochem Res 17: 877–885.
O'Connor JJ, Coogan AN (1999) Actions of the pro-inflammatory cytokine IL-1 beta on central synaptic transmission. Exp Physiol 84: 601–614.
O'Neill LAJ, Dinarello CA (2000) The IL-1 receptor/toll-like receptor superfamily: crucial receptors for inflammation and host defense. Immunol Today 21: 206–209.
O'Neill LAJ, Greene C (1998) Signal transduction pathways activated by the IL-1 receptor family: Ancient signaling machinery in mamals, insects and plants. J Leukoc Biol 63: 650–657.
Okamura H, Tsutsi H, Komatsu T, Yutsudo M, Hakura A, Tanimoto T, Torigoe K, Okura T, Nukada Y, Hattori K, Akita K, Namba M, Tanabe F, Konishi K, Fokuda S, Kurimoto M (1995) Cloning of a new cytokine that induces IFN-gamma production by T cells. Nature 378: 88–91.
Oppmann B, Lesley R, Blom B, Timans JC, Xu Y, Hunte B, Vega F, Yu N, Wang J, Singh K, Zonin F, Vaisberg E, Churakova T, Liu M, Gorman D, Wagner J, Zurawski S, Liu Y, Abrams JS, Moore KW, Rennick D, Waal-Malefyt R, Hannum C, Bazan JF, Kastelein RA (2000) Novel p19 protein engages IL-12p40 to form a cytokine, IL-23, with biological activities similar as well as distinct from IL-12. Immunity 13: 715–725.
Otero GC, Merrill JE (1994) Cytokine receptors on glial cells. Glia 11: 117–128.
Otero GC, Merrill JE (1997) Response of human oligodendrocytes to interleukin-2. Brain Behav Immun 11: 24-38.
Pan W, Banks WA, Kennedy MK, Gutierrez EG, Kastin AJ (1996) Differential permeability of the BBB in acute EAE: enhanced transport of TNT-alpha. Am J Physiol 271: E636–E642.
Petitto JM, Huang Z, Raizada MK, Rinker CM, McCarthy DB (1998) Molecular cloning of the cDNA coding sequence of IL-2 receptor-gamma (gamma$_c$) from human and murine forebrain: expression in the hippocampus in situ and by brain cells in vitro. Brain Res Mol Brain Res 53: 152–162.
Petitto JM, McNamara RK, Gendreau PL, Huang Z, Jackson AJ (1999) Impaired learning and memory and altered hippocampal neurodevelopment resulting from interleukin-2 gene deletion. J Neurosci Res 56: 441–446.
Prinz M, Hanisch UK (1999) Murine microglial cells produce and respond to interleukin-18. J Neurochem 72: 2215–2218.
Prinz M, Hanisch UK, Kettenmann H, Kirchhoff F (1998) Alternative splicing of mouse IL-15 is due to the use of an internal splice site in exon 5. Mol Brain Res 63: 155–162.

Prinz M, Kann O, Draheim HJ, Schumann RR, Kettenmann H, Weber JR, Hanisch UK (1999) Microglial activation by components of Gram-positive and -negative bacteria: Distinct and common routes to the induction of ion channels and cytokines. J Neuropathol Exp Neurol 58: 1078–1089.

Raber J, Sorg O, Horn TF, Yu N, Koob GF, Campbell IL, Bloom FE (1998) Inflammatory cytokines: putative regulators of neuronal and neuro- endocrine function. Brain Res Brain Res Rev 26: 320–326.

Raine CS, Bonetti B, Cannella B (1998) Multiple sclerosis: expression of molecules of the tumor necrosis factor ligand and receptor families in relationship to the demyelinated plaque. Rev Neurol (Paris) 154: 577–585.

Raivich G, Bluethmann H, Kreutzberg GW (1996) Signaling molecules and neuroglial activation in the injured central nervous system. Keio J Med 45: 239–247.

Raivich G, Bohatschek M, Kloss CU, Werner A, Jones LL, Kreutzberg GW (1999a) Neuroglial activation repertoire in the injured brain: graded response, molecular mechanisms and cues to physiological function. Brain Res Brain Res Rev 30: 77–105.

Raivich G, Jones LL, Werner A, Bluthmann H, Doetschmann T, Kreutzberg GW (1999b) Molecular signals for glial activation: pro- and anti-inflammatory cytokines in the injured brain. Acta Neurochir Suppl (Wien) 73: 21–30.

Ren L, Lubrich B, Biber K, Gebicke-haerter PJ (1999) Differential expression of inflammatory mediators in rat microglia cultured from different brain regions. Brain Res Mol Brain Res 65: 198–205.

Renno T, Krakowski M, Piccirillo C, Lin JY, Owens T (1995) TNF-alpha expression by resident microglia and infiltrating leukocytes in the central nervous system of mice with experimental allergic encephalomyelitis. Regulation by Th1 cytokines. J Immunol 154: 944–953.

Rogers J, Webster S, Lue LF, Brachova L , Civin WH, Emmerling M, Shivers B, Walker D, McGeer P (1996) Inflammation and Alzheimer's disease pathogenesis. Neurobiol Aging 17: 681–686.

Rostworowski M, Balasingam V, Chabot S, Owens T, Yong VW (1997) Astrogliosis in the neonatal and adult murine brain post-trauma: elevation of inflammatory cytokines and the lack of requirement for endogenous interferon-gamma. J Neurosci 17: 3664–3674.

Rothwell N, Allan S, Toulmond S (1997) Perspectives Series: Cytokines and the brain. J Clin Invest 100: 2648–2652.

Rothwell N, Allan S, Toulmond S (1997) The role of Interleukin 1 in acute neurodegeneration and stroke: Pathophysiological and therapeutic implications. J Clin Invest 100: 2648–2652.

Rothwell NJ, Hopkins SJ (1995) Cytokines and the nervous system II: Actions and mechanisms of action. Trends Neurosci 18: 130–136.

Rothwell NJ, Luheshi G, Toulmond S (1996) Cytokines and their receptors in the central nervous system: physiology, pharmacology, and pathology. Pharmacol Ther 69: 85–95.

Rothwell NJ, Luheshi GN (2000) Interleukin 1 in the brain: biology, pathology and therapeutic target. Trends Neurosci 23: 618–625.

Sanz JM, Di Virgilio F (2000) Kinetics and mechanism of ATP-dependent IL-1β release from microglial cells. J Immunol 164: 4893–4898.

Sawada M, Suzumura A, Hosoya H, Marunouchi T, Nagatsu T (1999) Interleukin-10 inhibits both production of cytokines and expression of cytokine receptors in microglia. J Neurochem 72: 1466–1471.

Schneider H, Pitossi F, Balschun D, Wagner A, del Rey A, Besedovsky HO (1998) A neuromodulatory role of interleukin-1β in the hippocampus. Proc Natl Acad Sci USA 95: 7778–7783.

Schubert P, Morino T, Miyazaki H, Ogata T, Nakamura Y, Marchini C, Ferroni S (2000) Cascading glia reactions: a common pathomechanism and its differentiated control by cyclic nucleotide signaling. Ann N Y Acad Sci 903: 24–33.

Schwaiger FW, Hager G, Raivich G, Kreutzberg GW (1998) Cellular activation in neuroregeneration. Prog Brain Res 117: 197–210.

Seilhean D, Kobayashi K, He Y, Uchihara T, Rosenblum O, Katlama C, Bricaire F, Duyckaerts C, Hauw JJ (1997) Tumor necrosis factor-alpha, microglia and astrocytes in AIDS dementia complex. Acta Neuropathol (Berl) 93: 508–517.

Selmaj K, Raine CS, Farooq M, Norton WT, Brosnan CF (1991b) Cytokine cytotoxicity against oligodendrocytes. Apoptosis induced by lymphotoxin. J Immunol 147: 1522–1529.

Selmaj K, Shafit-Zagardo B, Aquino DA, Farooq M, Raine CS, Norton WT, Brosnan CF (1991a) Tumor necrosis factor-induced proliferation of astrocytes from mature brain is associated with down-regulation of glial fibrillary acidic protein mRNA. J Neurochem 57: 823–830.

Selmaj KW, Farooq M, Norton WT, Raine CS, Brosnan CF (1990) Proliferation of astrocytes in vitro in response to cytokines. A primary role for tumor necrosis factor. J Immunol 144: 129–135.

Seto D, Kar S, Quirion R (1997) Evidence for direct and indirect mechanisms in the potent modulatory action of interleukin-2 on the release of acetylcholine in rat hippocampal slices. Br J Pharmacol 120: 1151–1157.

Shadiack AM, Hart RP, Carlson CD, Jonakait GM (1993) Interleukin-1 induces substance P in sympathetic ganglia through the induction of leukemia inhibitory factor (LIF). J Neurosci 13: 2601–2609.

Shadiack AM, Vaccariello SA, Sun Y, Zigmond RE (1998) Nerve growth factor inhibits sympathetic neurons' response to an injury cytokine. Proc Natl Acad Sci USA 95: 7727–7730.

Sheng JG, Griffin WS, Royston MC, Mrak RE (1998) Distribution of interleukin-1-immunoreactive microglia in cerebral cortical layers: implications for neuritic plaque formation in Alzheimer's disease. Neuropathol Appl Neurobiol 24: 278–283.

Sheng JG, Ito K, Skinner RD, Mrak RE, Rovnaghi CR, Van Eldik LJ, Griffin WS (1996) In vivo and in vitro evidence supporting a role for the inflammatory cytokine interleukin-1 as a driving force in Alzheimer pathogenesis. Neurobiol Aging 17: 761–766.

Sheng JG, Mrak RE, Griffin WS (1995) Microglial interleukin-1 alpha expression in brain regions in Alzheimer's disease: correlation with neuritic plaque distribution. Neuropathol Appl Neurobiol 21: 290–301.

Smith DE, Renshaw BR, Ketchem RR, Kubin M, Garka KE, Sims JE (2000) Four new members expand the interleukin-1 superfamily. J Biol Chem 275: 1169–1175.

Smith EM, Cadet P, Stefano GB, Opp MR, Hughes TK, Jr. (1999) IL-10 as a mediator in the HPA axis and brain. J Neuroimmunol 100: 140–148.

Smith ME, van-der MK, Somera FP (1998) Macrophage and microglial responses to cytokines in vitro: phagocytic activity, proteolytic enzyme release, and free radical production. J Neurosci Res 54: 68–78.

Stalder AK, Pagenstecher A, Yu NC, Kincaid C, Chiang CS, Hobbs MV, Bloom FE, Campbell IL (1997) Lipopolysaccharide-induced IL-12 expression in the central nervous system and cultured astrocytes and microglia. J Immunol 159: 1344–1351.

Sternberg EM (1997) Neural-immune interactions in health and disease. J Clin Invest 100: 2641–2647.

Stohwasser R, Giesebrecht J, Kraft R, Müller E-C, Häusler KG, Kettenmann H, Hanisch UK, Kloetzel P-M (2000) Biochemical analysis of proteasomes from microglia: Induction of immunoproteasomes by interferon-gamma and lipopolysaccharide. Glia 29: 355–365.

Stoll G, Jander S (1999) The role of microglia and macrophages in the pathophysiology of the CNS. Prog Neurobiol 58: 233–247.

Stoll G, Jander S, Schroeter M (1998) Inflammation and glial responses in ischemic brain lesions. Prog Neurobiol 56: 149–171.

Stoll G, Jander S, Schroeter M (2000) Cytokines in CNS disorders: neurotoxicity versus neuroprotection. J Neural Transm Suppl 59: 81–89.

Streit WJ (1994) The role of microglia in regeneration. Eur Arch Otorhinolaryngol Suppl S69–S70.

Streit WJ (2000) Microglial response to brain injury: a brief synopsis. Toxicol Pathol 28: 28–30.

Streit WJ, Hurley SD, McGraw TS, Semple-Rowland SL (2000b) Comparative evaluation of cytokine profiles and reactive gliosis supports a critical role for interleukin-6 in neuron-glia signaling during regeneration. J Neurosci Res 61: 10–20.

Streit WJ, Walter SA, Pennel NA (2000a) Reactive microgliosis. Prog Neurobiol 57: 563–581.

Suzumura A, Sawada M, Itoh Y, Marunouchi T (1994) Interleukin-4 induces proliferation and activation of microglia but suppresses their induction of class II major histocompatibility complex antigen expression. J Neuroimmunol 53: 209–218.

Suzumura A, Sawada M, Takayanagi T (1998) Production of interleukin-12 and expression of its receptors by murine microglia. Brain Res 787: 139–142.

Szczepanik AM, Funes S, Petko W, Ringheim GE (2001) IL-4, IL-10 and IL-13 modulate Abeta(1-42)-induced cytokine and chemokine production in primary murine microglia and a human monocyte cell line. J Neuroimmunol 113: 49–62.

Takahashi S, Kapas L, Fang J, Krueger JM (1999) Somnogenic relationships between tumor necrosis factor and interleukin-1. Am J Physiol 276: R1132–R1140.

Tan J, Town T, Paris D, Mori T, Suo Z, Crawford F, Mattson MP, Flavell RA, Mullan M (1999) Microglial activation resulting from CD40-CD40L interaction after beta-amyloid stimulation. Science 286: 2352–2355.

Thomas WE (1999) Brain macrophages: on the role of pericytes and perivascular cells. Brain Res Brain Res Rev 31: 42–57.

Thomassen E, Bird TA, Renshaw BR, Kennedy MK, Sims JE (1998) Binding of interleukin-18 to the interleukin-1 receptor homologous receptor IL-1Rrp1 leads to activation of signaling pathways similar to those used by interleukin-1. J Interferon Cytokine Res 18: 1077–1088.

Torres C, Aranguez I, Rubio N (1995) Expression of interferon-gamma receptors on murine oligodendrocytes and its regulation by cytokines and mitogens. Immunology 86: 250–255.

Tran EH, Prince EN, Owens T (2000) IFN-gamma shapes immune invasion of the central nervous system via regulation of chemokines. J Immunol 164: 2759–2768.

Tringali G, Dello-Russo C, Preziosi P, Navarra P (1998) Hypothalamic interleukin-1 in physiology and pathology. Toxicol Lett 102: 295–299.

Tsuji-Takayama K, Aizawa Y, Okamoto I, Kojima H, Koide K, Takeuchi M, Ikegami H, Ohta T, Kurimoto M (1999) Interleukin-18 induces interferon-gamma production through NF-kappaB and NFAT activation in murine T helper type 1 cells. Cell Immunol 196: 41–50.

Tsuji-Takayama K, Matsumoto S, Koide K, Takeuchi M, Ikeda M, Ohta T, Kurimoto M (1997) Interleukin-18 induces activation and association of p56(lck) and MAPK in a murine TH1 clone. Biochem Biophys Res Commun 237: 126–130.

Vandenabeele P, Fiers W (1991) Is amyloidogenesis during Alzheimer's disease due to an IL-1-/IL-6-mediated 'acute phase response' in the brain? Immunol Today 12: 217–219.

Vilcek J (1998) The cytokines: an overview. In: The cytokine handbook (Thomson A, ed), pp 1–20. San Diego: Academic Press.

Vitkovic L, Bockaert J, Jacque C (2000) "Inflammatory" cytokines: Neuromodulators in normal brain? J Neurochem 74: 457–471.

Waguespack PJ, Banks WA, Kastin AJ (1994) Interleukin-2 does not cross the blood-brain barrier by a saturable transport system. Brain Res Bull 34: 103–109.

Ware CF, Santee S, Glass A (1998) Tumor necrosis factor-related ligands and receptors. In: The cytokine handbook (Thomson A, ed), pp 549–592. San Diego: Academic Press.

Wei R, Jonakait GM (1999) Neurotrophins and the anti-inflammatory agents interleukin-4 (IL-4), IL- 10, IL-11 and transforming growth factor-beta1 (TGF-beta1) down- regulate T cell costimulatory molecules B7 and CD40 on cultured rat microglia. J Neuroimmunol 95: 8–18.

Wewers MD, Herzyk DJ (1989) Alveolar macrophages differ from blood monocytes in human IL-1b release. J Immunol 143: 1635–1641.

Wheeler, R.D., Culhane AC, Hall MD, Pickering-Brown S, Rothwell N, Luheshi GN (2000) Detection of the interleukin 18 family in rat brain by RT-PCR. Mol Brain Res 77: 290–293.

Wildbaum G, Youssef S, Grabie N, Karin N (1998) Neutralizing antibodies to IFN-gamma-inducing factor prevent experimental autoimmune encephalomyelitis. J Immunol 161: 6368–6374.

Woodroofe MN, Sarna GS, Wawda M, Hayes GM, Loughlin AJ, Tinker A, Curzner ML (1991) Detection of interleukin-1 and interleukin-6 in adult rat brain following mechnical injury by in vivo microdialysis: Evidence of a role for microglia in cytokine production. J Immunol 33: 227–236.

Xia MQ, Hyman BT (1999) Chemokines/chemokine receptors in the central nervous system and Alzheimer's disease. J Neurovirol 5: 32-41.

Xia MQ, Qin SX, Wu LJ, Mackay CR, Hyman BT (1998) Immunohistochemical study of the beta-chemokine receptors CCR3 and CCR5 and their ligands in normal and Alzheimer's disease brains. Am J Pathol 153: 31–37.

Yang RB, Mark MR, Gray A, Huang A, Xie MH, Zhang M, Goddard A, Wood WI, Gurney AL, Godowski PJ (1998) Toll-like receptor-2 mediates lipopolysaccharide-induced cellular signalling. Nature 395: 284–288.

Yao J, Keri JE, Taffs RE, Colton CA (1992) Characterization of interleukin-1 production by microglia in culture. Brain Res 591: 88–93.

Yates SL, Burgess LH, Kocsis-Angle J, Antal JM, Dority MD, Embury PB, Piotrkowski AM, Brunden KR (2000) Amyloid beta and amylin fibrils induce increases in proinflammatory cytokine and chemokine production by THP-1 cells and murine microglia. J Neurochem 74: 1017–1025.

Yoshimura A, Lien E, Ingalls RR, Tuomanen E, Dziarski R, Golenbock D (1999) Cutting edge: Recognition of gram-positive bacterial cell wall components by the innate immune system occurs via Toll-like receptor 2. J Immunol 163: 1–5.

Zhang M., Tracey KJ (1998) Tumor necrosis factor. In: The cytokine handbook (Thomson A, ed), pp 517–548. San Diego: Academic Press.

Zhang GX, Baker CM, Kolson DL, Rostami AM (2000) Chemokines and chemokine receptors in the pathogenesis of multiple sclerosis. Mult Scler 6: 3–13.

Zujovic V, Benavides J, Vige X, Carter C, Taupin V (2000) Fractalkine modulates TNF-alpha secretion and neurotoxicity induced by microglial activation. Glia 29: 305–315.

6

Microglia and Macrophage Responses in Cerebral Ischemia

GUIDO STOLL, SEBASTIAN JANDER AND MICHAEL SCHROETER

1. Pathophysiology of cerebral ischemia

Focal impairment, or cessation of blood flow to the brain, restricts the delivery of substrates, most importantly oxygen and glucose, and thereby impairs maintenance of ionic gradients. This is followed by depolarization of neurons and glia that release excitatory amino acids (glutamate) into the extracellular space and accumulate Ca2+ (reviewed in Dirnagl et al. (1999)). Ca2+ is a universal second messenger leading to production of proteolytic enzymes and free-radical species, and activation of glutamate receptors. In the center of the ischemic territory, where the flow reduction is most severe, these processes induce rapid cell death. A significant proportion of neurons, however, dies by an internal program of self-destruction, designated apoptosis or programmed cell death (Bredesen (1995)). Apoptotic neurons are intermingled with necrotic neurons in the core of infarctions. In the boundary zone, apoptotic cell death is ongoing during the first week after focal ischemia (Li et al. (1995); Braun et al. (1996); Isenmann et al. (1998)). Accordingly, several studies using modern imaging techniques provided evidence for infarct growth during the first few days after cerebral ischemia (Marchal et al. (1996); Beaulieu et al. (1999)). In experimental animals, mediators of the immune system appear to play an essential role in this secondary infarct growth. Mice lacking interferon regulatory factor (IRF), a nuclear transcription factor, developed similar infarct volumes at 24 hours, but significant differences in favor of the knock-out animals became evident at day 3 (Iadecola et al. (1999)). Similarly, partial neuroprotection was observed in mice lacking inducible nitric oxide synthase (Iadecola et al. (1997)).

Several animal models have been developed that more or less reflect the variable conditions of cerebral ischemia in humans (reviewed in McAuley (1995)). Surgical occlusion of major arteries is widely used (Bolander et al. (1989)). Permanent occlusion of the middle cerebral artery (pMCAO) at proximal sites leads to infarctions of the basal ganglia and the neocortex, which start with necrosis of neuronal populations in the caudate nucleus and putamen between three and twelve hours and are followed by pannecrosis after 24 to 48 hours (Fig. 1A). Pannecrosis incorporating the entire area supplied by the MCA is apparent at 72 to 96 hours. Studies in Wistar rats have revealed that the infarct volume increases between 6 and 72 hours, but not thereafter (Garcia et al. (1993)). Focal infarcts are surrounded by a large ischemic penumbra that can be salvaged by reperfusion (Memezawa et al. (1992)). The penumbra refers to cortical areas where the blood flow is reduced, but is still above a threshold that maintains a reversible state of ischemic neuronal injury (Hossmann et al. (1995)). Thus, early reperfusion significantly modifies the chronology of the events described above. Withdrawal of an intraluminal thread advanced into the proximal MCA reconstitutes perfusion (transient MCAO) and leads to variable infarct extension depending on the duration of ischemia (Longa et al. (1989)). Short-lasting ischemia by 30 minutes leads to a restricted infarction with pannecrosis of neurons and glial cells in the caudate nucleus and putamen, but only partly affects the neocortex (Fig. 1B). In this setting, pannecrotic infarctions are surrounded by areas of selective neuronal death characterized by loss of large pyramidal neurons, but preservation of endothelial and glial structures (Chen et al. (1993)) (Fig. 1B).

As a different approach to study focal ischemic lesions, Watson and colleagues (1985) introduced the photothrombosis model. After systemic injection of the dye rose bengal, focal illumination of a brain area through the intact skull leads to endothelial alterations. These are followed by an early disruption of the blood-brain barrier, vasogenic edema, photochemically stimulated platelet aggregation with formation of clots, and thrombotic occlusion of small intracerebral vessels with ensuing infarction (Fig. 1D). Since areas of ischemic brain damage are highly reproducible in size and location, this model allows the precise distinction between cellular and molecular responses within the lesion, within secondarily degenerating fibre tracts, and in remote nonischemic cerebral cortex (Fig. 1C).

Neuronal death in transient global cerebral ischemia is restricted to vulnerable areas and delayed by several days. In this model, circulation to the entire brain ("four vessel" occlusion model) or the forebrain ("two vessel" occlusion model) is transiently interrupted inducing neuronal injury similar to the situation after cardiac arrest in humans. After short periods of global ischemia CA1 neurons in the hippocampus die, while CA3 neurons are spared (Pulsinelli et al. (1982); Ordy et al. (1993)). Global ischemia allows evaluation of pathogenetic mechanisms of delayed neuronal death in the absence of pannecrosis, and of the intrinsic mechanisms of neuroprotection in surviving neurons within the same brain region.

2. Microglia and macrophage responses in focal ischemia

2.1. Areas of necrosis

2.1.1. Basic observations. Focal cerebral ischemia leads to local activation of microglia and astrocytes and provokes a massive invasion of leukocytes (reviewed

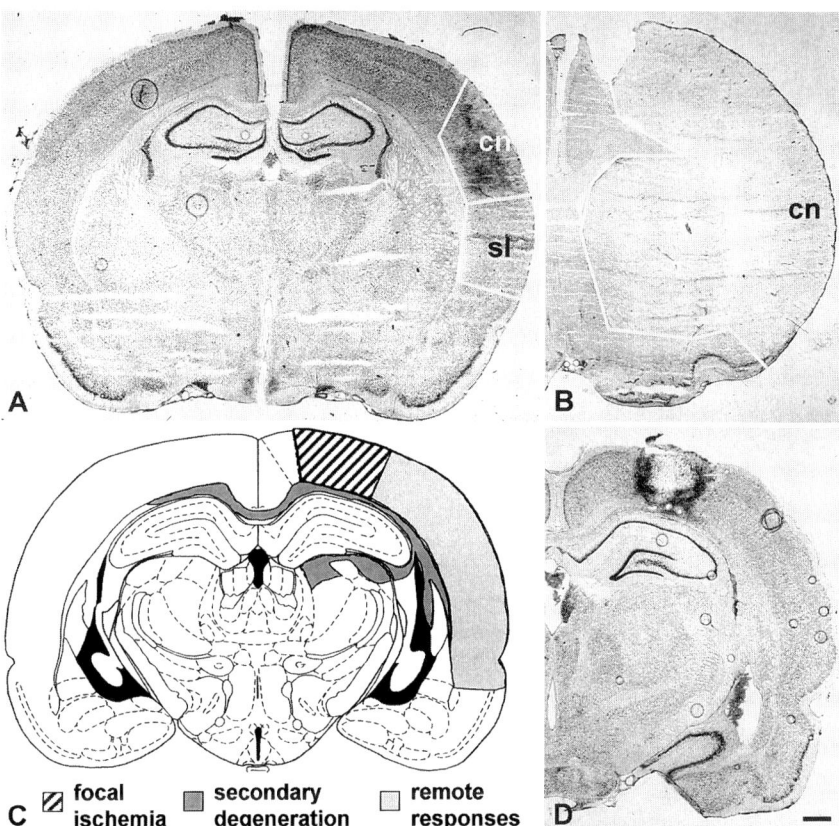

FIGURE 1. Cresyl-violet-stained coronal sections of ischemic lesions at the level of the sensorimotor cortex. Infarcts were induced in rats by transient (A) or by permanent (B) occlusion of the middle cerebral artery (MCAO) or by photothrombosis (D). Notice the small pannecrotic area within the neocortex (cn= cortical necrosis) and the neighboring region with selective neuronal death (sl) 7 days after transient (30 min) MCAO. The concomitant pannecrosis of the basal ganglia is out of this particular plane at the sensorimotor level. (B) shows a large pannecrotic lesion involving the basal ganglia and neocortex 3 days after permanent MCAO. Photothrombotic (PT) lesions are smaller and highly reproducible in size and location (D). Notice the hypercellular rim around the necrotic core 7 days after PT. As indicated in the schematic drawing (C), photothrombotic lesions allow a distinction between cellular and molecular reactions at the level of the ischemic focus (hatched area), in fiber tracts undergoing secondary Wallerian degeneration (dark gray), and in remote ipsilateral areas covered by spreading depression, but lacking neuronal death (light gray) (see also Fig. 6). Black areas mark ventricles. Bar represents 1 mm in A–D.

in Kochanek and Hallenbeck (1992); Stoll et al. (1998)). Overall, there are no qualitative differences in the composition of the cellular infiltrates within the ischemic lesions after permanent, transient MCAO or photothrombosis (Stoll et al. (1998)). Granulocytes are the first hematogenous cells that appear in ischemic brain tissue. In experimental animals and in humans, granulocytes accumulate in cerebral vessels within hours of ischemia and further invade the infarct and its boundary zone. This process peaks 24 hours after infarction; thereafter, the number of granulocytes rapidly declines (Kochanek and Hallenbeck (1992); Garcia et al. (1994); Linsberg et al. (1996); Stoll et al. (1998)). Accumulation of phagocytes in the boundary zone and infarct region usually starts within 12 hours, but numbers continuously increase during the first two weeks after infarction (Clark et al. (1993); Garcia et al. (1994); Schroeter et al. (1994)). Inflammation also involves a considerable number of T-cells, which are preferentially located in the perivascular space in the border zones and prevail between days 3 and 7 (Schroeter et al. (1994); Jander et al. (1995); Becker et al. (1997); Stoll et al. (1998)).

The population of phagocytes involves both activated resident microglia and hematogenous macrophages. There is an intrinsic problem with the distinction between microglia and macrophages after cerebral ischemia. Microglia, the resident "macrophages" of the brain, are activated after ischemia, and then become indistinguishable from hematogenous macrophages because they lose their characteristic ramified shape and become round ameboid cells containing cell debris. Likewise, activated microglia and macrophages share surface and cytoplasmic markers (Flaris et al. (1993); Stoll and Jander (1999)). Similar to monocytes/macrophages of hematogenous origin, resident microglia express the complement type 3 receptor (CR3 or CD11b/CD18 complex) recognized by the monoclonal antibody OX42 in the rat. Upon activation, both cell types upregulate MHC class I and II, and CD4 molecules. During the stage of phagocytic transformation, resident microglia and blood-derived monocytes contain phagolysosomes that can be visualized by the monoclonal antibody ED1 in the rat. Therefore, in CNS injury the pool of phagocytes contains activated microglia and hematogenous monocytes/macrophages in uncertain and variable proportions (Stoll and Jander (1999); see 2.1.2).

The dynamics of the microglial and phagocytic response depend upon the duration and local extension of ischemia. Transient MCAO in rats leads to ischemic pannecrosis of the lateral part of the striatum and adjacent neocortex, and to selective neuronal necrosis in the border zones. With increasing delay of reperfusion, additional surrounding tissue becomes pannecrotic. These different patterns in the progression of tissue destruction are mirrored by differential microglial/macrophage responses (Lehrmann et al. (1997); Zhang et al. (1997); Schroeter et al. (2001b)). In the most severely ischemic regions, no intact microglial cells, but only CR3+ microglial processes, were identifiable at 6 hours. In contrast, microglia in adjacent areas containing scattered, shrunken neurons residing in otherwise normal-appearing tissue showed increased CR3-immunoreactivity, and were less ramified, an indication of their activation. Ameboid cells were rare. At day 1, CR3-immunoreactivity further increased in areas surrounding the infarct in the striatum and neocortex, whereas the core of the infarct was almost devoid of CR3+ cells. In the boundary zone, round and ameboid cells, either representing transformed microglia or hematogenous monocytes, appeared and were mingled with highly ramified microglia. At two days after reperfusion, ischemic lesions became pannecrotic and ameboid cells preferentially localized to the inner boundary of the ischemic lesions with a further increase at day 3. At this stage, the hypocellularity within the ischemic focus was contrasted by a mildly hypercellular boundary zone, that contained numerous round CR3+ and ED1+ phagocytes. At day 7, the ischemic focus became hypercellular. Immunocytochemistry confirmed infiltration of the entire infarct region by CR3+/ED1+ phagocytes, which persisted for several weeks until debris removal was completed, and brain atrophy with a glial scar had developed.

Permanent MCAO leads to an infarction of the entire territory supplied by the MCA. (Fig. 1B). Pannecrosis first develops in the basal ganglia and with some delay in the neocortex. In contrast to transient MCAO, the inflammatory response is more vigorous and appears to be accelerated (Clark et al. (1993); Schroeter et al. (1994)). Using diffusion-weighted magnetic resonance imaging, Rupall and colleagues (1998) identified the periphery of the ischemic lesion, and by staining matched tissue sections found activated microglia with hypertrophic cell bodies and stout processes in this region as early as 30 minutes after pMCAO. Within the first 24 hours, a considerable number of ED1+ phagocytes encompassing activated microglia, and probably monocytes, already infiltrated the infarct core. Within two days, the entire infarct area was covered by ED1+ phagocytes. The number of phagocytes further increased during the next two weeks. By day 30, most of the infarcted tissue had been cleared by ED1+ phagocytes, and the cortex showed profound atrophy at the site of infarction. In this region, ED1+ phagocytes persisted.

In the photothrombosis model, ischemic lesions are smaller (Fig. 1C, D) and evoke a more circumscribed phagocytic response (Fig. 2) that can reliably be differentiated from microglial reactions to fiber tract degeneration occuring secondarily to the ischemic assult (see 2.2) and from remote responses in the intact ipsilateral cortex (see 2.3) (Jander et al. (1995); Schroeter et al. (1999)). In the infarct core, signs of microglial activation were visible as early as four hours post-event. Microglia showed shortening of cellular processes and more intense OX42 staining, which was lost within the next 24 hours because of ongoing pannecrosis. At 24 hours, only some ameboid microglia with rounded cell bodies were present in the pannecrotic infarct core. At three days after photothrombosis, ED1+ phagocytes were located in an annular fashion around the ischemic core of necrotic tissue (Fig. 2A). During the following week, the number of ED1+ phagocytes strongly increased in the peri-infarct region (Fig. 2B). The core of the infarct was infiltrated with a significant delay by ED1+ phagocytes, and became hypercellular during the second week after ischemia (Fig. 6B). At two weeks after photothrombosis, ED1+ phagocytes covered the entire infarct area (Jander et al.

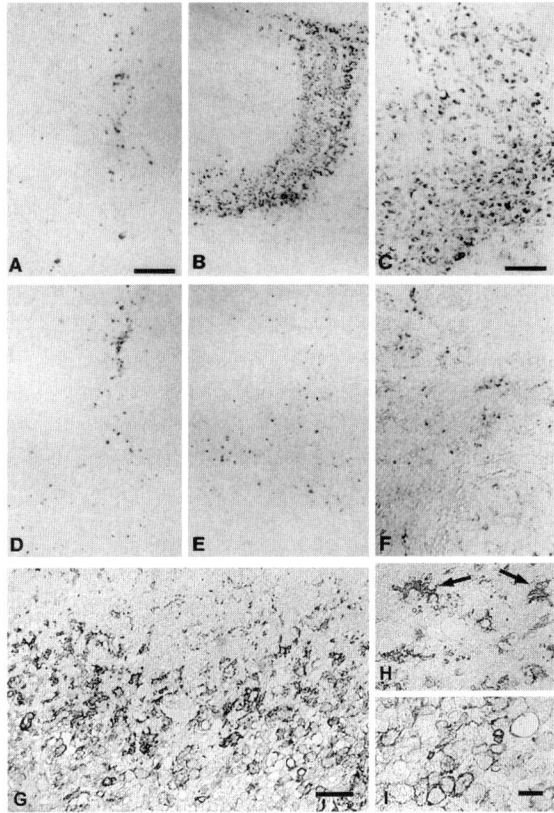

FIGURE 2. The relative contribution of microglia and macrophages to the pool of phagocytes in focal ischemia: coronal sections showing the right border zone of photothrombotic lesions at day 3 (A,D) and day 7 (B,E) stained for ED1, a marker for phagocytic microglia/macrophages. Notice the increase of ED1+ phagocytes between day 3 (A) and 7 in (B). Systemic depletion of hematogenous macrophages induces no change in the number of phagocytes at day 3 (D), but a dramatic reduction at day 7 (E). Thus, ED1+ phagocytes at day 3 mainly represent activated microglia whereas hematogenous macrophages infiltrate the lesion with a delay between day 3 and 7. (C) represents the normal ED1+ macrophage population in the spleen (C), and (F) shows successful depletion after liposome treatment (F). (G,H,I) show microglial/macrophage activation at the border zone of a human infarct 5 days after MCAO, as revealed by staining for MHC class II molecules. Normal brain tissue is on top, the infarct border zone at the bottom, of (G). Notice the gradual transition of ramified MHC class II positive microglia (G) into stout microglia (arrows in H), and finally into round phagocytes (G, bottom; I) towards the infarct border. These round MHC class II+ cells can no longer be separated into activated microglia or hematogenous macrophages by morphological criteria. (A–F) are reproduced from Schroeter et al. (1997) *Stroke* 28:382–386, with permission. Bars represent 250μm in A,B,D,F; 100μm in C,F; 100μm in G; and 20μm in H,I.

(1995); Schroeter et al. (1999)) (Fig. 6B). High-resolution magnetic resonance imaging at a field strength of 7T could depict distinct stages of tissue remodeling in photothrombotic lesions, but failed to differentiate between discrete infarct subareas preferentially covered by ED1+ phagocytes, or reactive astrocytes (Schroeter et al. (2001a)).

2.1.2 Differentiating between microglial and macrophagic responses: the liposome paradigm. To assess the relative contribution of resident microglia and hematogenous macrophages to the population of ED1+ phagocytes after cerebral ischemia, we depleted blood-derived macrophages in rats by systemic application of dichloromethylene diphosphonate containing liposomes (Van Roijen et al. (1989)). This treatment almost completely destroys monocytes/macrophages in lymphoid organs (Fig. 2F) and the blood stream for a few days, but leaves microglia unaffected (Bauer et al. (1995)). We then induced ischemic lesions by photothrombosis because, in this model, phagocytic responses as described above (2.1.1.) are highly reproducible in time course and location. In MCAO models, areas of ischemic damage are variable owing to differences in collateral blood supply. When the phagocytic response between sham-treated and macrophage-depleted animals was assessed by immunocytochemistry, there was no difference in the number and distribution of ED1+ phagocytes at day 3 after photothrombosis (Schroeter et al. (1997)) (Fig. 2A,D). In contrast, a dramatic difference was seen at day 7 (Fig. 2B,E). In sham-treated rats, ED1+ phagocytes largely outnumbered those in macrophage-depleted rats. This indicates that, in the initial days after photochemically induced infarction, phagocytes at the border zone were mainly derived from resident microglia, whereas blood-derived monocytes/macrophages were recruited with a delay of several days.

A possible mechanistic explanation for the great profusion of hematogenous macrophages after focal ischemia is that the chemokine monocyte chemoattractant protein-1 (MCP-1) is locally induced after MCAO (Kim et al. (1995); Wang et al. (1995)). MCP-1 mRNA was increased as early as six hours after MCAO and remained elevated up to day 5. MCP-1 mRNA was translated into protein and expressed by endothelial cells and "macrophage-like cells" (Kim et al. 1995), while another study localized MCP-1 mRNA and protein at early stages to astrocytes surrounding the ischemic tissue, and after day 4 to macrophages and reactive microglia in the infarcted tissue (Gourmala et al. (1997)). Moroever, intercellular and vascular cell adhesion molecules serving as counterreceptors for CD11b and VLA-4 involved in transendothelial migration of monocytes/macrophages (Springer (1994)) are induced in intraparenchymal blood vessels shortly after focal ischemia (Okada et al. (1994); Schroeter et al. (1994); Wang and Feuerstein (1995); Jander et al. (1996)). These findings suggest that a coordinated local expression of chemokines and cell adhesion molecules leads to the recruitment of hematogenous macrophages to ischemic brain lesions.

In macrophage-depleted rats, the morphological transition of resting ramified microglia into phagocytes could be selectively studied. Changes included upregulation of CR3, retraction of processes ("stellate microglia"), and later rounding of the cell body ("ameboid microglia"). According to the findings in rat photothrombosis, we found different morphological stages of microglial transition into phagocytes with increased numbers of stellate and ameboid microglia towards the border zone in human infarcts of recent age (Fig. 2G). It is an intriguing question what signal activates resident microglia to transform into phagocytes after cerebral infarction. In light of the potential neurotoxicity of activated microglia, this issue may have important therapeutic implications (of which, see below).

2.1.3 CD8+ microglia/macrophages as indicators of pannecrosis. Ischemic lesions in the rat brain contain a large number of CD8+ cells, which are most abundant during the second half of the first postischemic week (Schroeter et al. (1994; 2001b); Jander et al. (1995; 1998)) (Fig. 3,4). CD8+ cells were of similar size and codistributed with the population of perilesional phagocytes (Fig. 3). The only two established CD8+ cell populations in the rat immune system to date had been T-cells and natural killer (NK) cells (Lawetzky et al. (1990)). Most of the CD8+ cells in ischemic brain lesions, however, were negative for T-cell markers, and no NK cells were present (Schroeter et al. (1994); Jander et al. (1995; 1998)). By using double labeling immunofluorescence in combination with confocal laser microscopy, we could show that a subpopulation of phagocytes in the lesions expressed the CD8 molecule (Fig. 3) (Jander et al. (1998)). This unexpected finding could be further substantiated by PCR amplification of mRNA for the α and β chains of the CD8 molecule, essentially excluding NK-cells, which are always CD8α-positive (but CD8β-negative), as a cellular source of CD8. Besides CD8+ phagocytes, a second population of perilesional phagocytes could be identified based on the expression of the CD4 molecule. The kinetics of these two phagocyte populations differed fundamentally. Although CD8+ phagocytes were most prevalent between day 3 and 7 (Fig. 4), the number of CD4+ phagocytes gradually increased from day 2 on, and peaked at day 14 when they covered the entire infarct region. Double labeling with CD4 and CD8 antibodies revealed that perilesional CD4+ and CD8+ cells represented two distinct, largely nonoverlapping populations. The newly recognized population of CD8+ phagocytes appeared to be relatively unique to ischemic brain lesions, since no such cells were seen in secondarily degenerating fiber tracts after ischemia, or optic nerve transsection (Jander et al. (1998), Schroeter et al. (1999)).

To assess the relative contributions of infiltrating macrophages and resident microglia to the CD8+ phagocyte response, we depleted peripheral macrophages, as described above (2.1.2). Macrophage depletion led to a dramatic reduction, but not complete abolishment, of CD8+ cells in the ensuing infarcts (Schroeter et al.

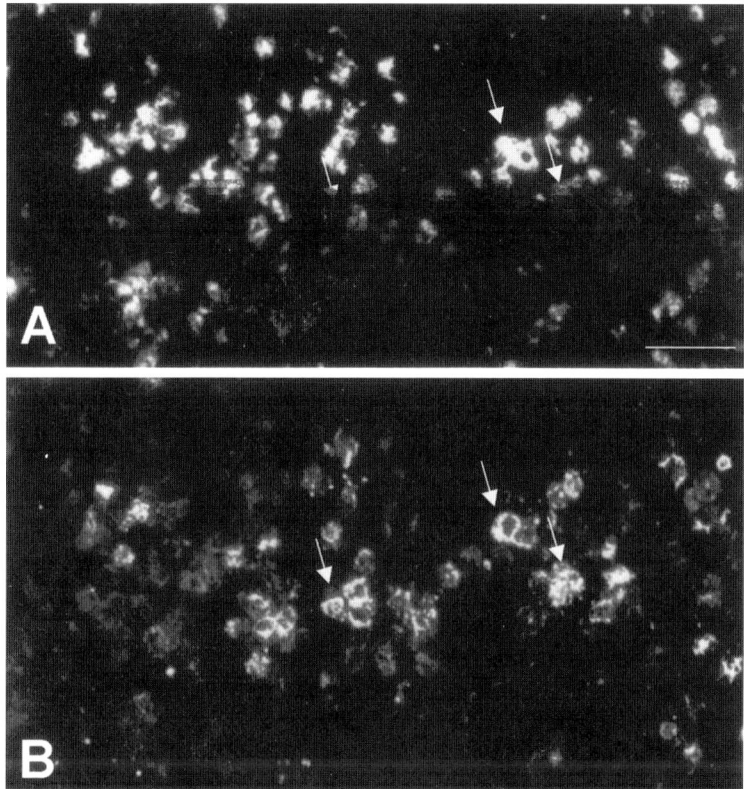

FIGURE 3. Identification of most CD8+ cells (B) as ED1+ microglia/macrophages (A) by double immunofluorescence in focal cerebral infarcts. Examples of double positive cells are marked by arrows. Bar represents 50μm.

(2001b)). This suggests that the majority of CD8+ cells was derived from hematogenous macrophages, but microglia to some extent also upregulated CD8. Accordingly, CD8+ cells resembling shape ramified microglia could be detected in the boundary zone (Fig. 4B, inset).

To further characterize the lesion conditions facilitating recruitment of CD8+ phagocytes, mild focal ischemia was induced by transient MCAO. As shown in Fig. 1A, this model leads to a core infarction with ischemic pannecrosis, which is surrounded by areas with delayed and selective neuronal death. In transient MCAO, recruitment of CD8+ phagocytes was restricted to areas of ischemic pannecrosis. In areas undergoing selective neuronal loss, microglia upregulated CR3, showed a limited degree of phagocytic activity as indicated by ED1 staining, and to some extent expressed CD4, but not CD8, molecules (Schroeter et al. (2001b)). Recently, a similar CD8+ microglia/

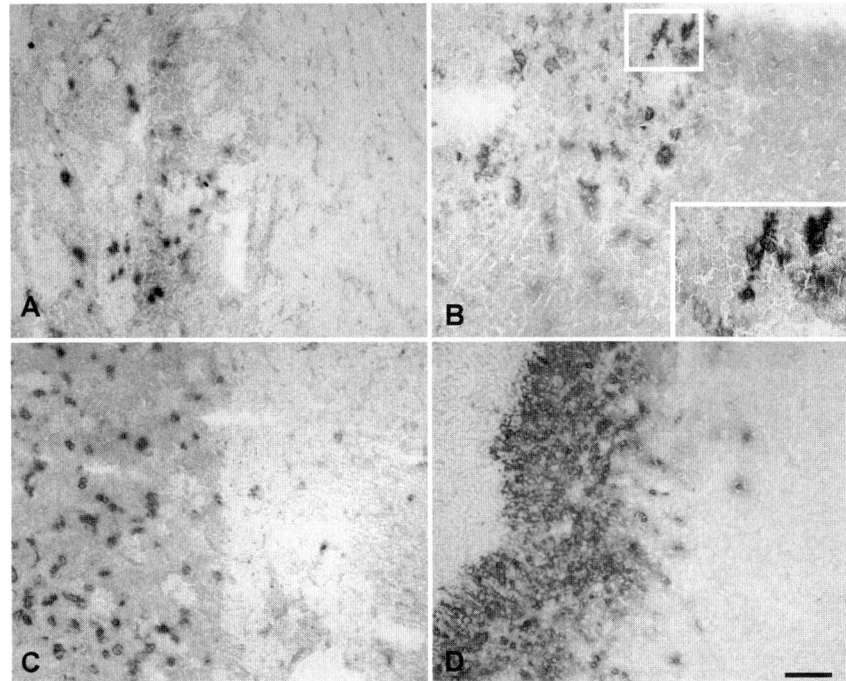

FIGURE 4. Selective targeting of CD8+ microglia/macrophages to areas of ischemic pannecrosis at day 3 (A,B) and day 7 (C,D) in the striatum after transient MCAO (A,C) and in the neocortex after photothrombosis (B,D). Notice the increase of CD8+ infiltrates between day 3 and 7. Inset in (B) shows stout microglia expressing the CD8 molecule. Bar represents 100μm for (A–D).

macrophage population was detected in Bornavirus-infected neonatal Lewis rats and found to be restricted to regions with prominent dropout of neurons or neuronal processes (Weissenböck et al. (2000)). In conclusion, it appears that the presence of CD8+ microglia/macrophages in ischemic CNS lesions indicates pannecrosis.

The functional role of CD8+ microglia/macrophages in brain ischemia is unclear. CD8+ cells could contribute to exacerbation of ischemic brain damage, as well as to tissue remodeling and healing processes. *In vitro* studies done independently on alveolar macrophages showed that signalling via the CD8 molecule led to activation of TNF-α, IL-1β and iNOS expression, with concomitant enhanced cytotoxicity toward Leishmania major (Hirji et al. (1998, 1999)). NO derived from neurons and inflammatory cells has been implicated as a major pathogenic factor in secondary infarct growth after focal cerebral ischemia (Iadecola (1997); Iadecola et al. (1997)).

2.1.4 Induction of inflammatory effector molecules. Focal cerebral ischemia elicits a strong cytokine response in the CNS (Arvin et al. (1996); Barone et al. (1997); Iadecola (1997); Shohami et al. (1999); Touzani et al. (1999); Jander et al. (2000); Stoll et al. (2000)) (Fig. 5). Cytokine induction is an early event and accompanied by signs of glial activation. Transcripts for IL-1β and TNF-α were detectable as early as four hours in the infarct core and border zone with a maximum expression during the first two days in focal cerebral ischemia. Thus, in contrast to autoimmune disorders such as experimental autoimmune encephalomyelitis, cytokine expression in focal ischemia preceded the step-wise infiltration of the infarct and its border zone by inflammatory cells (Fig. 5). Based on cellular identification by morphological criteria, putative microglia in ischemic lesions and surrounding tissue expressed IL-1β protein by six hours (Davis et al. (1999)). By 24 and 48 hours after ischemia, the number and spread of IL-1β immunoreactive cells had increased and included activated microglia and infiltrating macrophages. The cellular source of TNF-α in cerebral ischemia is under debate. In the study by Liu et al. (1994), TNF-α immunoreactivity was confined to neurons in the evolving infarct at six and twelve hours, and was primarily associated with macrophages located within the infarcted tissue at five days. Botchkina et al. (1997) described a more diffuse cellular distribution of TNF in neurons, astrocytes, microglia and infiltrating granulocytes. Conversely, Gregersen and colleagues (2000) recently colocalized TNF-mRNA exclusively with MAC-1+ microglia/macrophages in mice after pMCAO. In this study, granulocytes and astrocytes were TNF-negative when examined at six and twenty-four hours after ischemia.

Both IL-1β and TNF-α have been implicated in infarct growth after focal ischemia, probably through the induction of iNOS (reviewed in Shohami et al. (1999)); Touzani et al. (1999)). In culture systems, IL-1β and TNF-α synergistically induce iNOS activity in microglia (Minghetti and Levi (1998)). In focal ischemic lesions, IL-1β and TNF-α expression was followed by strong induction of iNOS-mRNA, which reached peak levels during the first 24 to 48 hours after ischemia (Iadecola et al. (1997)) (Fig. 5). After disruption of the iNOS gene, mice developed smaller infarcts and fewer motor deficits. Most importantly, such reduction in ischemic damage and neurological deficit was observed 96 hours after ischemia, but not at 24 hours, providing strong evidence that iNOS expression is one of the critical factors that contribute to the delayed expansion of brain damage.

2.2 Microglial activation in secondarily degenerating fiber tracts and nuclei

A restricted pattern of microglial activation occurs in fiber tracts undergoing Wallerian degeneration after focal ischemia (Schroeter et al. (1999)). In the photothrombosis model, microglia were activated in commissural, thalamocortical

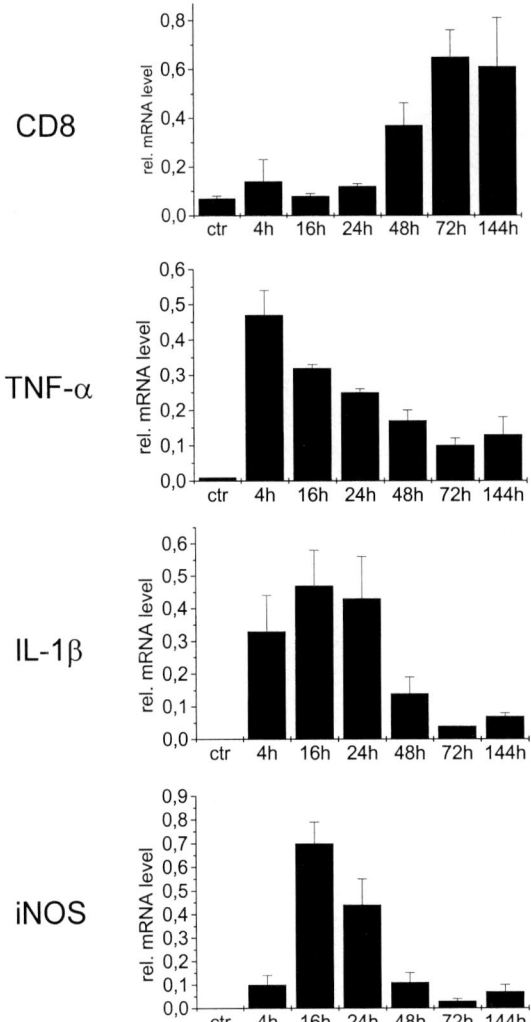

FIGURE 5. Temporal relation between cytokine/iNOS induction and inflammation as examplified by CD8 expression in focal ischemic lesions after permanent MCAO. Notice that transcripts for TNF-α and IL-1 β are upregulated as early as 4 hours after induction of cerebral ischemia, but that iNOS induction is delayed to 16 hours. In accordance with the immunocytochemical data, CD8-mRNA expression and macrophage invasion starts at day 2 and reaches peak levels between day 3 and 7. Notice that cytokine transcripts are already downregulated at the peak of CD8 expression. This indicates that cytokine expression is an intrinsic response of the CNS to ischemic injury probably involving microglia as a major cellular source.

and corticospinal fiber tracts, and in the posterior thalamic nuclei and lateral geniculate body when the primary ischemic lesion was localized in the sensorimotor and occipital corteces (Fig. 6). Ramified microglia transformed into stout microglia and expressed increased CR3-immunoreactivity after day 3, an activation that was fully developed at day 14. In contrast to the necrotic lesion (star in Fig. 6A), microglia only partially rounded and expressed ED1 immunoreactivity indicating limited phagocytic transformation in degenerating thalamic nuclei. (inset Fig. 6B). The vast majority of OX42+ microglia in degenerating fiber tracts and thalamus still showed cellular processes and exhibited strong MHC class II (Fig. 6C) and CD4 (Fig. 6D) expression. Although CD8+ microglia/macrophages constituted a substantial proportion of perilesional phagocytes (2.1.3), they were completely absent from secondarily degenerating fiber tracts and nuclei. This limited microglial response reflects the usual slow phagocytic reaction seen after fiber tract injury in the CNS (Perry and Gordon (1987); Stoll et al. (1989); Perry et al. (1993); George and Griffin (1994); Kreutzberg (1996); Stoll and Jander (1999); Jander et al. (2001a)).

2.3. Remote microglial responses

Microglia not only respond in neocortical areas with primary or secondary structural damage, but also in the remote ipsilateral cortex lacking infiltration of hematogenous leukocytes (Fig. 7). In areas remote from the focal ischemic lesion, no detectable tissue damage occurred and, in particular, there was no indication of ongoing apoptosis (Braun et al. (1996); Isenmann et al. (1998); Schroeter et al. (1999)).

Focal ischemic lesions in rodents provoke cortical spreading depression (CSD) (Dietrich et al. (1994); Schroeter et al. (1995)). CSD is characterized by a transient suppression of all neuronal activity that extends from sites of increased extracellular potassium concentrations to the entire ipsilateral, but not the contralateral, hemisphere (reviewed in Martins-Ferreira et al. (2000)). In the normal brain, CSD does not induce neuronal injury, and indeed facilitates neuronal tolerance against subsequent lethal ischemic challenge (Kobayashi et al. (1995); Yanamoto et al. (1998)). CSD can be blocked by pretreatment with the noncompetitive N-methyl-D-aspartate (NMDA) receptor antagonist MK-801, which suggests glutamate-signaling through the NMDA receptor (Lauritzen and Hansen (1992)). In the photothrombosis model, microglia of the ipsilateral cortex stained more intensely with OX42 antibodies at day 3, indicating upregulation of CR3 (Schroeter et al. (1999)) (Fig. 7A, B). Moreover, microglial cell density increased. These changes were less prominent at day 7, and returned to baseline levels at day 14. This remote microglial response was not accompanied by upregulation of MHC class II molecules, CD4 or CD8. MK-801-mediated suppression of CSD partly reduced, but did not abolish, microglial activation in the remote ipsilateral cortex. Similar to microglia, cortical astrocytes of the entire ipsilateral cortex showed

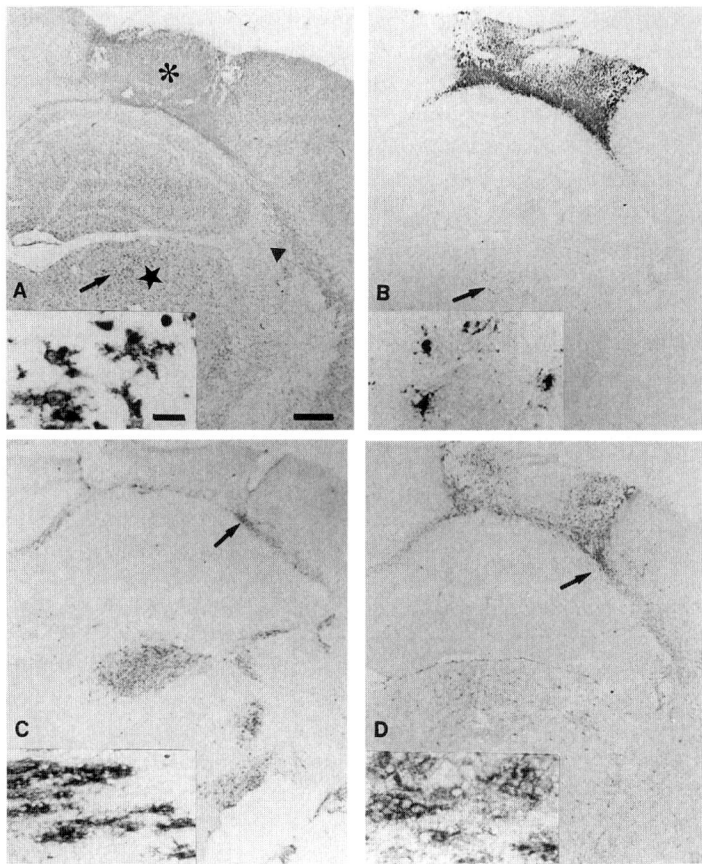

FIGURE 6. Microglial responses in the ischemic core (asterisk in A) compared with the fiber tracts undergoing Wallerian degeneration (arrowhead in A) and the thalamus (star in A) at day 14 after photothrombosis (see also fig. 1C). Serial sections stained for OX42 (A), ED1 (B), MHC class II (C) and CD4 molecules (D). Arrows in (A–D) mark areas where the inset is taken from to show labelled microglia at higher magnification. Notice the widespread and intense microglia staining in the ischemic lesion (where additionally hematogenous macrophages are present and stained), the fibre tracts and the thalamus (inset) with mab OX42 in (A). (B) is a serial section stained with mab ED1 showing that phagocytic activity is predominant in the ischemic lesion, and present to some extent in the thalamus (inset), but almost absent in degenerating fiber tracts. Contrastingly, MHC class II expression is low in the ischemic core at this late stage of infarct development, but strong in degenerating fiber tracts (inset in C) and the thalamus (C). CD4 is predominantly upregulated in the periphery of the ischemic core, degenerating fiber tracts (inset) and to lesser extent in the thalamus (D). Scale bars: (A–D) = 800μm, (insets) = 40μm. Reproduced from Schroeter et al. (1999) *Neuroscience* 89:1367–1377, with permission.

6. Microglia and Macrophage Responses in Cerebral Ischemia   139

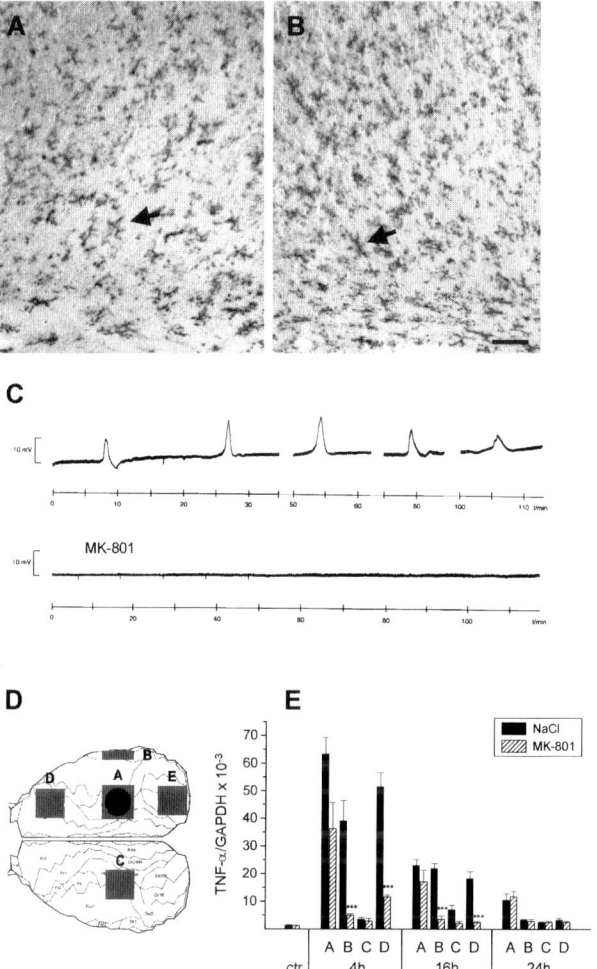

FIGURE 7. Microglial activation remote from an ischemic lesion (light gray area in Fig. 1C) as assessed by immunostaining with mab OX42. (A) shows OX42 labelling of microglia in the normal neocortex. Notice that 3 days after focal ischemia microglia have increased in number and show more intense OX42 staining indicating upregulation of CR3 (B). Single microglial cells are marked by arrows. Bar in (B) represents 50μm. Immediately after infarction spreading depression (SD) "travel" over the entire ipsilateral hemisphere (C), which can be completely blocked by the NMDA-antagonist MK-801. (D), (E): In the ischemic focus (area A), TNF-α-mRNA is induced with peak levels at 4 hours. TNF-α induction is not restricted to the ischemic focus, but also occurs in remote ipsilateral (areas B,D,E), but not contralaterally (area C). Notice that blocking of SD by MK-801 almost completely abolished remote cytokine induction (hatched bars in (E)).

signs of activation, as revealed by increased expression of glial fibrillary acidic protein with peak levels between days 3 and 7, which could also be blocked by MK-801 (Schroeter et al. (1995)). The pathophysiological consequences of this widespread activation of microglia and astrocytes are unknown, but could be part of a stress response leading to the development of ischemic tolerance and perilesional plasticity. Accordingly, microglial and astrocytic activation was preceded by induction of the proinflammatory cytokines TNF-α and IL-1β (Jander et al. (2000; 2001b)) which have been inflicted in ischemic preconditioning (Wang et al. (2000)) and increased long-term potentiation in the vicinity of ischemic foci (Hagemann et al. (1998); Schneider et al. (1998)). In contrast to the ischemic focus, TNF-α and IL-1β expression in remote brain regions was not accompanied by iNOS induction. Since iNOS plays an essential role in NMDA-mediated neuronal injury (Hewett et al. (1994)), this may explain why TNF-α and IL-1β expression in remote brain areas is neuroprotective rather than neurotoxic.

3. Microglial activation in global ischemia and areas of selective neuronal loss

Global cerebral ischemia refers to transient interruption of the blood flow to the brain by four-vessel occlusion, or bilateral occlusion of the carotid arteries. After brief periods of global cerebral ischemia (up to 15 minutes), selective loss of hippocampal CA1 pyramidal cells occurs and is typically delayed to two to four days, whereas CA3 neurons survive because they are more resistant to hypoxia (Ordy et al. (1993)). Thirty minutes of global ischemia produces more severe CA1 damage and less severe, but definite, involvement of CA3.

Morioka and colleagues (1991) described the microglial reaction in the dorsal hippocampus following 25 minutes of transient forebrain ischemia using the microglia/macrophage marker *Griffonia simplicifolia* B4 isolectin. Increased staining of microglial cells was detected in the dentate hilus and CA1 area as early as twenty minutes after reperfusion, and more intensely at 24 hours. The strongest microglial reaction was observed four to six days after reperfusion. At all time points examined, microglial reactivity in the CA3 pyramidal and dentate granule cell layers was lower than in CA1. Similar findings were reported by Jorgensen et al. (1993) using histochemical nucleoside diphosphatase (NDPase) and immunocytochemical CR3 stains. With both techniques, an early but transient microglial reaction with upregulation of the above markers and retraction of cellular processes in all hippocampal subfields was found, which included areas devoid of subsequent neuronal degeneration. Furthermore, MHC class I, but not MHC class II, molecules were expressed by microglia (Morioka et al. (1992); Finsen et al. (1993)) and an increase in microglial C1q biosynthesis, the recognition subcomponent of the classical complement-activation pathway was described by Schäfer and colleagues (2000). From days 4 to 7, the shape and staining characteristics of microglial cells returned to normal in areas without neuronal death. In the CA1 region, internucleosomal DNA fragmentation and apoptotic bodies were observed between days 3 and 4 after ischemia, a period when neuronal death is maximal (Bhat et al. (1996); Friedlander et al. (1997)). Accordingly, the number of pyramidal neurons was significantly reduced at one week after ischemia (Lin et al. (1998)). Neuronal death was associated with the appearance of large numbers of intensely NDPase/CR3-positive microglial cells, which were now ameboid (Jorgensen et al. (1993)). Ameboid microglia expressed increased levels of nitric oxide synthase (Yamashita et al. (1995, 2000)) and of IL-1-converting enzyme (ICE, also designated caspase 1). ICE cleaves corresponding precursor molecules into bioactive proinflammatory cytokines IL-1 and IL-18, and furthermore is a key enzyme in the molecular cascade leading to apoptotic cell death (Bhat et al. (1996)). In the dentate hilus, another hippocampal area undergoing neuronal cell loss, microglia actually clustered around degenerating neurons. In addition to MHC class I molecules, these cells upregulated leukocyte common antigen and, to a variable degree, MHC class II molecules (Morioka et al. (1992); Finsen et al. (1993)). At day 21, CA1 was the only hippocampal region where these microglial activation markers were still upregulated. In contrast to focal cerebral ischemia, global ischemia does not attract many hematogenous inflammatory cells. T-cells were scarce (Morioka et al. (1992)), and macrophage and granulocyte infiltration was modest, although the endothelial adhesion molecule ICAM-1 was upregulated early in CA1 (Clark et al. (1995)). Treatment with anti-ICAM-1 antibodies, accordingly, did not significantly affect appearance of macrophage-like cells at day 4 (Clark et al. (1995)).

As described above (1.), areas with selective neuronal death are also present after transient focal ischemia, and are characterized by loss of pyramidal neurons, but preservation of endothelial cells and glia similar to global ischemia (Chen et al. (1993)) (Fig.1B). Accordingly, these regions showed a similar restricted microglial response as CA1 in global ischemia (Schroeter et al. (2000b)). Microglia were less ramified, upregulated CR3 and CD4 molecules and exhibited weak ED1 immunoreactivity, an indication of partial phagocytic transformation. Selective neuronal death in these regions can be attenuated by treatment with caspase inhibitors (Endres et al. (1998)). In contrast to pannecrotic ischemic lesions, CD8+ cells were virtually absent from areas of selective neuronal death (Schroeter et al. (2000b)).

4. Microglia activation and macrophage recruitment: beneficial or harmful?

Activated microglia have the capacity to synthesize and secrete a large number of molecules, which potentially cause brain damage, induce repair processes, or orchestrate immune responses. Molecules produced by microglia include neurotoxins, cytokines, coagulation and complement factors, prostanoids, free radicals, extracellular matrix components, enzymes and growth factors (Piani et al. (1991); Chao et al. (1992); Banati et al. (1993); Giulian et al. (1993); Zielasek and Hartung (1996)). In general, our knowledge about the circumstances under which each of these molecules is produced, and the regulatory elements involved, is limited. This is mainly due to the complexity of possible interactions and the unknown contribution of other glial cells and infiltrating leukocytes *in vivo*. Most of the information on the cytotoxic properties of activated microglia pertain to *in vitro* observations, and require *in vivo* confirmation. Thus, the functional role of microglial activation in the pathophysiology of cerebral infarction awaits further clarification.

Evidence for direct neurotoxic effects is derived from culture studies. Giulian and colleagues reported on a neurotoxic activity secreted by cultured rat microglia and macrophages obtained from ischemic brain tissue two days after transient MCAO (Giulian et al. (1993)). This neurotoxic factor of unknown identity was resistant to proteases, and its neurotoxic effects could be blocked by NMDA antagonists. In an independent study, Piani and colleagues showed that murine macrophages secrete glutamate, and thereby damage neurons *in vitro* (Piani et al. (1991)), further supporting the notion that microglia/macrophages participate in NMDA-mediated neuronal damage. Moreover, as yet unknown factors secreted by macrophages triggered apoptosis in cultured hippocampal neurons (Flavin et al. (1997)). Peters and colleagues (1998) found increased formation of reactive oxygen species in ischemic lesions after transient and permanent MCAO. Microglia and macrophages are potential sources of these neurotoxic radicals (Banati et al. (1993)). *In vivo*, pharmacological treatment of rats with minocycline, a second generation tetracycline with profound anti-inflammatory activity, diminished neuronal loss in focal and global ischemia models (Yrjänheikki et al. (1998; 1999)). On the molecular level, minocycline treatment led to reduced levels of interleukin-1 converting enzyme (ICE), cyclooxygenase-2 and partly iNOS-mRNA expression in ischemic lesions. This was paralleled by attenuated microglial responses, as revealed by less pronounced CR3 upregulation and lower degrees of ameboid transformation. Inhibition of ICE proteases reduced ischemic neuronal damage (Hara et al. (1997)). Whether reduced microglial activation was the cause, or a consequence, of reduced infarct volume is unclear.

Microglia are involved in cytokine networks, both as cellular sources and as target cells. As outlined above (2.1.4), TNF-α and IL-1β are rapidly induced in ischemic brain lesions and contribute to lesion expansion (Arvin et al. (1996); Jander et al. (2000)). Microglia and macrophages at least partly contribute to this TNF-α and IL-1β production in ischemic brain lesions (Botchina et al. (1997); Davis et al. (1999); Gregersen et al. (2000)). *In vitro*, TNF-α and IL-1β exacerbated neuronal cell death by excitotoxins, an effect dependent on the coinduction of iNOS (Hewett et al. (1994)). Rodent microglia respond to stimulation by TNF-α and IL-1β with iNOS induction, and subsequent production of large amounts of NO (Minghetti and Levi (1998)). NO is a *pleiotropic* molecule engaged in multiple molecular interactions. Upon reaction with oxygen, the reactive nitrogen oxygen species peroxynitrite is generated, a cytotoxic agent. The important pathophysiological role of NO and its derivates in stroke development was underscored by *in vivo* experiments. iNOS-knock-out mice and mice treated with selective iNOS inhibitors developed smaller infarcts (Iadecola et al. (1997); Iadecola, (1997)). Taken together, it appears that activated microglia within the first four days contribute to ischemic brain damage by release of proinflammatory cytokines and NO. However, within the same time frame the cytokine transforming growth factor-β (TGF-β) is induced in focal ischemic lesions (Lehrmann et al. (1995)). Besides exerting anti-inflammatory effects, TGF-β can rescue cultured neurons from excitotoxic and hypoxic cell death, which has been attributed to its antioxidative and antiapoptotic properties (Prehn et al. (1993); Henrich-Noack et al. (1996)). Microglia and macrophages express TGF-β in ischemic brain lesions (Lehrmann et al. (1998)). The functional consequences of the local interaction between pro- and anti-inflammatory cytokines in ischemic lesions is unclear. The coordinated expression of counteracting cytokines may serve as an autocrine mechanism to control microglial/macrophage actions in the infarct region.

Is the inflammmatory response beneficial at later stages of infarct development? Most evidence suggests that infarcts no longer grow beyond day 4, when macrophage infiltration still is at its early stage. Phagocyte numbers further increase during the first two weeks after cerebral infarction. As a consequence of this strong combined microglial/macrophage response, intact and viable neurons are separated from pannecrotic tissue (infarct demarcation) and cellular debris is rapidly removed. This rapid phagocytic response is in sharp contrast to the limited and delayed microglial activation in CNS lesions with pure fiber tract degeneration undergoing Wallerian degeneration. These lesions lack significant macrophagic migration, and myelin remnants impeding nerve regeneration persist for months (Stoll et al. (1989); Schwab (1990); Stoll and Jander (1999)). In contrast, during focal cerebral ischemia microglia appear to be rapidly and appropriately activated to act as phagocytes and to aid hematogenous macrophages in debris removal and subsequent tissue remodeling. Elucidation of the underlying signaling cascades of this rapid microglial activation in focal ischemia may help to develop new strategies to foster regeneration after CNS injuries.

In vitro, microglia produce neurotrophins such as nerve growth factor, neurotrophin 3, and basic fibroblast growth factor upon stimulation with cytokines (Elkabes et al. (1996); Heese et al. (1998)), the neurite-outgrowth-promoting extracellular matrix molecule thrombospondin (Chamak et al. (1994)), and other as yet unidentified growth-promoting activities (Rabchevsky and Streit (1997)).

Neurotrophins are upregulated in the environment of ischemic lesions, but immunoreactivity was mostly localized in neurons (Comelli et al. (1993); Kokaia et al. (1995; 1998). Currently it is unclear whether microglia contribute to neurotrophin synthesis after focal ischemia. Osteopontin (OPN) is a multifunctional glycoprotein that is transiently expressed at high levels during wound healing in many tissues. OPN is significantly upregulated in the surround of ischemic brain lesions. OPN could be localized in peri-infarct microglia beginning at 24 hours, and in microglia/macrophages at 48 hours, within the infarct core after permanent MCAO (Ellison et al. (1998)). OPN was secreted into the extracellur matrix and reached peak levels at five days (Wang et al. (1998)). Functionally, OPN is chemotactic for macrophages and astrocytes that bear the corresponding OPN receptor integrin alpha(v)beta 3 on their surface. Thus, OPN has been implicated in the recruitment of additional hematogenous macrophages and in the formation of the glial scar after focal ischemia. Moreover, OPN inhibits NO production in activated macrophages by suppression of iNOS activity (Rollo et al. (1996)) that might mediate protective effects.

Summary

Microglial responses to cerebral ischemia are tightly and differentially regulated. After focal cerebral ischemia, three principal patterns of microglial activation can be distinguished: (i) rapid microglial transformation into phagocytes in the border zone of infarction accompanied by a unusual CD8+ phagocyte population in the rat; (ii) a slowly evolving response in fiber tracts undergoing Wallerian degeneration, and in nuclei with retrograde neuronal degeneration; and (iii) a transient ipsilateral activation remote from the infarction that is partly due to NMDA-receptor signaling. Hematogenous macrophages and other leukocytes are *recruited* with a delay preferentially to sites of ischemic pannecrosis and the immediate border zone. Global cerebral ischemia lacks a significant inflammatory response from the circulatory system. Microglial activation is most prominent and persistent in areas of ongoing neuronal death and axonal degeneration, but rapidly reversible in brain areas without neuronal damage.

In focal ischemic brain lesions, morphological signs of microglial activation are accompanied by upregulation of cytokines, iNOS, and immunological cell adhesion molecules that precede inflammation by hematogenous cells. Activated microglia potentially contribute to neuronal death at early stages of focal ischemia by secretion of TNF-α, IL-1β with ensuing NO production, and other neurotoxic molecules. Beyond day 4 after focal ischemia, activated microglia in concert with infiltrating macrophages probably assist in wound healing and tissue remodelling.

Acknowledgement

The authors' work cited in this chapter was supported by the Deutsche Forschungsgemeinschaft (SFB 194, B6). Guido Stoll held a Hermann and Lilly Schilling professorship.

References

Arvin B, Neville LF, Barone FC, Feuerstein GZ (1996). The role of inflammation and cytokines in brain injury. *Neurosci Biobehav Reviews* 20:445–452.

Banati RB, Gehrmann J, Schubert P, Kreutzberg GW (1993). Cytotoxicity of microglia. *Glia* 7:111–118.

Barone FC, Arvin B, White RF, Miller A, Web CL, Willette RN, Lysko PG, Feuerstein GZ (1997). Tumor necrosis factor-alpha. A mediator of focal ischemic brain injury. *Stroke* 28:1233–1244.

Bauer J, Huitinga I, Zhao W, Lassmann H, Hickey WF, Dijkstra CD (1995). The role of macrophages, perivascular cells, and microglial cells in the pathogenesis of experimental autoimmune encephalomyelitis. *Glia* 15:437–446.

Beaulieu C, De Crespigny A, Tong DC, Moseley ME, Albers GW, Marks MP (1999). Longitudinal magnetic resonance imaging study of perfusion and diffusion in stroke: evolution of lesion volume and correlation with clinical outcome. *Ann Neurol* 46:568–578.

Becker KJ, Mc Carron RM, Ruetzler C, Laban O, Sternberg E, Flanders KJ, Hallenbeck JM (1997). Immunologic tolerance to myelin basic protein decreases stroke size after transient focal cerebral ischemia. *Proc Natl Acad Sci* (USA) 94:10873–10878.

Bhat RV, DiRocco R, Marcy VR, Flood DG, Zhu Y, Dobrzanski P, Siman R, Scott R, Contreras PC, Miller M (1996). Increased expression of IL1-β converting enzyme in hippocampus after ischemia: selective localization in microglia. *J Neurosci* 16:4146–4154.

Bolander HG, Persson L, Hillered L, D'Argy R, Ponten U, Olsson Y (1989). Regional cerebral blood flow and histopathological changes after middle cerebral artery occlusion in rats. *Stroke* 20:930–937.

Botchina GI, Meistrell ME, Botchina IL, Tracey KJ (1997). Expression of TNF and TNF receptors (p55 and p75) in the rat brain after focal cerebral ischemia. *Mol Med* 3:765–781.

Braun JS, Jander S, Schroeter M, Witte OW, Stoll G (1996). Spatiotemporal relationship of apoptotic cell death to lymphomonocytic infiltration in photochemically induced focal ischemia of the rat cerebral cortex. *Acta Neuropath* 92:255–263.

Bredesen DE (1995). Neural apoptosis. *Ann Neurol* 38:839–851.

Chao CC, Hu S, Molitor TW, Shaskan EG, Peterson PK (1992). Activated microglia mediate neuronal cell injury via a nitric oxide mechanism. *J Immunol* 149:2736–2741.

Chamak B, Morandi V, Mallat M (1994). Brain macrophages stimulate neurite outgrowth and regeneration by secreting thrombospondin. *J Neurosci Res* 38:221–233.

Chen H, Chopp M, Schultz L, Bodzin G, Garcia JH (1993). Sequential neuronal and astrocytic changes after transient middle cerebral artery occlusion in the rat. *J Neurol Sci* 118:109–116.

Clark RK, Lee EV, Fish CJ, White RF, Price WJ, Jonak ZL, Feuerstein GZ, Barone FC (1993). Development of tissue damage, inflammation and resolution following stroke: an immunohistochemical and quantitative planimetric study. *Brain Res Bull* 31:565–572.

Clark WM, Lauten JD, Lessov N, Woodward W, Coull BM (1995). Time course of ICAM-1 expression and leukocyte subset infiltration in rat forebrain ischemia. *Mol Chem Neuropathol* 26:213–230.

Comelli MC Guidolin D, Seren MS, Zanoni R, Canella R, Rubini R, Manev H (1993). Time course, localisation and pharmacological modulation of immediate early inducible genes, brain-derived neurotrophic factor and trkB messenger RNAs in the rat brain following photochemical stroke. *Neuroscience* 55:473–490.

Davis CA, Loddick SA, Toulmond S, Stroemer RP, Hunt J, Rothwell NJ (1999). The progression and topographic distribution of interleukin-1β expression after permanent middle cerebral artery occlusion in the rat. *J Cereb Blood Flow Metab* 19:87–98.

Dietrich D, Feng ZC, Leistra H, Watson BD, Rosenthal M (1994). Photothrombotic infarction triggers multiple episodes of cortical spreading depression in distant brain regions. *J Cereb Blood Flow Metab* 14:20–28.

Dirnagl U, Iadecola C, Moskowitz MA (1999). Pathobiology of ischaemic stroke: an integrated view. *Trends Neurosci* 22:391-397.

Elkabes S, DiCicco-Bloom EM, Black IB (1996). Brain microglia/macrophages express neurotrophins that selectively regulate microglial proliferation and function. *J Neurosci* 16:2508–2521.

Ellison JA, Velier JJ, Spear P, Jonak ZL, Wang X, Barone FC, Feuerstein GZ (1998). Osteopontin and its integrin receptor alpha(v)beta3 are upregulated during formation of the glial scar after focal stroke. *Stroke* 29:1698–1706.

Endres M, Namura S, Shimizu-Sasamata M, Waeber C, Zhang L, Gomez-Isla T, Hyman BT, Moskowitz MA (1998). Attenuation of delayed neuronal death after mild focal ischemia in mice by inhibition of the caspase family. *J Cereb Blood Flow Metabol* 18:238–247.

Finsen BR, Jorgensen MB, Diemer NH, Zimmer J (1993). Microglial MHC antigen expression after ischemic and kainic acid lesions of the adult rat hippocampus. *Glia* 7:41–49.

Flaris NA, Densmore TL, Molleston MC, Hickey WF (1993). Characterization of microglia and macrophages in the central nervous system of rats: definition of the differential expression of molecules using standard and novel monoclonal antibodies in normal CNS and in four models of parenchymal reaction. *Glia* 7:34–40.

Flavin MP, Coughlin K, Ho LT (1997). Soluble macrophage factors trigger apoptosis in cultured hippocampal neurons. *Neurosci* 80:437–448.

Friedlander RM, Gagliardini V, Hara H, Fink KB, Li W, MacDonald G, Fishman MC, Greenberg AH, Moskowitz MA, Yuan J (1997). Expression of a dominant negative mutant of interleukin-1 beta converting enzyme in transgenic mice prevents neuronal cell death induced by trophic factor withdrawl and ischemic brain injury. *J Exp Med* 185:933–940.

Garcia JH, Yoshida Y, Chen H, Li Y, Zhang ZG, Lian J, Chen S, Chopp M (1993). Progression from ischemic injury to infarct following middle cerebral artery occlusion in the rat. *Am J Pathol* 142:623–635.

Garcia JH, Liu KF, Yoshida Y, Lian J, Chen S, del Zoppo G (1994). Influx of leukocytes and platelets in an evolving brain infarct (Wistar rat). *Am J Pathol* 144:188–199.

George R, Griffin JW (1994). Delayed macrophage responses and myelin clearance during Wallerian degeneration in the central nervous system: the dorsal radiculotomy model. *Exp Neurol* 129:225–236.

Giulian D, Corpuz M, Chapman S, Mansouri M, Robertson C (1993). Reactive mononuclear phagocytes release neurotoxins after ischemic and traumatic injury to the central nervous system. *J Neurosci Res* 36:681–693.

Gourmala NG, Buttini M, Limonta S, Sauter A, Boddeke HW (1997). Differential and time-dependent expression of monocyte chemoattractant protein-1 mRNA by astrocytes and macrophages in rat brain: effects of ischemia and peripheral lipopolysaccharide administration. *J Neuroimmunol* 74:35–44.

Gregersen R, Lambertsen K, Finsen B (2000). Microglia and macrophages are the major source of tumor necrosis factor in permanent middle cerebral artery occlusion in mice. *J Cereb Blood* Flow Metab 20:53–65

Hagemann G, Redecker C, Neumann-Haefelin T, Freund HJ, Witte OW (1998). Increased long-term potentiation in the surround of experimentally induced focal cortical infarction. *Ann Neurol* 44:255–258.

Hara H, Friedlander RM, Gagliardini V, Ayata C, Ayata G, Huang Z, Shimizu-Sasamata M, Yuan J, Moskowitz M (1997). Inhibition of ICE family proteases reduces ischemic and excitotoxic neuronal damage. *Proc Natl Acad Sci USA* 94:2007–2012.

Heese K, Hock C, Otten U (1998). Inflammatory signals induce neurotrophin expression in human microglial cells. *J Neurochem* 70:699–707.

Henrich-Noack P, Prehn JH, Krieglstein J (1996). TGF-beta 1 protects hippocampal neurons against degeneration caused by transient global ischemia. Dose-response relationship and potential neuroprotective mechanisms. *Stroke* 27:1609–1614.

Hewett SJ, Csernansky CA, Choi DW (1994). Selective potentiation of NMDA-induced neuronal injury following induction of astrocytic iNOS. *Neuron* 13:487–494.

Hirji N, Lin TJ, Bissonnette E, Belosevic M, Befus AD (1998). Mechanisms of macrophage stimulation through CD8: macrophage CD8 alpha and CD8 beta induce nitric oxide production and associate killing of the parasite Leishmani major. *J Immunol* 160:6004–6011.

Hirji N, Lin TJ, Gichrist M, Nault G, Nohara O, Grill BJ, Belosevic M, Stenton GR, Schreiber AD, Befus AD (1999). Novel CD8 molecule on macrophages and mast cells: expression, function and signaling. *Int Arch Allergy Immunol* 118:180–182.

Hossmann KA (1995). Viability thresholds and the penumbra of focal ischemia. *Ann Neurol* 36:557–565.

Iadecola C (1997). Bright and dark sides of nitric oxide in ischemic brain injury. *Trends Neurosci* 20:132–139.

Iadecola C, Zhang F, Casey R, Nagayama M, Ross ME (1997). Delayed reduction of ischemic brain injury and neurological deficits in mice lacking the inducible nitric oxide synthase gene. *J Neurosci* 17:9157–9164.

Iadecola C, Salkowski CA, Zhang F, Aber T, Nagayama M, Vogel SN, Ross ME (1999). The transcription factor interferon regulatory factor 1 is expressed after cerebral ischemia and contributed to ischemic brain injury. *J Exp Med* 189:719–727.

Isenmann S, Stoll G, Schroeter M, Krajewski S, Reed JC, Bähr M (1998). Differential regulation of bax, bcl-2, and bcl-x proteins in focal cortical ischemia in the rat. *Brain Pathol* 8:49–63.

Jander S, Kraemer M, Schroeter M, Witte OW, Stoll G (1995). Lymphocytic infiltration and expression of intercellular adhesion molecule-1 in photochemically induced ischemia of the rat cortex. *J Cereb Blood Flow Metab* 15:42–51.

Jander S, Pohl J, Gillen C, Schroter M, Stoll G (1996). Vascular cell adhesion molecule-1mRNA is expressed in immune-mediated and ischemic injury of the rat nervous system. *J Neuroimmunol* 70:75–80.

Jander S, Schroeter M, D'Urso D, Gillen C, Witte OW, Stoll G (1998). Focal cerebral ischemia of the rat brain elicits an unusual inflammatory response: early appearance of CD8+ macrophages/microglia. *Eur J Neurosci* 10:680–688.

Jander S, Schroeter M, Stoll G (2000). Role of NMDA receptor signalling in the regulation of inflammatory gene expression after focal brain ischemia. *J Neuroimmunol* 109:181–187.

Jander S, Lausberg F, Stoll G (2001a). Differential recruitment of CD8+ macrophages during Wallerian degeneration in the peripheral and central nervous system. *Brain Pathol* 11:27–38.

Jander S, Schroeter M, Peters O, Witte OW, Stoll G (2001b). Cortical spreading expression induces proinflammatory cytokine gene expression in the rat brain. *J Cereb Blood Flow Metab* 21:218–225.

Jorgensen MB, Finsen BR, Jensen MB, Castellano B, Diemer NH, Zimmer J (1993). Microglial and astroglial reactions to ischemic and kainic acid-induced lesions of the adult rat hippocampus. *Exp Neurol* 120:70–88.

Kim JS, Gautam SC, Chopp M, Zalonga C, Jones ML, Ward PA, Welch KM (1995). Expression of monocyte chemoattractant protein-1 and macrophage inflammatory protein-1 after focal cerebral ischemia in the rat. *J Neuroimmunol* 56:127–134.

Kobayashi S, Harris VA, Welsh FA (1995). Spreading depression induces tolerance of cortical neurons to ischemia in rat brain. *J Cereb Blood Flow Metab* 15:721–727.

Kochanek PM, Hallenbeck JM (1992). Polymorphonuclear leukocytes and monocytes/macrophages in the pathogenesis of cerebral ischemia and stroke. *Stroke* 23:1367–1379.

Kokaia Z, Zhao Q, Kokaia M, Elmer E, Metsis M, Smith ML, Siesjö BK, Lindvall O (1995). Regulation of brain-derived neurotrophic factor gene expression after transient middle cerebral artery occlusion with and without brain damage. *Exp Neurol* 136:73–88.

Kokaia Z, Andsberg G, Yan Q, Lindvall O (1998). Rapid alterations of BDNF protein levels in the rat brain after focal ischemia: evidence for increased synthesis and anterograde axonal transport. *Exp Neurol* 154:289–301.

Kreutzberg GW (1996). Microglia: a sensor for pathological events in the CNS. *Trends Neurosci* 19:312–318.

Lauritzen M, Hansen AJ (1992). The effect of glutamate receptor blockade on anoxic depolarization and cortical spreading depression. *J Cereb Blood Flow Metab* 12:223–229.

Lawetzky A, Tiefenthaler G, Kubo R, Hünig T (1990). Identification and characterization or rat T cell populations expressing T cell receptors α/β and γ/δ. *Eur J Immunol* 20:343–349.

Lehrmann E, Kiefer R, Finsen B, Diemer NH, Zimmer J, Hartung HP (1995). Cytokines in cerebral ischemia: expression of transforming growth factor beta-1 mRNA in the postischemic adult hippocampus. *Exp Neurol* 131:1–10.

Lehrmann E, Christensen T, Zimmer J, Diemer NH, Finsen B (1997). Microglial and macrophage reactions mark progressive changes and define the penumbra in the rat neocortex and striatum after transient middle cerebral artery occlusion. *J Comp Neurol* 386:461–476.

Lehrmann E, Kiefer R, Christensen T, Toyka KV, Zimmer J, Diemer NH, Hartung HP, Finsen B (1998). Microglia and macrophages are major sources of locally produced transforming growth factor-beta1 after transient middle cerebral artery occlusion in rats. *Glia* 24:437–448.

Li Y, Sharov VG, Jiang N, Zalonga C, Sabbah HN, Chopp M (1995). Ultrastructural and light microscopic evidence of apoptosis after middle cerebral artery occlusion in the rat. *Am J Pathol* 146:1045–1051.

Lin B, Ginsberg MD, Busto R, Dietrich WD (1998). Sequential analysis of subacute and chronic neuronal, astrocytic and microglial alterations after transient global ischemia in rats. *Acta Neuropath* 95:511–523.

Lindsberg PJ, Carpen O, Paetau A, Karjalainen-Lindsberg ML, Kaste M (1996). Endothelial ICAM-1 expression associated with inflammatory cell response in human ischemic stroke. *Circulation* 94:939–945.

Liu T, Clark RK, McDonell PC, Young PR, White RF, Barone FC, Feuerstein GZ (1994). Tumor necrosis factor-α expression in ischemic neurons. *Stroke* 25:1481–1488.

Longa EZ, Weinstein PR, Carlson S, Cummins R (1989). Reversible middle cerebral artery occlusion without craniectomy in rats. *Stroke* 20:84–91.

Marchal G, Beaudouin V, Rioux P, De la Sayette V, Le Doze F, Viader F, Derlon JM, Baron JC (1996). Prolonged persistence of substantial volumes of potentially viable brain tissue after stroke: a correlative PET-CT study with voxwel-based data analysis. *Stroke* 27:599–606.

Martins-Ferreira H, Nedergaard M, Nicholson C (2000). Perspectives on spreading depression. *Brain Res Rev* 32:215–234.

McAuley MA (1995). Rodent models of focal ischemia. *Cerebrovasc Brain Metabol Rev* 7:153–180.

Memezawa H, Smith ML, Siesjö BK (1992). Penumbral tissues salvaged by reperfusion following middle cerebral artery occlusion in rats. *Stroke* 23:552–559.

Minghetti L, Levi G (1998). Microglia as effector cells in brain damage and repair: focus on prostanoids and nitric oxide. *Prog Neurobiol* 54:99–125.

Morioka T, Kalehua AN, Streit WJ (1991). The microglial reaction in the rat dorsal hippocampus following transient forebrain ischemia. *J Cereb Blood Flow* Metabol 11:966–973.

Morioka T, Kalehua AN, Streit WJ (1992). Progressive expression of immunomolecules on microglial cells in the rat dorsal hippocampus following transient forebrain ischemia. *Acta Neuropath* (Berl) 83:149–157.

Okada Y, Copeland BR, Mori E, Tung MM, Thomas WS, Del Zoppo GJ (1994). P-selektin and intercellular adhesion molecule-1 expression after focal brain ischemia and reperfusion. *Stroke* 25:202–211.

Ordy JM, Wengenack TM, Bialobok P, Rodier P, Baggs RB, Dunlap WP, Kates B (1993). Selective vulnerability and early progression of hippocampal CA1 pyramidal cell degeneration and GFAP-positive astrocytes reactivity in the rat four-vessel occlusion model of transient global ischemia. *Exp Neurol* 119:128–139.

Perry VH, Gordon S (1987). Modulation of CD4 antigen on macrophages and microglia in rat brain. *J Exp Med* 166:1138–1143.

Perry VH, Andersson PB, Gordon S (1993). Macrophages and inflammation in the central nervous system. *Trends Neurosci* 16:268–273.

Peters O, Back T, Lindauer U, Busch C, Megow D, Dreier J, Dirnagl U (1998). Increased formation of reactive oxygen species after permanent and reversible middle cerebral artery occlusion in the rat. *J Cereb Blood Flow Metab* 18:196–205.

Piani D, Frei K, Do KQ, Cuenod M, Fontana A (1991). Murine brain macrophages induce NMDA receptor mediated neurotoxicity in vitro by secreting glutamate. *Neurosci Lett* 133:159–162.

Prehn JH, Backhauss C, Krieglstein J (1993). Transforming growth factor-beta 1 prevents glutamate neurotoxicity in rat neocortical cultures and protects mouse neocortex from ischemic injury in vivo. *J Cereb Blood Flow Metab* 13:521–525.

Pulsinelli WA, Brierly JB, Plum F (1982). Temporal profile of neuronal damage in a model of transient forebrain ischemia. *Ann Neurol* 11:491–498.

Rabchevsky AG, Streit WJ (1997). Grafting of cultured microglial cells into the lesioned spinal cord of adult rats enhances neurite outgrowth. *J Neurosci Res* 47:34–48.

Rollo EE, Laskin DL, Denhardt DT (1996). Osteopontin inhibits nitric oxide production and cytotoxicity by activated RAW264.7 macropahges. *J Leukoc Biol* 60:397–404.

Rupalla K, Allegrini PR, Sauer D, Wiesner C (1998). Time course of microglia activation and apoptosis in various brain regions after permanent focal cerebral ischemia in mice. *Acta Neuropath* 96:172–178.

Schäfer MK, Schwaeble WJ, Post C, Salvati P, Calabresi M, Sim RB, Petry F, Loos M, Weihe E (2000). Complement C1q is dramatically up-regulated in brain microglia in response to transient global cerebral ischemia. *J Immunol* 164:544–5452.

Schneider H, Pitossi F, Balschun D, Wagner A, Del Rey A, Besedovsky HO (1998). A neuromodulatory role of interleukin-1β in the hippocampus. *Proc Natl Acad Sci* USA 95:7778–7783.

Schroeter M, Jander S, Witte OW, Stoll G (1994). Local immune responses in the rat cerebral cortex after middle cerebral artery occlusion. *J Neuroimmunol* 55:195–203.

Schroeter M, Schiene K, Kraemer M, Hagemann G, Weigel H, Eysel UT, Witte OW, Stoll G (1995). Astroglial responses in photochemically induced focal ischemia of the rat cortex. *Exp Brain Res* 106:1–6.

Schroeter M, Jander S, Huitinga I, Witte OW, Stoll G (1997). Phagocytic response in photochemically induced infarction of the rat cerebral cortex: the role of resident microglia. *Stroke* 28:382–386.

Schroeter M, Jander S, Witte OW, Stoll G (1999). Heterogeneity of the microglial response in photochemically induced focal ischemia of the rat cerebral cortex. *Neuroscience* 89:1367–1377.

Schroeter M, Franke C, Stoll G, Hoehn M (2001a). Dynamic changes of MRI abnormalities in relation to inflammation and glial responses after photothrombotic cerebral infarction in the rat brain. *Acta Neuropathol* 101:114–122.

Schroeter M, Jander S, Huitinga I, Stoll G (2001b). CD8+ phagocytes in focal ischemia of the rat brain: predominant origin from hematogenous macrophages and targeting to areas of pannecrosis. *Acta Neuropathol* 101:440–448.

Schwab ME (1990). Myelin-associated inhibitors of neurite outgrowth and regeneration in the CNS. *Trends Neurosci* 13:452–456.

Shohami E, Ginis I, Hallenbeck JM (1999). Dual role of tumor necrosis factor alpha in brain injury. *Cytokine Growth Factor Rev* 10:119–130.

Springer TA (1994). Traffic signals for lymphocyte recirculation and leukocyte emigration: the multistep paradigm. *Cell* 76:301–314.

Stoll G, Trapp BD, Griffin JW (1989). Macrophage function during Wallerian degeneration of the rat optic nerve: clearance of degenerating myelin and Ia expression. *J Neurosci* 9:2327–2335.

Stoll G, Jander S, Schroeter M (1998). Inflammation and glial responses in ischemic brain lesions. *Prog Neurobiol* 56:149–161.
Stoll G, Jander S (1999). The role of microglia and macrophages in the pathophysiology of the CNS. *Prog Neurobiol* 58:233–247.
Stoll G, Jander S, Schroeter M (2000). Cytokines in CNS disorders: neurotoxicity versus neuroprotection. *J Neural Transm* 59 (suppl):81–89.
Touzani O, Boutin H, Chuquet J, Rothwell N (1999). Potential mechanisms of interleukin-1 involvement in cerebral ischemia. *J Neuroimmunol* 100:203–215.
Van Roojien N (1989). The liposome-mediated macrophage "suicide" technique. *J Immunol Meth* 124:1–6.
Wang X, Feuerstein GZ (1995). Induced expression of adhesion molecules following brain ischemia. *J Neurotrauma* 12:825–832.
Wang X, Yue TL, White RF, Barone FC, Feuerstein GZ (1995). Monocyte chemoattractant protein-1 messenger RNA expression in rat ischemic cortex. *Stroke* 26:661–665.
Wang X, Louden C, Yue TL, Ellison JA, Barone FC, Solleveld HA, Feuerstein GZ (1998). Delayed expression of osteopontin after focal stroke in the rat. *J Neurosci* 18:2075–2083.
Wang X, Li X, Erhardt JA, Barone FC, Feuerstein GZ (2000). Detection of tumor necrosis factor-alpha mRNA induction in ischemic brain tolerance by means of real-time polymerase chain reaction. *J Cereb Blood Flow Metab* 20:15–20.
Watson BD, Dietrich D, Busto R, Wachtel MS, Ginsberg MD (1985). Induction of reproducible brain infarction by photochemically initiated thrombosis. *Ann Neurol* 17:497–504.
Weissenböck H, Hornig M, Hickey WF, Lipkin WI (2000). Microglial activation and neuronal apoptosis in Bornavirus infected neonatal Lewis rats. *Brain Pathol* 10:260–272.
Yamashita K, Kataoka Y, Yamashita YS, Himeno A, Tsutsumi K, Niwa M, Taniyama K (1995). Glial endothelia/nitric oxide system participates in hippocampus CA1 neuronal death of SHRPS following transient forebrain ischemia. *Clin Exp Pharmacol Physiol* 22(suppl. 1):S227–278.
Yamashita K, Kataoka Y, Sakurai-Yamashita Y, Shigematsu K, Himeno A, Niwa M, Taniyama K (2000). Involvement of glial endothelin/nitric oxide in delayed neuronal death of rat hippocampus after transient forebrain ischemia. *Cell Mol Neurobiol* 20:541–551.
Yanamoto H, Hashimoto N, Nagata I, Kikuchi H (1998). Infarct tolerance against temporary focal ischemia following spreading depression in rat brain. *Brain Res* 784:239–249.
Yrjanheikki J, Keinanen R, Pellikka M, Hokfelt T, Koistinaho J (1998). Tetracyclines inhibit microglial activation and are neuroprotective in global brain ischemia. *Proc Natl Acad Sci USA* 95:15769–15774.
Yrjanheikki J, Tikka T, Keinanen R, Goldsteins G, Chan PH, Koistinaho J (1999). A tetracycline derivative, minocycline, reduces inflammation and protects against focal cerebral ischemia with a wide therapeutic window. *Proc Natl Acad Sci USA* 96:13496–13500.
Zhang ZG, Chopp M, Powers C (1997). Temporal profile of microglial response following transient (2h) middle cerebral artery occlusion. *Brain Res* 744:189–198.
Zielasek J, Hartung HP (1996). Molecular mechanisms of microglial activation. *Adv Neuroimmunol* 6:191–222.

7

Role of Microglia and Macrophages in Secondary Injury of the Traumatized Spinal Cord: Troublemakers or Scapegoats?

PHILLIP G. POPOVICH

Introduction

Spinal cord injury (SCI) initiates a cascade of cellular and biochemical reactions that propagate tissue damage beyond the original site of trauma. In theory, circumventing this destructive secondary pathology will lead to increased preservation of neurons and glia, and presumably to functional recovery. However, the actual onset, duration and mechanisms of secondary neuronal injury are poorly understood. To date, numerous mechanisms of secondary injury have been proposed including the possibility that microglia and macrophages play a role in lesion expansion. In 1985, Blight saw in a cat model of SCI that axons surviving the initial mechanical trauma did not undergo significant demyelination until at least two days post-injury. Because delayed demyelination correlated with the timing of significant macrophage influx, and large lipid-filled macrophages were closely apposed to denuded axons, Blight postulated a role for macrophages in secondary injury (Blight (1985); Blight (1992)). Later independent studies in guinea pig, rabbit and rodent models of SCI supported this notion that inflammation in general, and macrophages in particular, contribute to delayed secondary injury (Giulian and Robertson (1990); Blight (1994); Blight et al. (1995); Popovich et al. (1999)) (Guth et al. (1994a); Guth et al. (1994b); Zhang et al. (1997)). Owing to their ability to release prodigious amounts of noxious chemicals and enzymes involved in host defense, it is reasonable to assume a role for macrophages in acute neuronal injury, microvascular damage and delayed demyelination. Indeed, macrophage-mediated injury to the CNS is implicated in a variety of neurodegenerative and neuroinflammatory conditions, including Alzheimer's disease, cerebral ischemia, multiple sclerosis and HIV encephalomyelitis (Giulian et al. (1990); Dickson et al. (1993); Lees (1993); Mrak and Griffin (1997)). However, given the pivotal role played by macrophages in wound repair, the designation of these cells as militant effectors of tissue damage is somewhat perplexing. Aside from their destructive potential, it is also clear that macrophages can carry out protective functions in the pathological CNS. This apparent dichotomy of macrophage function is likely precipitated by several factors, including marked macrophage functional heterogeneity and diverse microenvironmental cues at the injury site. This chapter will provide a critical appraisal of the nebulous concept of immune or inflammatory-mediated secondary injury in the context of spinal trauma. Specifically, I shall focus on the potential role of microglia as *the* central mediator of the cellular and biochemical cascades elicited by SCI and, I hope absolve this mostly maligned glial cell of its neurotoxic, demyelinating reputation.

Spinal cord trauma and secondary injury

Soon (minutes) after traumatic injury to the mammalian spinal cord, light microscopy reveals little overt pathology. However, within a few hours, there is notable hemorrhage, periaxonal swelling and neuronal chromatolysis. Ultrastructural analyses reveal the onset of myelin vesiculation between two and four hours post-injury, with progressive demyelination occurring over the first week (Gledhill et al. (1973); Wakefield and Eidelberg (1975); Blight (1985); Rosenberg and Wrathall (1997)). This delay between the mechanical trauma and the first visible signs of neuropathology is the defining characteristic of "secondary injury." Of interest in the present chapter is what role, if any, do microglial activation and neuroinflammation play in propagating tissue injury? While neutrophils may play a role, their numbers are too few and their distribution such that they are not distributed throughout all regions undergoing secondary degeneration (Means and Anderson (1983); Zhang et al. (1997); Popovich et al. (1997a)). It is more likely that the earliest phases of secondary injury result from vascular injury and ischemia (Tator and Fehlings (1991)). However, after the first 24 to 48 hours, the temporal and spatial correlation between the onset of marked demyelination and rapid macrophage accumulation suggests that these cells are somehow involved in the delayed degenerative pathology so characteristic of spinal trauma (Blight (1992)). It is also possible that we are simply reading too much into what we see through the microscope. What is the evidence that activated macrophages, specifically those derived from microglia, contribute to secondary injury? To begin such a discussion requires an understanding of the biology of microglia and macrophages.

Microglia and secondary injury: guilty by association?

Fundamental microglial biology. Of all cells in the CNS that react to disruption of the brain and spinal cord, whether it is as a result of trauma, infection or ischemia, microglia are arguably the most responsive (Banati and Graeber (1994); Kreutzberg (1996)). These cells constitute as much as 20% of all glial cells and once activated, they envelop more surface area than astrocytes (Banati and Graeber (1994)). Unfortunately, because convenient histological techniques were lacking until the mid 1980s, contributions of microglia to post-injury sequelae were largely unappreciated. However, just as astrocyte research was speeded by the development of GFAP antibodies, inexpensive and reproducible lectin and monoclonal antibody techniques have heightened our awareness of microglial biology (Hume et al. (1984); Perry and Hume (1985)). Application of these techniques in models of neural development and neuropathology has facilitated the notion of "functional plasticity," i.e., microglial morphology and immunophenotype correlate with unique effector functions (Streit et al. (1988); Streit et al. (1999)). Since "functional plasticity" has been described in detail elsewhere (Streit et al. (1988); Streit et al. (1999)), only an overview of the basic features of phenotypic microglial activation will be provided.

Resting or ramified microglia can be visualized as cells with elaborate cellular processes. Although originally thought to be functionally quiescent, ramified microglia may provide various maintenance functions in the normal CNS. For example, the large surface area of resting microglia is believed to facilitate pinocytosis and surveillance of the extracellular milieu (e.g. shifts in ions or extracellular metabolites) (Ranson and Thomas (1991)). Following injury or infection, microglia retract their cellular processes (hypertrophy) and increase the expression of cell surface molecules, including major histocompatibility complex (MHC) antigens. These cells exist either alone or in clusters of "activated" or "hyper-ramified" microglia (Streit et al. (1999)). Although this morphology implies a heightened state of function, the biological impact that these cells have on the surrounding milieu is as ill-defined as the effector potential of the ramified stage. Nevertheless, given the increased expression of co-stimulatory molecules (e.g. B7.1), MHC I and II, complement receptors (CR3) and cytokine mRNA (e.g., TGFβ, TNF), activated microglia are certainly poised for action. Further activation renders microglia phagocytic, at which point they are referred to as "brain macrophages." Microglial-derived brain macrophages are phagocytic and capable of secreting a variety of cytokines, growth factors, oxygen and nitrogen free radicals and neurotrophins (Giulian and Baker (1986); Banati et al. (1993)). Although most of these secretory products have been associated with the brain macrophage phenotype *in vitro,* increased expression of inducible nitric oxide, matrix metalloproteinases (MMPs), cytokines, quinolinic acid and neurotrophic factors in tissue sections suggests that these structure-function relations hold up *in vivo* (Maeda and Sobel (1996); Moffett et al. (1997); Vezzani et al. (1999); Possel et al. (2000)). However, a recent study revealed discrete cytokine mRNA profiles between cultured microglia and brain macrophages derived from tissue samples (Streit et al. (2000)). These data emphasize the need to be conservative in our interpretation of *in vivo* morphological/function correlates and challenge us to question what the evolutionary advantage would be for blanketing the entire CNS with a population of cells capable of killing neurons and glia. Indeed, the default functional program for activated microglia is likely to be one of repair–a hypothesis supported by the absence of any *in vivo* data that indisputably demonstrates microglial-mediated tissue damage. During the remainder of this chapter, I shall explore the breadth of microglial responses in the injured spinal cord and propose potential mechanisms for skewing normal microglial-mediated CNS repair to a cellular response in which secondary injury prevails.

Evidence against microglial-mediated secondary injury If microglial activation, as determined by the in vivo morphological criteria described above, is inherently destructive then anywhere we "see" activated microglia, focal and bystander damage should occur (i.e., secondary injury). Through cell-cell contact with target cells and through the elaboration of enzymes, cytokines, nitric oxide and/or superoxide, microglia should be able to injure neurons and glia within a defined radius of their activation. Following contusion/compression trauma to the spinal cord, activated microglia demarcate the gray matter that will eventually degenerate (Popovich et al. (1997a); Carlson et al. (1998)). However, based on light microscopic analysis, microglia with similar morphology and phenotype can be found throughout white matter regions devoid of any apparent axonal injury or myelin degradation (Dusart and Schwab (1994); Popovich et al. (1997a); Frei et al. (2000)). For example, activated microglia delineate fasciculus gracilis rostral to the site of a spinal contusion injury, but there is little or no apparent pathology of adjacent fasciculus cuneatus or the corticospinal tract (Fig. 1). If the activated morphology were predictive of secondary injury, one would expect diffuse axonal pathology within the environs of the activated microglia. In areas remote from the primary focus of trauma, this does not appear to be the case, since axon pathology is clearly focused to regions of Wallerian degeneration (Fig. 1). Similarly, using an overhemisection spinal injury model, it has been shown that the spread of secondary injury is unrelated to neutrophil influx, microgliosis and the inflammatory reaction that accompanies that type of injury (Dusart and Schwab (1994)). Thus, morphological indicators of microglial activation may not signal impending tissue injury and the assumption that these cells are "killer" cells may be misleading. In fact, transplantation of activated microglia into the lesioned spinal cord promotes neurite growth, demonstrating the pro-regenerative properties of these cells (Rabchevsky and Streit (1997)).

Throughout the CNS, there are numerous examples of microglial activation without concomitant tissue injury. Axotomy of the facial nerve triggers robust microglial activation without neuronal cell death in the facial nucleus. Quite the contrary, microglial activation precedes regeneration of facial motorneurons in this model (Streit and Kreutzberg (1987)). Transient cerebral ischemia induces a rapid microglial response within the CA1 region of the hippocampus, but not within the adjacent CA3 region (Gehrmann et al. (1992); Morioka et al. (1992)). Although microglial activation precedes the degeneration of pyramidal CA1 neurons, if microglia were involved in secondary neuronal extirpation by the release of toxic mediators, one would expect graded injury throughout the penumbra, especially in nearby CA3 neurons. Following fluid percussion injury to the cerebral cortex, there

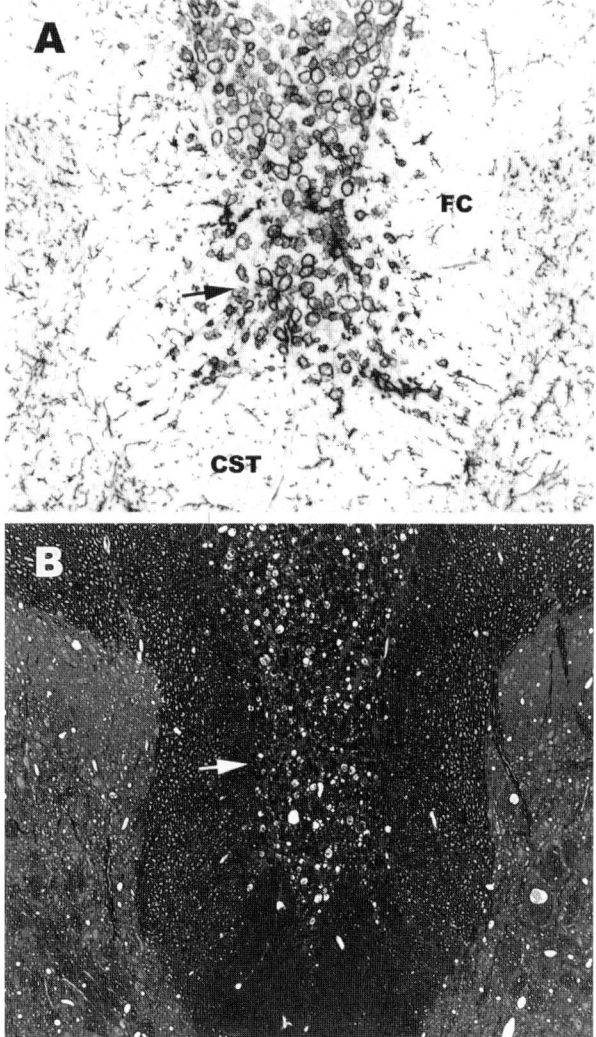

FIGURE 1. Secondary injury or bystander damage is not associated with activated microglia in the dorsal funiculus following spinal cord injury (SCI). Microglial activation (OX42 immunohistochemistry) is pronounced within fasciculus gracilis rostral to the site of a SCI (A), but is minimal in the adjacent corticospinal tract (CST) and fasciculus cuneatus (FC). Axon and myelin pathology is restricted to fasciculus gracilis, a region containing axons undergoing Wallerian degeneration (B). Despite the presence of activated microglia at the boundary region between the gracile and cuneate fasciculus (arrow, A), there is no discernible axon pathology in this region (arrow, B). The same is true for corticospinal axons.

is no obvious pathological change in thalamus or cerebellum despite robust microglial activation in these regions (Mautes et al. (1996); Fukuda et al. (1996)). Although microglial-derived TNF has been suggested to contribute to oligodendrocytic cell death in multiple sclerosis (MS), cerebral ischemia and SCI (Zajicek et al. (1992); Shuman et al. (1997); Gregersen et al. (2000)), neuronal degeneration is enhanced following brain injury in TNF receptor deficient mice, presumably because of the absence of microglial-derived TNF in promoting neuronal survival (Bruce et al. (1996); Liu et al. (1998)). Further evidence for microglial-mediated neuroprotection was provided in the osteopetrotic (*op/op*) mouse (Berezovskaya et al. (1995)). As a result of an autosomal recessive mutation, these mice lack colony-stimulating factor-1 (CSF-1)—a well characterized monocyte/macrophage growth factor. In *op/op* mice subjected to cerebral ischemia, microglial activation is severely impaired and corresponds with decreased survival of cortical neurons (Berezovskaya et al. (1995)). Microglial activation and neuronal survival are restored if CSF-1 is provided to the *op/op* mouse.

Still, the extensive literature implicating microglia in neuronal killing and glial injury make it difficult to exclude the possibility that microglia contribute to secondary degeneration. It is well established *in vitro* that microglia phenotype and function can be influenced by neurons, astrocytes and leukocytes (Giulian (1993); Zhang and Fedoroff (1996a); Dobbertin et al. (1997)). *In vivo,* these same types of intercellular reactions are likely to occur and will undoubtedly shape the eventual effector profile of activated microglia, i.e., their neurotoxic or neurotrophic potential.

Intercellular reactions affecting microglia

After spinal cord injury, inflammation and microglial activation are most notable at the site beneath the area of direct mechanical trauma. Whether neuroinflammation produces cavitation of injured spinal tissues is disputed. Still, a number of studies suggest that certain aspects of acute inflammation do more harm than good. For example, depletion of hematogenous macrophages during the first week post-injury improves functional recovery and preserves myelinated axons at the injury site (Popovich et al. (1999)). Furthermore, a decrease in the spread of the cavitating lesion to nearby spinal levels suggests that the blood-derived macrophage population contributes to secondary injury. Interestingly, despite the reduction of macrophage influx, activated microglia persisted at and nearby the injury site. Thus, it is intriguing to consider that in the absence of interactions with the infiltrating macrophages and their cytokines, microglia carry out repair and maintenance programs. The robust axon growth and deposition of a fibrous tissue matrix throughout the lesion center of macrophage-depleted animals would support this theory (Popovich et al. (1999)).

Following spinal trauma, it is believed that hematogenous macrophages are largely restricted to the impact site, i.e., they do not infiltrate proximal and distal zones of tissue degeneration (Stenwig (1972)). Consequently, it is conceivable that microglial-mediated repair could proceed in these regions (e.g. at the dorsal funiculus) without the confounding influences of infiltrating macrophages. Indeed, in these regions where subpopulations of axons are undergoing Wallerian degeneration, microgliosis occurs in parallel with pronounced revascularization and neurite growth (Stenwig (1972); Koshinaga and Whittemore (1995); Zhang and Guth (1997)). That infiltrating macrophages could induce cytotoxic properties in microglia is plausible, since macrophage depletion significantly reduced microglial activation in the animal model for multiple sclerosis (i.e., experimental autoimmune encephalomyelitis—EAE) (Bauer et al. (1995)). In EAE, neurological and histopathological changes in the CNS result from infiltration of lymphocytes and hematogenous macrophages. Morphological and phenotypic activation of microglia seems secondary to the infiltration of neuroantigen-specific T-cells and elaboration of pro-inflammatory cytokines (Dijkstra et al. (1992); Popovich et al. (1997b)). Consequently, in the absence of the appropriate signals from infiltrating leukocytes, microglia may be innocuous, or even beneficial, in the pathological CNS.

Even non-traumatic *in vivo* activation of microglia with pro-inflammatory cytokines or bacterial products (e.g. LPS) fails to elicit consistent or characteristic pathological changes in the brain or spinal cord. Indeed, intraparenchymal or intrathecal injection of LPS, TNFα or IFNγ are sufficient to induce microglial activation and upregulation of cytokine mRNA without notable damage to axons, myelin or neurons (Vass and Lassmann (1990); Andersson et al. (1992); Fitch et al. (1999); Kim et al. (2000); Stern et al. (2000)). However, if a specific threshold is reached, perhaps as a result of injecting pharmacological concentrations of the activating agents, blood-brain barrier injury and primary demyelination are seen in regions containing both blood monocytes and activated microglia (Andersson et al. (1992); Kim et al. (2000)). Although more studies are needed to define the potential of infiltrating leukocytes to induce microglial neurotoxic properties, it is already clear that microglia and peripheral macrophages have unique functional properties. Indeed, *in vivo* and *in vitro* models have demonstrated marked differences in the quantity and types of inflammatory cytokines, enzymes and reactive oxygen/nitrogen intermediates produced by microglia and macrophages derived from peripheral sources (e.g. blood, peritoneal cavity) (Giulian (1987); Giulian et al. (1989); Banati et al. (1991); Giulian et al. (1993); Alberati-Giani et al. (1996); Mosley and Cuzner (1996)). Thus, the balance between neurotoxicity and neural repair could depend on which macrophage subset is preferentially activated.

In addition to the inductive or modulatory effect exerted by infiltrating leukocytes, microglial cells are influenced by neurons and astrocytes. Not surprisingly, given the inherent difficulty in controlling these cellular reactions or interpreting their meaning *in vivo,* microglial-astrocyte and microglial-neuron interactions have largely been explored *in vitro*. Astrocytes can induce microglial proliferation through the release of M-CSF and GM-CSF (Giulian and Ingeman (1988); Sawada et al. (1990)) and may control microglial synthesis of extracellular matrix molecules (ECMs) (Rabchevsky and Streit (1997)). Microglia reciprocate by producing cytokines and growth factors (e.g. IL-1 and TGFβ) that trigger astroglial proliferation and production of ECMs (Giulian and Lachman (1985); Giulian et al. (1986); Smith and Hale (1997)). Given the co-localization and direct cell-cell contact between microglia and astrocytes surrounding the site of a spinal cord injury, as well as within the degenerating funiculi rostral and caudal to the injury site, it is reasonable to assume that this glial feedback loop also operates *in vivo* (Popovich et al. (1997a)). However, as discussed above, one must take care not to draw too many conclusions from *in vivo* morphological data.

Neuronal signaling of microglia has been assumed, largely from data obtained in the facial nerve axotomy model—a model of reversible neuronal injury. Using that model, it is well established that numerous microglia abut the affected neurons. Whether microglia participate in delivering trophic or neurotoxic signals is not clear, but is likely to be determined by the degree of trauma experienced by the neuron in question (Streit et al. (1999)). Recently, the neuronal chemokine fractalkine has been implicated in regulating microglial activation through the CX3CR1 receptor (Harrison et al. (1998)). Neurotoxic or neurotrophic microglial function may be dictated in large part by whether fractalkine remains membrane-bound, or is cleaved from the neuronal surface (Chapman et al. (2000)). It is interesting that, in mixed culture systems containing neurons and microglia, neuron survival was supported only when neurons and microglia were in direct contact. When neurons and microglia were co-cultured but physically separated, or when neurons were treated with microglial-conditioned medium, neuronal survival was decreased (Zhang and Fedoroff (1996b); Zietlow et al. (1999)). If neuron-microglial interactions are pivotal in determining whether microglia contribute to secondary injury or neural repair, it is reasonable to assume a distinct functional repertoire in gray and white matter microglia throughout the CNS. Moreover, the degree to which microglia contribute to injury or repair following spinal cord injury could be drastically influenced by the level of the injury. For example, injury to the cervical spinal cord, compared with thoracic cord, could trigger distinct populations of activated microglia, owing to larger gray-to-white matter ratios.

The pathological contributions of microglia and brain macrophages need not be restricted to secondary injury. Through their interactions with surrounding glia, microglia and macrophages may adversely affect axonal regeneration. For example, cytokines can influence astrocyte production of growth factors (Lindholm et al. (1990); Lindholm et al. (1992); Yoshida et al. (1992)). Macrophages and blood-brain barrier injury also have been associated with changes in the composition of CNS extracellular matrix molecules (ECMs). For example, a strong increase in ED1 immunoreactivity at the site of a penetrating spinal injury in rats uniformly co-localizes with deposition of chondroitin sulfate proteoglycans (CSPG)—

potent inhibitors of axon growth (Fitch and Silver (1997)). Away from the site of injury (e.g. dorsal funiculus), microglial activation is observed without concomitant changes in CSPG deposition. This latter observation suggests that microenvironmental factors (e.g. serum proteins) at the injury site could differentially influence the secretory potential of microglia and macrophages (also see above). Astrocyte production of tenascin, another ECM molecule capable of influencing axonal regeneration, is also promoted by macrophage-derived cytokines (Smith and Hale (1997)).

Microglia and macrophages could indirectly influence neuronal survival through activation of endogenous or exogenously administered neuroantigen-specific T-cells. It is well established in EAE that myelin-specific T-lymphocytes induce marked CNS pathology and neurological deficits. However, recently it was shown that these same cells may have neuroprotective effects in the traumatized nervous system. For instance, within the facial nucleus, T-lymphocytes co-localize and interact with activated microglia that surround regenerating facial motor neurons (Raivich et al. (1998)). Following SCI, injected myelin-reactive T-lymphocytes were shown to drastically reduce secondary injury and improve functional recovery, as demonstrated by increased survival of rubrospinal neurons and decreased lesion cavitation (i.e., decreased secondary injury) (Hauben et al. (2000)). Although the exact mechanism for the neuroprotective effective of myelin-reactive lymphocytes is unknown, the authors predict that the lymphocytes need to be reactivated within the CNS by antigen presenting cells to exert their neuroprotective potential (Hauben et al. (2000)). Given the neuroprotection afforded to the rubrospinal neurons, a region of the CNS well removed from the injury site and without infiltrating leukocytes, it is tempting to speculate that lymphocytes may be engaged and activated by microglia within or surrounding the red nucleus.

Conclusions

Whether neurotoxic properties can be attributed to activated microglia in the traumatically injured CNS is still open to debate. It is even possible that, once activated, microglia display both neurotrophic and neurotoxic properties. The net influence of this activity may be entirely dependent on whether cells in the surrounding tissue (e.g. astrocytes, neurons) can buffer the neurotoxic molecules. In this way, microglial activation would serve to protect the neural parenchyma from infection (release bactericidal/neurotoxic molecules), while simultaneously supplying neurons and surrounding glia with neurotrophic support (e.g. via the release of neurotrophins). Thus, under normal circumstances, microglia should be considered sentries-always on the lookout for possible perturbations in the normal structure and function of the nervous tissue, whether by injury or infection. When challenged with minor disturbances, they initiate a predetermined pattern of activation, i.e., they "awaken" from their dormant state by expressing cell surface molecules and altering their shape. This heightened state of awareness places the microglia in an ideal situation to interact with other cells in the microenvironment, including astrocytes, neurons, and endothelia or infiltrating leukocytes. If sufficient inflammatory signals are not delivered (e.g., infectious agent, cytokines, etc.), microglia will eventually retire to their posts as sentinels, or they will be *"relieved of duty"* via programmed cell death (Gehrmann and Banati (1995)). However, if the disturbance sufficiently threatened the chronic survival or function of nearby neurons and glia, a distinct series of intracellular signaling pathways could create both neurotoxic and neurotrophic microglia. Although distinct in phenotype and morphology, these unique microglial populations would elicit different functional profiles governed in large part by the degree of trauma, the composition of the microenvironment and the type of intercellular reactions. However, the complexities of these cellular interactions accompanying a pathological challenge to the CNS makes it difficult to ascertain what the molecular basis is that controls microglial activation. Revealing these secrets of microglial cell biology are likely to provide us with unique therapeutic targets for treating neurodegenerative and neuroinflammatory diseases.

References

Alberati-Giani D, Ricciardi-Castagnoli P, Kohler C, Cesura AM (1996). Regulation of the kynurenine metabolic pathway by interferon-gamma in murine cloned macrophages and microglial cells. *J Neurochem* 66: 996–1004.

Andersson PB, Perry VH, Gordon S (1992). The acute inflammatory response to lipopolysaccharide in CNS. *Neuroscience* 48: 169–186.

Banati RB, Gehrmann J, Schubert P, Kreutzberg GW (1993). Cytotoxicity of microglia. *Glia* 7: 111–118.

Banati RB, Graeber MB (1994). Surveillance, intervention and cytotoxicity: is there a protective role of microglia? *Dev Neurosci* 16: 114–127.

Banati RB, Rothe G, Valet G, Kreutzberg GW (1991). Respiratory burst activity in brain macrophages: a flow cytometric study on cultured rat microglia. *Neuropathol Appl Neurobiol* 17: 223–230.

Bauer J, Huitinga I, Zhao W, Lassmann H, Hickey WF, Dijkstra CD (1995). The role of macrophages, perivascular cells, and microglial cells in the pathogenesis of experimental autoimmune encephalomyelitis. *Glia* 15: 437–446.

Berezovskaya O, Maysinger D, Fedoroff S (1995). The hematopoietic cytokine, colony-stimulating factor 1, is also a growth factor in the CNS: congenital absence of CSF-1 in mice results in abnormal microglial response and increased neuron vulnerability to injury. *Int J Dev Neurosci* 13: 285–299.

Blight AR (1985). Delayed demyelination and macrophage invasion: a candidate for secondary cell damage in spinal cord injury. *CNS Trauma* 2: 299–315.

Blight AR (1992). Macrophages and inflammatory damage in spinal cord injury. *J Neurotrauma* 9 Suppl 1: S83–S91.

Blight AR (1994). Effects of silica on the outcome from experimental spinal cord injury: implication of macrophages in secondary tissue damage. *Neuroscience* 60: 263–273.

Blight AR, Cohen TI, Saito K, Heyes MP (1995). Quinolinic acid accumulation and functional deficits following experimental spinal cord injury. *Brain* 118: 735–752.

Bruce AJ, Boling W, Kindy MS, Peschon J, Kraemer PJ, Carpenter MK, Holtsberg FW, Mattson MP (1996). Altered neuronal and microglial responses to excitotoxic and ischemic brain injury in mice lacking TNF receptors. *Nat Med* 2: 788–794.

Carlson SL, Parrish ME, Springer JE, Doty K, Dossett L (1998). Acute inflammatory response in spinal cord following impact injury. *Exp Neurol* 151: 77–88.

Chapman GA, Moores K, Harrison D, Campbell CA, Stewart BR, Strijbos PJ (2000). Fractalkine cleavage from neuronal membranes represents an acute event in the inflammatory response to excitotoxic brain damage. *J Neurosci* (online) 20:RC87.

Dickson DW, Lee SC, Mattiace LA, Yen S-HC, Brosnan C (1993). Microglia and cytokines in neurological disease, with special reference to AIDS and Alzheimer's disease. *Glia* 7: 75–83.

Dijkstra CD, De Groot CJ, Huitinga I (1992). The role of macrophages in demyelination. *J Neuroimmunol* 40: 183–188.

Dobbertin A, Schmid P, Gelman M, Glowinski J, Mallat M (1997). Neurons promote macrophage proliferation by producing transforming growth factor-$\beta 2$. *J Neurosci* 17: 5305–5315.

Dusart I, Schwab ME (1994). Secondary cell death and the inflammatory reaction after dorsal hemisection of the rat spinal cord. *Eur J Neurosci* 6: 712–724.

Fitch MT, Doller C, Combs CK, Landreth GE, Silver J (1999). Cellular and molecular mechanisms of glial scarring and progressive cavitation: In vivo and in vitro analysis of inflammation-induced secondary injury after CNS trauma. *J Neurosci* 19: 8182–8198.

Fitch MT, Silver J (1997). Activated macrophages and the blood-brain barrier: inflammation after CNS injury leads to increases in putative inhibitory molecules. *Exp Neurol* 148:587–603.

Frei E, Klusman I, Schnell L, Schwab ME (2000). Reactions of oligodendrocytes to spinal cord injury: cell survival and myelin repair. *Exp Neurol* 163: 373–380.

Fukuda K, Aihara N, Sagar SM, Sharp FR, Pitts LH, Honkaniemi J, Noble LJ (1996). Purkinje cell vulnerability to mild traumatic brain injury. *J Neurotrauma* 13: 255–266.

Gehrmann J, Banati RB (1995). Microglial turnover in the injured CNS: activated microglia undergo delayed DNA fragmentation following peripheral nerve injury. *J Neuropathol Exp Neurol* 54: 680–688.

Gehrmann J, Bonnekoh P, Miyazawa T, Hossmann KA, Kreutzberg GW (1992). Immunocytochemical study of an early microglial activation in ischemia. *J Cereb Blood Flow Metab* 12: 257–269.

Giulian D (1987). Ameboid microglia as effectors of inflammation in the central nervous system. *J Neurosci Res* 18: 155–71, 132–3.

Giulian D (1993). Reactive glia as rivals in regulating neuronal survival. *Glia* 7:102–110.

Giulian D, Baker TJ (1986). Characterization of ameboid microglia isolated from developing mammalian brain. *J Neurosci* 6: 2163–2178.

Giulian D, Baker TJ, Shih LN, Lachman LB (1986). Interleukin 1 of the central nervous system is produced by ameboid microglia. *J Exp Med* 164: 594–604.

Giulian D, Chen J, Ingeman JE, George JK, Noponen M (1989). The role of mononuclear phagocytes in wound healing after traumatic injury to adult mammalian brain. *J Neurosci* 9: 4416–4429.

Giulian D, Corpuz M, Chapman S, Mansouri M, Robertson C (1993). Reactive mononuclear phagocytes release neurotoxins after ischemic and traumatic injury to the central nervous system. *J Neurosci Res* 36: 681–693.

Giulian D, Ingeman JE (1988). Colony-stimulating factors as promoters of ameboid microglia. *J Neurosci* 8: 4707–4717.

Giulian D, Lachman LB (1985). Interleukin-1 stimulation of astroglial proliferation after brain injury. *Science* 228: 497–499.

Giulian D, Robertson C (1990). Inhibition of mononuclear phagocytes reduces ischemic injury in the spinal cord. *Ann Neurol* 27: 33–42.

Giulian D, Vaca K, Noonan CA (1990). Secretion of neurotoxins by mononuclear phagocytes infected with HIV-1. *Science* 250: 1593–1596.

Gledhill RF, Harrison BM, McDonald WI (1973). Demyelination and remyelination after acute spinal cord compression. *Exp Neurol* 38: 472–487.

Gregersen R, Lambertsen K, Finsen B (2000). Microglia and macrophages are the major source of tumor necrosis factor in permanent middle cerebral artery occlusion in mice. *J Cereb Blood Flow Metab* 20: 53–65.

Guth L, Zhang Z, DiProspero NA, Joubin K, Fitch MT (1994a). Spinal cord injury in the rat: treatment with bacterial lipopolysaccharide and indomethacin enhances cellular repair and locomotor function. *Exp Neurol* 126: 76–87.

Guth L, Zhang Z, Roberts E (1994b). Key role for pregnenolone in combination therapy that promotes recovery after spinal cord injury. *Proc Natl Acad Sci USA* 91: 12308–12312.

Harrison JK, Jiang Y, Chen S, Xia Y, Maciejewski D, McNamara RK, Streit WJ, Salafranca MN, Adhikari S, Thompson DA, Botti P, Bacon KB, Feng L (1998). Role for neuronally derived fractalkine in mediating interactions between neurons and CX3CR1-expressing microglia. *Proc Natl Acad Sci USA* 95: 10896–10901.

Hauben E, Butovsky O, Nevo U, Yoles E, Moalem G, Agranov E, Mor F, Leibowitz-Amit R, Pevsner E, Akselrod S, Neeman M, Cohen IR, Schwartz M (2000). Passive or active immunization with myelin basic protein promotes recovery from spinal cord contusion. *J Neurosci* 20: 6421–6430.

Hume DA, Perry VH, Gordon S (1984). The mononuclear phagocyte system of the mouse defined by immunohistochemical localisation of antigen F4/80: macrophages associated with epithelia. *Anat Rec* 210: 503–512.

Kim WG, Mohney RP, Wilson B, Jeohn GH, Liu B, Hong JS (2000). Regional difference in susceptibility to lipopolysaccharide-induced neurotoxicity in the rat brain: role of microglia. *J Neurosci* 20: 6309–6316.

Koshinaga M, Whittemore SR (1995). The temporal and spatial activation of microglia in fiber tracts undergoing anterograde and retrograde degeneration following spinal cord lesion. *J Neurotrauma* 12: 209–222.

Kreutzberg GW (1996). Microglia: A sensor for pathological events in the CNS. *Trends Neurosci* 19: 312–318.

Lees GJ (1993). The possible contribution of microglia and macrophages to delayed neuronal death after ischemia. *J Neurol Sci* 114: 119–122.

Lindholm D, Castren E, Kiefer R, Zafra F, Thoenen H (1992). Transforming growth factor-β1 in the rat brain: increase after injury and inhibition of astrocyte proliferation. *J Cell Biol* 117: 395–400.

Lindholm D, Hengerer B, Zafra F, Thoenen H (1990). Transforming growth factor-β1 stimulates expression of nerve growth factor in the rat CNS. *Neuroreport* 1: 9–12.

Liu J, Marino MW, Wong G, Grail D, Dunn A, Bettadapura J, Slavin AJ, Old L, Bernard CC (1998). TNF is a potent anti-inflammatory cytokine in autoimmune-mediated demyelination. *Nat Med* 4: 78–83.

Maeda A, Sobel RA (1996). Matrix metalloproteinases in the normal human central nervous system, microglial nodules, and multiple sclerosis lesions. *J Neuropathol Exp Neurol* 55: 300–309.

Mautes AE, Fukuda K, Noble LJ (1996). Cellular response in the cerebellum after midline traumatic brain injury in the rat. *Neurosci Lett* 214: 95–98.

Means ED, Anderson DK (1983). Neuronophagia by leukocytes in experimental spinal cord injury. *J Neuropathol Exp Neurol* 42: 707–719.

Moffett JR, Els T, Espey MG, Walter SA, Streit WJ, Namboodiri MA (1997). Quinolinate immunoreactivity in experimental rat brain tumors is present in macrophages but not in astrocytes. *Exp Neurol* 144: 287–301.

Morioka T, Kalehua AN, Streit WJ (1992). Progressive expression of immunomolecules on microglial cells in rat dorsal hippocampus following transient forebrain ischemia. *Acta Neuropathol* (Berl) 83: 149–157.

Mrak RE, Griffin WS (1997). The role of chronic self-propagating glial responses in neurodegeneration: implications for long-lived survivors of human immunodeficiency virus. *J Neurovirol* 3: 241–246.

Perry VH, Hume DA (1985). Immunohistochemical localization of macrophages and microglia in the adult and developing mouse brain. *Neuroscience* 15: 313–326.

Popovich PG, Guan Z, Wei P, Huitinga I, Van Rooijen N, Stokes BT (1999). Depletion of hematogenous macrophages promotes partial hindlimb recovery and neuroanatomical repair after experimental spinal cord injury. *Exp Neurol* 158: 351–365.

Popovich PG, Wei P, Stokes BT (1997a) The cellular inflammatory response after spinal cord injury in Sprague-Dawley and Lewis rats. *J Comp Neurol* 377: 443–464.

Popovich PG, Yu JY, Whitacre CC (1997b). Spinal cord neuropathology in rat experimental autoimmune encephalomyelitis: modulation by oral administration of myelin basic protein. *J Neuropathol Exp Neurol* 56: 1323–1338.

Possel H, Noack H, Putzke J, Wolf G, Sies H (2000). Selective upregulation of inducible nitric oxide synthase (iNOS) by lipopolysaccharide (LPS) and cytokines in microglia: In vitro and in vivo studies. *GLIA* 32: 51–59.

Rabchevsky AG, Streit WJ (1997). Grafting of cultured microglial cells into the lesioned spinal cord of adult rats enhances neurite outgrowth. *J Neurosci Res* 47: 34–48.

Raivich G, Jones LL, Kloss CUA, Werner A, Neumann H, Kreutzberg GW (1998). Immune surveillance in the injured nervous system: T-lymphocytes invade the axotomized mouse facial motor nucleus and aggregate around sites of neuronal degeneration. *J Neurosci* 18: 5804–5816.

Ranson PA, Thomas WE (1991). Pinocytosis as a select marker of ramified microglia in vivo and in vitro. *J Histochem Cytochem* 39: 853–858.

Rosenberg LJ, Wrathall JR (1997). Quantitative analysis of acute axonal pathology in experimental spinal cord contusion. *J Neurotrauma* 14: 823–838.

Sawada M, Suzumura A, Yamamoto HAUM (1990). Activation and proliferation of the isolated microglia by colony stimulating factor-1 and possible involvement of protein kinase C. *Brain Res* 509: 119–124.

Shuman SL, Bresnahan JC, and Beattie MS (1997). Apoptosis of microglia and oligodendrocytes after spinal cord contusion in rats. *J Neurosci Res* 50: 798–808.

Smith GM, Hale JH (1997). Macrophage/Microglia regulation of astrocytic tenascin: synergistic action of transforming growth factor-beta and basic fibroblast growth factor. *J Neurosci* 17: 9624–9633.

Stenwig AE (1972). The origin of brain macrophages in traumatic lesions, Wallerian degeneration, and retrograde degeneration. *J Neuropathol Exp Neurol* 31: 696–704.

Stern EL, Quan N, Proescholdt MG, Herkenham M (2000). Spatiotemporal induction patterns of cytokine and related immune signal molecule mRNAs in response to intrastriatal injection of lipopolysaccharide. *J Neuroimmunol* 106: 114–129.

Streit WJ, Graeber MB, Kreutzberg GW (1988). Functional plasticity of microglia: A review. *GLIA* 1: 301–307.

Streit WJ, Hurley SD, McGraw TS, Semple-Rowland SL (2000). Comparative evaluation of cytokine profiles and reactive gliosis supports a critical role for interleukin-6 in neuron-glia signaling during regeneration. *J Neurosci Res* 61: 10–20.

Streit WJ, Kreutzberg GW (1987). Lectin binding by resting and reactive microglia. *J Neurocytol* 16: 249–260.

Streit WJ, Walter SA, Pennell NA (1999). Reactive microgliosis. *Prog Neurobiol* 57: 563–581.

Tator CH, Fehlings MG (1991). Review of the secondary injury theory of acute spinal cord trauma with emphasis on vascular mechanisms. *J Neurosurg* 75: 15–26.

Vass K, Lassmann H (1990). Intrathecal application of interferon gamma. Progressive appearance of MHC antigens within the rat nervous system. *Am J Pathol* 137: 789–800.

Vezzani A, Conti M, De Luigi A, Ravizza T, Moneta D, Marchesi F, De Simoni MG (1999). Interleukin-1 beta immunoreactivity and microglia are enhanced in the rat hippocampus by focal kainate application: functional evidence for enhancement of electrographic seizures. *J Neurosci* 19: 5054–5065.

Wakefield CL, Eidelberg E (1975). Electron microscopic observations of the delayed effects of spinal cord compression. *Exp Neurol* 48: 637–646.

Yoshida K, Kakihana M, Chen LS, Ong M, Baird A, Gage FH (1992). Cytokine regulation of nerve growth factor-mediated cholinergic neurotrophic activity synthesized by astrocytes and fibroblasts. *J Neurochem* 59: 919–931.

Zajicek JP, Wing M, Scolding NJ, Compston DA (1992). Interactions between oligodendrocytes and microglia. A major role for complement and tumour necrosis factor in oligodendrocyte adherence and killing. *Brain* 115: 1611–1631.

Zhang SC, Fedoroff S (1996a). Neuron-microglia interactions in vitro. *Acta Neuropathol* (Berl) 91: 385–395.

Zhang Z, Guth L (1997). Experimental spinal cord injury: Wallerian degeneration in the dorsal column is followed by revascularization, glial proliferation, and nerve regeneration. *Exp Neurol* 147: 159–171.

Zhang Z, Krebs CJ, Guth L (1997). Experimental analysis of progressive necrosis after spinal cord trauma in the rat: etiological role of the inflammatory response. *Exp Neurol* 143: 141–152.

Zietlow R, Dunnett SB, Fawcett JW (1999). The effect of microglia on embryonic dopaminergic neuronal survival in vitro: diffusible signals from neurons and glia change microglia from neurotoxic to neuroprotective. *Eur J Neurosci* 11: 1657–1667.

8

Microglial Response in the Axotomized Facial Motor Nucleus

GENNADIJ RAIVICH

Introduction

The activation of microglial cells in the central nervous system following peripheral facial nerve injury is in many ways similar to that following other forms of trauma and brain pathology. Injury to the nervous system triggers a variety of morphologic and metabolic changes that appear to play a key role in two crucial physiological processes: protection against infectious agents and repair of the damaged nervous tissue. The activation of microglial cells, a macrophage- and monocyte-related, brain-resident cell, is a highly cellular response. The axotomized facial motor nucleus has turned into one of the key *in vivo* models to dissect the molecular mechanisms involved in neuronal survival, axonal regeneration and the activation of microglia and their role in the overall network of posttraumatic response. This chapter will describe the microglial response in the facial motor nucleus in terms of morphology and activation markers, summarize the current data on the associated molecular mechanisms, and discuss the biological function subserved by this highly conserved cellular reaction.

The facial axotomy model

The facial nucleus is the largest brainstem motor nucleus and contains, in the mouse, approximately 2000 motoneurons innervating the facial, platysmal and auricular muscles. The motoneurons are organized into several muscle- and region-specific subnuclei, producing a horseshoe-like pattern with a narrow ventral opening on frontal sections. Overall, the mouse facial motor nucleus is a

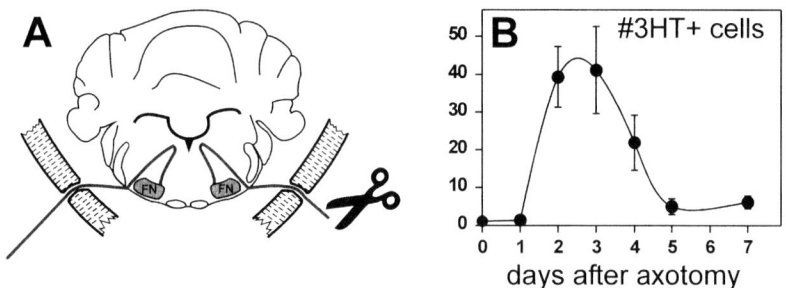

FIGURE 1. **A:** A schematic summary of the facial axotomy model. The facial motor nerve is cut at the stylomastoid foramen and the cellular changes in the facial motor nucleus (FN) examined at defined time points after transection. The facial nucleus on the contralateral side serves as an uninjured control. B: Time course of cell proliferation in the axotomized facial motor nucleus. The Y-axis shows the number of proliferating cells per 20 μm tissue section using autoradiography for [^{3}H]-thymidine (mean ± SEM). Notice the sudden onset of proliferation between day 1 and day 2, and the rapid downturn after day 3. Modified from Raivich et al., *Eur J Neurosci* 6:1515, 1994.

compact nucleus of 0.3mm^3, bulging at the ventral brainstem surface, and extending 0.7 mm in the rostrocaudal and mediolateral, and 0.5mm in the dorsoventral, direction. These dimensions are larger in larger animals (in the rat, for example) by a factor of 1.5-2, with a motoneuron number of 3,5-4,000 and a total volume of approximately 1.5mm^3, but the overall shape and relation between different dimensions is generally well maintained. After leaving the nucleus, the facial axons converge in the mediodorsal part of the brainstem, at the *genu nervi facialis* or inner knee of the facial nerve, and then turn 180° around the abducens nucleus to exit the brainstem at its ventrolateral surface, slightly rostral and lateral of the facial nucleus (Fig. 1A). With the exception of a tiny nerve branch innervating the stapedius muscle with an exit in the middle ear, all other facial nerve branches stay together during their passage through the internal acoustic meatus and the tympanic bulla bone, and radiate to their target muscles only after leaving the stylomastoid foramen. Thus, a complete transection of the facial nerve close to the stylomastoid foramen, including the main and the auricular branch, will lead to a uniform lesion that affects all facial motoneurons equally. This situation is quite different from direct trauma or cerebral ischemia, where all stages of brain pathology ranging from healthy to completely necrotic tissue may be present. Owing to the distance between the site of the peripheral axotomy and the facial motor nucleus in the central nervous system—around 6 mm in mice, 10 mm in rats—the cellular response in the affected facial motor nucleus is largely restricted to CNS—resident cells: the axotomized motoneurons, astrocytes and microglia. Studies using systemic injection of horseradish peroxidase (HRP) or human serum albumin also show an intact blood brain barrier to soluble molecules (Raivich et al. (1998a); Bohatschek et al. (2001)). Thus, the glial response is probably mediated by locally synthesized signals, and not by circulating molecules, hence reducing the number of potential activating pathways. From a technical standpoint, the size of the facial nucleus also provides enough tissue for detailed biochemical analysis to identify the presence and regulation of potential signaling molecules. Under normal conditions, there is also no infiltration by circulating granulocytes (Raivich et al. (1998a)), or by macrophages using the rhodamine isothiocyanate labeling (Bohatschek et al. (2001)). This is particularly important in the case of macrophages that express the same markers as the activated microglial cells. If infiltration does occur, it becomes very difficult to differentiate between resident microglia and the invading monocytes on the basis of the immunohistochemical markers. In fact, this is a common problem in interpreting the results of direct trauma, infection or autoimmune disease with respect to resident microglia versus extravasating macrophages. The absence of monocytic influx in the facial motor nucleus clearly simplifies the interpretation of the acquired data.

Microglial response to facial axotomy

The microglial cells form a substantial part of the total glial population that is functionally related to peripheral tissue macrophages, and other cells of monocytic lineage. Like macrophages in other tissues, the normal, resting microglia appear to participate in the immune surveillance of the nervous system. They express receptors for key immune molecules such as immunoglobulin (FcγR) and complement fragments (αMβ2 integrin, complement receptor type 3), and their surface is covered with moderate levels of endogenous immunoglobulin and thrombospondin, which may enable microglia to monitor their local environment for foreign antigens, opsonized particles and apoptotic cells (Fishman and Savitt (1989); Perry et al. (1985); Raivich et al. (1994); Liu et al. (1995); Möller et al. (1996); Engelhardt et al. (1998); Werner et al. (1998). The resting microglia have long, ramified processes: in white matter, they are oriented in parallel to the nerve fibers; in gray matter, they display a stellate morphology (Perry et al. (1985); Compston et al. (1997); Raivich et al. (1998b)). The highly ramified form of resting microglia, extending over territories 30–40 μm in diameter, probably supports their immune-surveillance function.

The rapid transformation of ramified microglia from a resting to an activated state has been clearly recognized for more than a century (Nissl (1899); Merzbacher (1909); Rio Hortega (1932)). As in other forms of injury, reactive microglia in the axotomized facial motor nucleus show an increase in cell body size, a thickening of proximal processes and a decrease in the ramification of distal branches. On the molecular level, this activation appears to proceed through a series of steps, which differ in their expression of molecules for cell adhesion, cytoskeletal organization and antigen presentation, as summarized in Figure 2.

Early activation/stage 1.

Early microglial activation in the axotomized facial motor nucleus is observed within the first 24 hours and associated with an increase in molecules with immune function. There is an increase in microglial IgG-immunoreactivity, in thrombospondin, in complement C1q, clusterin and in αMβ2-integrin, the receptor for complement 3bi fragment (Liu et al. (1995); Möller et al. (1996); Kloss et al. (1999); Schafer et al. (2000)). Mouse microglia also increase in ICAM1 (Werner et al. (1998); Fig. 3A-D). ICAM1 is an important cell-surface ligand of αMβ2 and αLβ2 integrins that mediate adhesion to different leukocyte lineages and related cells including granulocytes, microglia and lymphocytes (Hynes (1992)).

Adhesion and proliferation/stage 2.

The second stage is characterized by homing and adhesion to the axotomized motoneurons. It is associated with an increase in α5β1 and α6β1 integrins in mice and α4β1 and αLβ2 integrins in rats (Moneta et al. (1993); Hailer et al. (1997); Kloss et al. (1999)), cytoskeletal proteins like vimentin (Graeber et al. (1988); Raivich et al. (1993)), and a further reduction in ramification (Raivich et al. (1999)). These changes apparently pave the way for enhanced post-traumatic mobility and changes in the adhesion properties (Angelov et al. (1995); Schiefer et al. (1999)). Acute phase markers, such as αMβ2 and ICAM1 are downregulated during stage 2 (Raivich et al. (1998b); Werner et al. (1998); Kloss et al. (1999)), further documenting the progression of microglia to a new, and biochemically different phase.

The molecular basis of this adhesive behavior is currently subject of intense investigation. Studies using animals deficient for different inflammation-associated cytokines suggest that this is also a multi-step process. It begins with a microglial cell touching all the neuronal surface (Fig. 3E), followed by apposition of the microglial cell body (Fig. 3F), the close wrapping of the neuron with microglial processes, and lastly hypertrophy of the apposed microglial cell body. Microglia in mice lacking the macrophage-colony stimulating factor/MCSF will freeze during the first step: their processes will touch axotomized neurons, but they will not proceed to apposition, ensheatment or cell body swelling (Kalla et al. (2001)). Absence of TGFβ1 will interfere with ensheatment (Jones et al. (1998)), and absence of IL6 will show normal attachment, apposition and ensheatment, but no

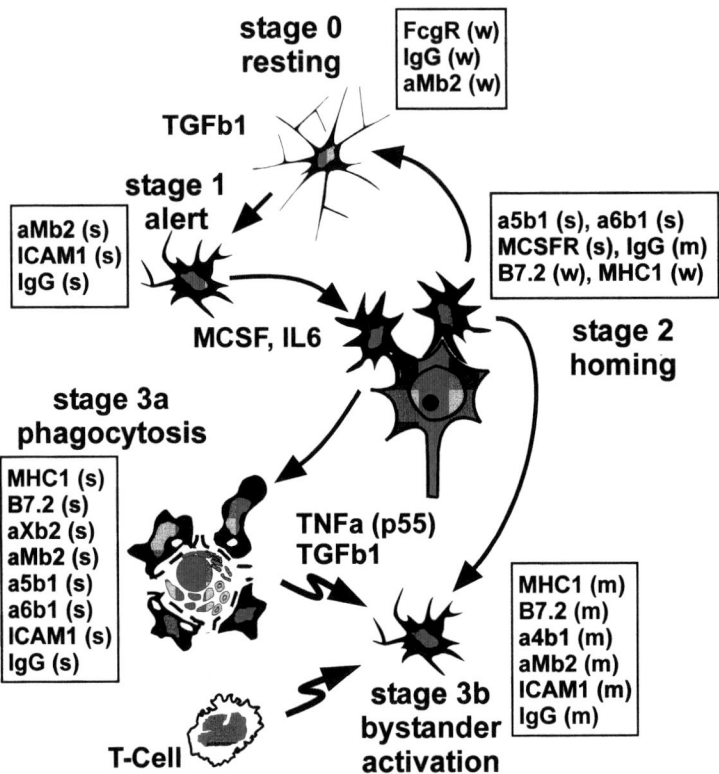

FIGURE 2. Microglial activation, a schematic summary. Cellular changes in activated microglia proceed through a series of steps, which differ in their morphology and molecular profile. Neural injury leads to a rapid transformation of the highly ramified resting microglia (stage 0) to more stout and deramified form (stage 1: state of alert), which home on damaged structures such as injured neurons (stage 2: homing), but without additional damage, gradually return to the normal, resting state (stage 0). Cell death leads to a further transformation of microglia into phagocytotic cells (stage 3a: phagocytosis). These foci of phagocytosis also lead to an activation of the surrounding, non-phagocytotic microglia (stage 3b: bystander activation). The transformation of microglia into phagocytotic, ameboid macrophages in the presence of cell debris appears to be cytokine-independent. The early stages of activation are affected by MCSF, IL6 and TGFβ1, the late bystander activation by TGFβ1, by TNF and its p55 receptor (TNFR1).

The boxes show the microglial activation markers specific for the activation stages in the mouse. Their staining intensity is shown in brackets: (w)-weak, (m)-moderate, (s)-strong. All integrins are encoded with standard nomenclature (a-alpha, b-beta). Modified from *Brain Res Rev*, Vol. 30, Raivich et al., Neurological activation repertoire in the injured brain, pp. 77–105, Copyright 1999, with permission from Elsevier Science.

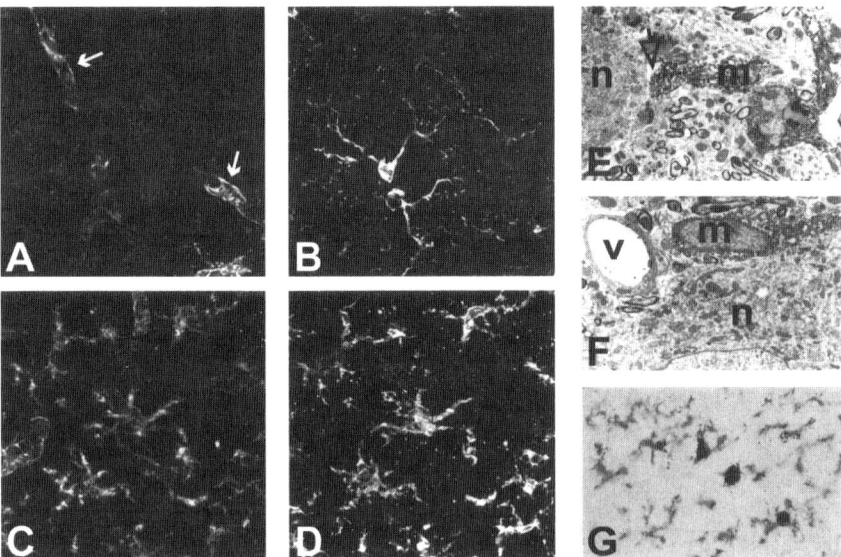

FIGURE 3. Microglial activation in the injured facial motor nucleus. changes in morphology and immunohistochemical markers. **A–D:** ICAM1 and IBA1 immunoreactivity in the facial axotomy model (**A–B:** normal facial motor nucleus, C, D: facial nucleus 3 days after transection of the facial nerve). A, C: High levels of ICAM1 are already present on vascular endothelia (A, arrows), but appear on activated microglia following injury (C). **B, D:** Immunoreactivity for IBA1, a constitutive microglial marker, also reveals changes in microglial morphology: Normal microglia are highly ramified, with long and slender processes (B), but this ramification decreases after injury (D). At day 3, there is also an overall increase in microglial cell number. **E, F:** Microglia adhere to damaged neurons in the facial motor nucleus 2 days after injury. Ultrastructural level, the microglial cytoplasm is labelled with MCSFR immunohistochemistry. **E:** Early stage of adhesion, with microglial process contacting neuronal surface (open arrow). **F:** Late stage, direct apposition of microglial cell body on the axotomized neuron. **G:** Microglial proliferation in the adult facial motor nucleus, 2 days after axotomy. Double labeling for microglial αMβ2-integrin immunohistochemical staining) and [^{3}H]-thymidine as a marker of cell proliferation (autoradiographic silver grains). In this model, cell proliferation is restricted to the αMβ2-positive microglia. (*continued*)

cell body swelling (Galiano et al. (submitted)). The neuronal expression of fractalkine and MCP1 in axotomized motoneurons (Harrison et al. (1998); Schwaiger et al. (2000)) and their cognate receptors CX3CR1 and CCR2, respectively, on activated microglia have also suggested chemokines as a source of chemical attraction between neurons and microglia. However, this attraction was not inhibited in single CX3CR1 or CCR2 knockouts (Jung et al. (2000)) pointing to a high degree of redundancy, at least at the level of chemokines.

As the activated microglia move into direct contact with injured neuronal cell bodies, they increase their expression of receptors for the microglial mitogens

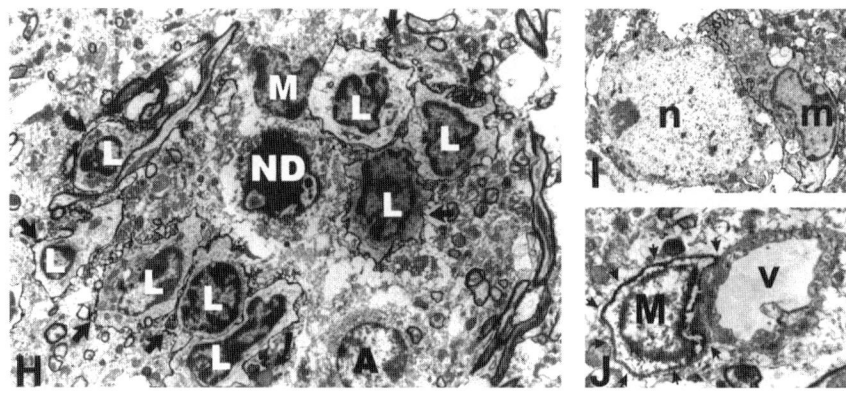

FIGURE 3. (*continued*). **H:** infiltrating T-lymphocytes (L) adhere to phagocytotic microglia (M) removing the debris of a dead motoneuron (ND). Immunoelectron microscopy of the facial motor nucleus in an adult mouse, 14 days after facial nerve cut. The lymphocyte cell surface carries large amounts of different adhesion molecules, including the αMβ2 integrin labeled in this immunoelectron microscopy. The arrow points to the phagosomes in an αMβ2-negative, microglial cell process adhering to a αMβ2-immunoreactive T-cell, a morphological prerequisite of antigen processing and presentation in the injured brain. These phagosomes are a common, characteristic feature in the phagocytotic microglial cells. (A) Adjacent astrocytes. **I:** Cerebral cortex 4 days after direct trauma (cortical hemisection). Note the extensive cell surface ruffling of an activated microglial cell adhering to an adjacent neuron. The microglial surface is labeled with an antibody against CD44. Microglial CD44 is an indicator of severe pathology such as direct trauma or disruption of the blood-brain barrier (Jones et al., 2000). CD44 it is an example of a microglial antigen that is not present in the normal or axotomized facial motor nucleus (Jones et al., 1997; 2000). **J:** MHC2-immunoreactive perivascular macrophage attached to a cerebral blood vessel. Unlike the parenchymal microglia, this cell is always surrounded by a perivascular basal membrane (arrows). Abbreviations: m-microglia, M-macrophage, n-neuron, v-vessel. A–D, 800x; E: 4,000x; F: 4,800x. G: 450x; H: 4400x; I: 4,000x L: 6,000x.

Figs. **A–D, G** and **J** are modified from Raivich et al., *Brain Res Rev,* 30: 77–105 and reproduced with permission from Elsevier Science; Figs. **E** and **F** are reproduced from Raivich et al., *J Comp Neurol,* 395:342–358; Fig. **H** from Raivich et al., *J Neurosci,* 18:5804–5816, Copyright 1998 by the Society for Neuroscience; and Fig. **I** from Jones et al., *J Comp Neurol,* 426:468–492, Wiley-Liss, Inc., a subsidiary of John Wiley & Sons, Inc.

MCSF and GMCSF (Raivich et al. (1991; 1998b)) and begin to proliferate (Kreutzberg (1966); Sjöstrand (1966); Graeber et al. (1988b)). Mitotic activity in these cells becomes maximal two to three days after axotomy (Raivich et al. (1994); see Figs. 1B and 3G, leading to an approximately four- to six-fold increase in the local microglial cell number (Klein et al. (1997)). Proliferative activity subsides after day 4 (Streit and Kreutzberg (1988); Raivich et al. (1994); Horvat et al. (in press)). This is accompanied by the onset of immunoreactivity for major histocompatibility complex (MHC) class I molecules (Raivich et al. (1993); Bohatschek et al. (1999)) and immunoaccessory glycoproteins such as

B7.2 (Bohatschek et al. (1999)). Damaged cells are frequently leaky and the tight adhesion of microglia could allow for an efficient uptake of diffusible molecules and their presentation to infiltrating lymphocytes.

Late changes/stage 3

Without additional damage, microglia gradually decrease in number, lose activation markers and revert to a resting state, characterized by a territorial distribution, highly ramified shape and moderate levels of complement receptors, Fcγ-receptors and IgG immunoreactivity. Neuronal cell death, on the other hand, leads to a further transformation of microglia into phagocytotic cells that remove neural debris (Streit and Kreutzberg (1988); Möller et al. (1996)). Both types of response can be observed in the axotomized facial motor nucleus, but they differ in the species studied. Adult rats normally show very little neuronal cell death after facial axotomy with a gradual, uneventful deactivation of the activated microglial cells (Streit et al. (1988)). In the adult mouse, axotomy will lead to a loss of approximately 40% of the neurons, which peaks two weeks after injury (Möller et al. (1996)) and results in widespread phagocytosis of dead motorneurons by microglia. Neuronal cell death induced by the retrograde transport of ricin in the rat facial nucleus will lead to a similarly prompt microglial phagocytosis (Streit and Kreutzberg (1988)).

Removal of large cellular structures such as the degenerating motorneurons frequently leads to the formation of microglial nodules, consisting of three to twenty microglial phagocytes (Streit and Kreutzberg (1988); Möller et al. (1996); Raivich et al. (1998a)). This state of phagocytosis causes a further increase in the expression of most activation markers already present during stages 1 and 2. It includes a long list of cell adhesion molecules: the $\alpha 5\beta 1$, $\alpha 6\beta 1$ and $\alpha M\beta 2$ integrins, thrombospondin and ICAM1 (Möller et al. (1996); Werner et al. (1998); Kloss et al. (1999)). All of these molecules could participate in the adhesion and internalization to the many different and diverse components of the cellular debris.

Microglial cells express strongly a number of immunologically important molecules, such as in Fcγ receptors, B7 and MHC class I (McGeer et al. (1993); De Simone (1995); Bohatschek et al. (1999)). Phagocytotic microglia in white matter express high levels of MHC class II (Butter et al. (1991); Kosel et al. (1997); Schmitt et al. (1998)). MHC Class II antigens are also present on gray matter microglia in the human brain (Köse et al. (1997); Schmitt et al. (1998)). On the other hand, phagocytotic microglia in rodent gray matter are generally not MHC class-II-positive, if phagocytosis is not associated with a strong, immune response-mediated activation (Streit et al. (1989); Angelov et al. (1996); Jungi et al. (1997); Bohatschek et al. (1999)). Surprisingly, the reverse is true for the speed of phagocytosis. In gray matter, phagocytotic microglia remove cellular debris within a few days, detach from the microglial nodules and downregulate their activation markers (Streit and Kreutzberg (1988)); Möller et al. (1996); Raivich et al. (1998a)). In white matter, phagocytosis and the degradation of myelin is an extremely slow process, with debris-laden macrophages in the pyramidal tract still present years after the original cerebral injury (Kösel et al. (1997)).

In addition to the antigen-presenting complexes MHCI and/or MHCII, there is also an upregulation of costimulatory molecules such as B7.2, ICAM1 and αXβ2 integrin (Werner et al. (1998); Bohatschek et al. (1999)), which may serve as potent stimuli of T-cell activation in the presence of specific antigen (Bretscher and Cohn (1970); Robey et al. (1995)). Of particular interest is the phagocyte-specific induction of αXβ2 integrin (Kloss et al. (1999)). In peripheral tissues, this integrin is a highly selective marker of dendritic cells (Brocker et al. (1997)). It is a coreceptor for the bacterial lipopolysaccharide (LPS) that mediates activation of macrophages and dendritic cells (Ingalls and Golenbock (1995)) and may strengthen the binding between lymphocytes and the antigen-presenting cells (Hynes (1992)). Its expression on phagocytotic microglia points to a similar function when they are contacted by infiltrating lymphocytes (Raivich et al. (1998a) Fig. 3H).

Bystander activation (stage 3b)

In addition to changes in the phagocytotic microglia, the process of phagocytosis also affects their immediate cellular environment. Adjacent microglia show an increase in MHC class I, B7.2, α4β1 and αMβ2 integrin (Raivich et al. (1998a); Bohatschek et al. (1999); Kloss et al. (1999)). In the case of MHC class I, this leads to the formation of blobs, 60–100μm in diameter, with an intensely immunoreactive phagocytotic nodule at the center, surrounded by gradually decreasing staining farther away from the Center (Raivich et al. (1998a)). A similar concentric distribution was also observed for lymphocyte aggregates around the microglial nodules, indicating the presence of a gradient for soluble cytokines produced by phagocytotic microglia. This coincidence with lymphocyte infiltration seems to be, under normal conditions, a unidirectional process, from microglia to lymphocytes and not in the other direction. Thus, a similar activation of bystander microglia, with the induction of MHCI, is also observed in mice with severe combined immunodeficiency (SCID), which lack differentiated T- and B-cells (Raivich et al. (1998a)).

What is the functional relevance of bystander activation? In normal mice, lymphocytes will also adhere to the activated, but non-phagocytotic, microglia (Raivich et al. (1998a)). Although these microglial cells express MHCI, the lower levels of costimulatory molecules ICAM1 and B7.2, and the lack of αXβ2-integrin, could induce anergy in the interacting T-cells. Selective absence of costimulatory molecules, in the presence of antigen and the appropriate MHC molecules, can cause the inactivation, or even the death, of the antigen-recognizing T-cell (Bretscher and Cohn (1970); Robey and Allison (1995)). In the context of immune surveillance in the injured brain, this may cause the inactivation or disappearance of T-cells that recognize self-antigens presented on the surface of the bystander-activated and non-phagocytotic microglial cells—a physiological form of a differential antigen display.

Perivascular macrophages

In addition to the parenchymal microglia, facial axotomy also leads to activation of perivascular macrophages, a second, monocyte-related cell type in the brain. However like the parenchymal microglia, they are long and flattened cells that are located in the space between neural parenchyma and the vascular endothelial cells, the Virchow-Robin space, and completely enclosed by a basal membrane (Fig. 3J). Studies in sublethal and lethal motor neuron injury models suggested that this activation proceeds in two consecutive stages, but there is considerable variation between different species (Streit et al. (1989b); Angelov et al. (1996)). In the first step, there is a simple increase in the number of MHC class-II-positive cells, peaking at around two to three weeks, a change observed both in the rat and mouse facial nucleus (Streit et al. (1989b); Bohatschek et al. (1999)). In the rat, starting two to three weeks after injury, some of these cells appear to leave their perivascular position, migrate into the neural parenchyma and differentiate into ramified, microglia-like cells. However, they do retain their strong MHC II immunoreactivity for more than half a year after injury (Streit et al. (1989b)). Hypothetically, these perivascular macrophages-turned-microglia may also contribute to the high number of ramified, MHC class-II-positive cells that accumulate in the gray matter of the aging human brain (Streit and Sparks (1997)). Although neuronal cell death enhanced the appearance of MHC class-II-positive, ramified cells in the motoneuron injury models, they did not participate in phagocytosis (Streit et al. (1989b); Angelov et al. (1996)), or, by imgilichin, in the presentation of internalized and processed antigen. MHC expression when not accompanied by additional immunoaccessory molecules can serve as a potent and antigen-specific inhibitory stimulus for the activated T-cells (Bretscher and Cohn (1970); Robey et al. (1995)). In this context, the long-term expression MHC class II after a traumatic event may be instrumental in turning down an immune response in the recovering brain, since most immunoaccessory molecules like B7.2, $\alpha X \beta 2$ or ICAM1 disappear rapidly after injury (Werner et al. (1998); Bohatschek et al. (1999); Kloss et al. (1999)).

Inflammatory cytokines: effects on microglial activation

Cytokine expression in the facial nucleus shows a graded response

The axotomized facial nucleus is a source of many inflammation-associated cytokines, but these are expressed in a tightly controlled and grade-specific expression pattern (Raivich et al. (1999)). Moderate but effective amounts of macrophage-colony stimulating factor (MCSF) and transforming growth factor $\beta 1$ are already present in the normal facial motor nucleus (Kiefer et al. (1993);

Jones et al. (1998); Raivich et al. (1998b); Streit et al. (1998)). Inside the first 24–48 hours, facial axotomy will lead to a rapid upregulation of interleukin-6/IL6 (Kiefer et al. (1993); Klein et al. (1997)), transforming-growth factor beta 1 (TGFβ1), its receptors (Kiefer et al. (1993) Jones et al. (1998) Streit et al. (1998)) and the receptor for MCSF (Raivich et al. (1991; 1998b); Streit et al. (1998)). The expression of MCSF, a key mitogen for microglia *in vivo,* however, will stay unchanged (Raivich et al. (1998b)). In the mouse, the onset of neuronal cell death (Torvik and Skörten (1971); Möller et al. (1996)), their phagocytosis by microglia and the massive influx by lymphocytes peaking at day 14 (Raivich et al. (1998a)) are associated with a strong induction of a second tier of cytokines: interferon gamma (IFNγ), tumor-necrosis factor alpha (TNFα) and interleukin 1 beta (IL1β). Here, recent data into the function of these cytokines *in vivo* are beginning to yield insights into the molecular mechanisms that determine the graded neuroglial response in the injured brain, as summarized in Figure 2.

Macrophage-Colony Stimulating Factor (MCSF)

MCSF is a homodimeric, 45–90 kD glycoprotein that is a potent mitogen for monocyte precursors (Roth and Stanley (1992)) and related cell types, including brain-derived ameboid macrophages (Giulian and Ingeman (1988); Hao et al. (1990); Sawada et al. (1990); Suzumura et al. (1990)) and ramified microglia (Kloss et al. (1997)). MCSF is constitutively expressed during brain development, and in the adult central nervous system. The data on the additional MCSF increase following injury are inconclusive, but all the different brain pathologies studied so far have shown a strong and selective induction of MCSF receptors (MCSFR) on activated microglia during the post-traumatic proliferation of this glial cell type (Raivich et al. (1991; 1998b); Akiyama et al. (1994); Hulkower et al. (1998); Streit et al. (1998)). Insights into the function of MCSF have come from *osteopetrotic* mice, which carry a natural, frameshift mutation in the coding region for MCSF (Yoshida et al. (1990)). In the homozygous animals (op/op), this mutation leads to a complete absence in the biological activity for this hematopoetic molecule (Wiktor-Jedrzejczak et al. (1990)), a 70–80% reduction in microglial proliferation (Raivich et al. (1994)), and a strongly curtailed expression of microglial-activation markers characteristic of stage 1 and stage 2, the cell-adhesion molecule thrombospondin, the α5β1, α6β1 and αMβ2 integrins, and the receptor for MCSF (Kalla et al. (2001)). This is also accompanied by a more than 60% reduction in the recruitment of CD3+ T-cells at day 1. The selective presence of MCSF receptors on microglia, but not on neighboring vessels, astrocytes, neurons or lymphocytes, clearly shows an important role for MCSF in this early phase of T-cell extravasation. Late stages of microglial activation—the phagocytosis of dead motoneurons, the bystander activation of the neighboring non-phagocytotic microglial cells—and the peak of CD3-positive lymphocytes, were not affected by the absence of MCSF (Kalla et al. (2001)). Altogether, this points to an early window of MCSF-mediated effects during the initial stages of microglial activation.

At the same time, the absence of MCSF and the ensuing massive reduction in microglial activation did not affect axonal outgrowth, astrocyte response, neuronal survival, or any of the biochemical changes in the injured motoneurons. Moreover, these data agree with the absence of non-microglial effects in the cytosine arabinoside model, wherein direct application of this antimitotic drug completely inhibited microglial proliferation, but had no effect on axonal regeneration (Svensson and Aldskogius (1993)).

Interleukin-6 (IL6)

The induction of IL6, a proinflammatory cytokine that belongs to a family of neurokines, is an almost ubiquitous, and early, marker of cerebral tissue damage. It occurs following direct trauma, ischemia, infection, in neurodegeneration (Alzheimer's, Parkinson), in autoimmune disease, or following indirect injury, e.g. after a peripheral axotomy in the facial motor nucleus (for reviews, see Hopkins and Rothwell (1995); Raivich et al. (1999)). The receptors for IL6 are localized on neurons and astrocytes, but not on microglia, and this expression pattern does not change after injury (Klein et al. (1997)). Absence of IL6 following transgenic deletion leads to striking changes in the pattern of neuroglial activation. In the axotomized facial motor nucleus model, all three major cell types, neurons, astrocytes and microglia, are affected. Compared with normal, wild-type control mice, the IL6-deficient animals show a much smaller, and sometimes completely absent, post-traumatic induction of GFAP-positive astrocytes. Similar but more moderate effects are also observed on microglial proliferation. There is a 30–40% decrease in the number of [3H]-thymidine labeled, microglial profiles (Klein et al. (1997)) and a reduction in early microglial activation markers (described above), including the early phase of MHCI increase (Galiano et al. 2001). Late microglial response is unaffected, and there is no apparent difference in neuronal survival. There is a 13% decrease in the speed of axonal regeneration. However, this is a mild effect and is probably direct, owing to neuronal receptors for IL6.

In contrast, lymphocyte recruitment is strongly curtailed (Galiano et al. (2001)), in a stage-specific way. Absence of IL6 leads to a ten-fold decrease at the early stage of lymphocyte entry, at day 1, but just to 2.3-fold decrease at day 14. Thus, as with MCSF, this cytokine appears to play a crucial role in the early phase of immune surveillance, activating antigen presentation and the entry of antigen-detecting lymphocytes in the presence of non-lethally injured neurons.

Transforming Growth Factor-Beta 1 (TGFβ1)

Unlike all the preceding and following cytokines mentioned in this chapter, TGFβ1 is an anti-inflammatory molecule. It is also a homodimeric protein that belongs to a family of pleiotrophic cytokines with potent neurotrophic and immunosuppressive properties. TGFβ1 is already present at low levels in the normal central nervous system (Kiefer et al. (1993)), but the expression of this cytokine is strongly upregulated in almost all cerebral pathologies, quite similar to the regulation pattern for the MCSF receptor and for IL6 (Morganti-Kossmann et al. (1992); Kiefer et al. (1995)). Although TGFβ receptors are present on many different cultured cell types, their expression in the brain appears to be restricted to neurons (Unsicker et al. (1996)). Thus, transection of the facial nerve led to a selective upregulation of the TGFβ receptor II (TR2) on axotomized motoneurons, but not on the adjacent astrocytes or microglia (Jones et al. (1998a)). This selective neuronal expression could offer one explanation for the striking differences in the glial response in cell culture and *in vivo* following transgenic deletion of TGFβ1: *in vitro,* the effects of TGFβ1 are direct (Jones et al. (1998b)); *in vivo,* they are probably mediated by the TGFβ1-responsive, TR2-positive neurons. The absence of TGFβ1 in the uninjured, adult brain leads to gliosis and inflammatory changes (Jones et al. (1998a)). The microglia exhibit reduction in ramification, cell-body attachment to neighboring neurons, and a high number of proliferating profiles. Approximately 5–10% of the microglia are also seen to transform into ameboid and phagocytotic cells, with a concomitant expression of late activation markers MHCI, B7.2 and aXb2 integrin, indicative of a severe form of brain pathology and the presence of cell debris. This absence of TGFβ1 also interferes with the normal microglial response to facial axotomy. Cell proliferation is reduced by 50–60% and there is a severe reduction in early activation markers ICAM1 and $\alpha M\beta 2$, and $\alpha 6\beta 1$ integrins. Late-stage activation markers such as $\alpha X\beta 2$ and B7.2 are already elevated in the uninjured facial nucleus, but there is no further increase after axotomy. Unlike IL6 or MCSF, the effects of TGFβ1 are not limited to the early stages of cellular response in the axotomized facial motor nucleus.

Interleukin 1(IL1)

Interleukin-1 (IL1) is an important player in inflammation and the immune response, and is strongly upregulated in severe brain injury (Hopkins and Rothwell (1995); Raivich et al. (1999)). The interleukin-1 family consists of three ligands, IL1α, IL1β and the IL1 receptor antagonist IL1ra, and two receptors, IL1R1 and IL1R2 (Alheim and Bartfai (1998)). Activated microglia/macrophages are the major source of IL1α, IL1β and IL1ra in the damaged brain (Hopkins and Rothwell (1995)). In the axotomized facial motor nucleus, however, the effects of IL1 appear relatively limited. Thus, transgenic deletion of IL1R1, the main IL1 signal-transducing cell surface receptor, leads to a pronounced reduction in lymphocytic migration. However, it does not affect neuronal survival, or late stages of microglial activation (Bohatschek et al. (1999)).

Interferon-gamma (IFNγ)

IFNγ is a critical regulator of T-cell mediated inflammation. It induces the production of cytotoxic oxygen radicals, phagocytes, and molecules involved in the presentation of antigen (MHCII) in cells derived from the central nervous system, particularly in microglia (Suzumura et al. (1987)). The effects in facial axotomy models appear very mild. Transgenic deletion of interferon gamma receptor type 1 (IFNγR1), the main IFNγ signal-transducing cell surface receptor, does not affect microglial activation, neuronal survival, or lymphocyte recruitment (Bohatschek and Raivich (1999); Raivich et al. (1999b)). Neither was an effect on microglial activation observed for mice with severe combined immunodeficiency/SCID (Raivich et al. (1998a)), which lack differentiated T- or B-cells. However, preliminary data do show a mild enhancing effect of IFNαR1 deficiency on the number of MHCII-positive, perivascular macrophages. Interestingly, ablation of the IFNγ-signaling pathway did not interfere with the onset of transferred experimental allergic encephalomyelitis/EAE (Willenborg et al. (1996)). However, it did inhibit the recovery, pointing to a suppressive effect of this cytokine in aggressive autoimmune disease (Krakowski and Owens (1996); Willenborg et al. (1996)). Although hypothetical, the downregulation in the number of MHCII-positive, perivascular brain macrophages may be one of the components mediating the inhibitory activity of IFNγ.

Tumor Necrosis Factor-α (TNFα)

TNFα is a proinflammatory cytokine, strongly induced following trauma, ischemia, infection or excitotoxic injury (Bruce et al. (1996); Seilhean et al. (1997); Uno et al. (1998); Yang et al. (1998)). However, it is absent during the early phase of indirect brain injury, as in the facial axotomy model (Kiefer et al. (1993); Raivich et al. (1998a)), suggesting that the function of TNFα is restricted to severe forms of brain pathology. *In vitro*, TNFα is a pleiotrophic molecule with direct effects on neurons, oligodendrocytes, astrocytes and microglia. Two different TNF receptor types, p55 and p75, have been identified, and both types are expressed on neurons and glial cells. Both receptors play important functional roles, either alone or in combination. Thus, the absence of p55 (TNFR1) interferes with the induction of macrophage-inducible nitrogen-oxide synthesizing enzyme iNOS, blocking their parasite killing potential in *T.gondii* encephalitis (Deckert-Schluter et al. (1998)). This molecule is apparently redundant in peripheral *T.gondii* infection, but needed to combat the intracerebral parasites (Scharton-Kesten et al. (1997)). On the other hand, removal of p75 (TNFR2) inhibits the induction of endothelial ICAM1, preventing a detrimental neurovascular response in cerebral malaria (Lucas et al. (1997)). Combined transgenic deletion of both TNF receptors also leads to drastic changes in the cellular response. In the murine facial axotomy model, absence of both TNF receptors almost completely abolished neuronal cell death (>75%) in the first 30 days after axotomy (Bohatschek et al. (1999)), and the ensuing formation of microglial nodules that express the long panel of late-activation markers typical of phagocytotic macrophages, summarized in Figure 2. In this case, the effect on the phagocytotic microglia appears to be indirect, due to a primary action on neuronal survival. In contrast, deletion of just TNFR1 or TNFR2 alone does not affect neuronal cell death, with a similar number of microglial nodules; however, these deletions affect the late non-neuronal response. Absence of TNFR1 leads to a significant reduction in microglial MHC and B7.2 expression, absence of TNFR2 to a 50% decrease in lymphocyte recruitment. While preventing neuronal cell death, the combined deletion of both receptors leads to only a mild increase in the non-neuronal response, pointing to different signaling requirements in the execution of the TNF-mediated effects, with the microglial bystander activation dependent on TNFR1, lymphocyte entry on TNFR2, and neuronal cell death or survival on the presence of both cell surface molecules.

Overview

The microglial changes in the damaged brain form a graded response, which is a consistent feature in almost all forms of brain pathology. In the axotomized facial motor nucleus, the microglia proceed from a resting stage, through early activation, adhesion to injured neurons, proliferation and phagocytosis, to interaction with other cell types, particularly with the non-lethally and lethally injured neurons and the invading T-lymphocytes. Recent studies on the inflammation-associated molecules have begun to uncover the mechanisms that underlie this post-traumatic cellular activation. Thus, early stages of microglial activation are dependent on MCSF and IL6, and the late bystander activation on the effects of TNF and TNFR1. The microglial effects of IL1 and IFNγ appear restricted to more severe forms of injury, such as direct trauma, ischemia, infection and autoimmune disease (for reviews, see Morganti-Kossman et al. (1992); Rothwell and Hopkins (1995); Raivich et al. (1999)), that of TGFβ1 are particularly broad, extending from resting microglia in uninjured brain, to those in particularly severe pathology. Currently, there are several lines of evidence that activated microglia may exert a trophic effect on injured neurons, particularly in cortical ischemia and in heavy-metal neurotoxicity, based of the effects of MCSF and TNFR1/2 deletions (Berezovskaya et al. (1995); Bruce et al. (1996); Bruccoleri and Harry (2000)). The effects of these deletions are very different after facial axotomy. Absence of MCSF does not affect neuronal survival or axonal outgrowth, while strongly curtailing microglial activation and lymphocyte recruitment (Kalla et al. (2000)). Deletion of both TNF receptors prevents neurotoxicity after facial axotomy (Bohatschek et al. (1999)). Thus, the microglial synthesis of TNF (Bruce et al. (1996)) could have a neurotoxic effect, at least in the facial injury model.

In discussing the biological effects of activated microglia, it is important to acknowledge that these cells may also have an another function, apart from a trophic or toxic action on the neighboring neurons. The microglia are the only component of the immune system with a continuous, ongoing residence in brain parenchyma, and thus form the first line of defense during neural infection. The different steps of microglial activation, from early explorative changes, through adhesion to damaged cells, phagocytosis, the activation of neighboring non-phagocytotic microglia and their interactions with lymphocytes, all appear to relate to a specific function in this anti-infection regime. That function also appears supported by astrocytes, cerebral blood vessels and the recruitment of circulating leukocytes (Raivich et al. (1999)). In this scheme, the elimination of some of the injured neurons struggling for survival may provide antigenic material from potentially infected cells, which will be processed by microglia and then assessed by the T-cells, the cellular component of the immune system. Indeed, this is what appears to happen in the axotomized facial motor nucleus model. It is clear that this evolutionarily conserved mechanism may be inadequate to deal with other, noninfectious forms of neurological disease, and in promoting neural repair. Studies on pro- and anti-inflammatory cytokines, adhesion molecules and bacteriocidic enzymes have begun to uncover the molecular mechanisms that animate this cellular regime. Its complete elucidation will greatly enhance our insight into the cellular and molecular activation cascade in the injured neural tissue, and could identify possible targets for pharmacological intervention, to help repair the damaged central nervous system.

References

Akiyama H, Nishimura T, Kondo H, Ikeda K, Hayashi Y, McGeer PL (1994) Expression of the receptor for macrophage colony stimulating factor by brain microglia and its upregulation in brains of patients with Alzheimer's disease and amyotrophic lateral sclerosis. *Brain Res* 639:171–174.

Alheim K, Bartfai T (1998). The interleukin-1 system: receptors, ligands, and ICE in the brain and their involvement in the fever response. *Ann NY Acad Sci.* 840:51–58.

Angelov DN, Gunkel A, Stennert E, Neiss WF (1995). Phagocytic microglia during delayed neuronal loss in the facial nucleus of the rat-time course of the neuronofugal migration of brain macrophages. *GLIA.* 13:113–129.

Angelov DN, Neiss WF, Streppel M, Walther M, Guntinas-Lichius O, Stennert E (1996). ED2-positive perivascular cells act as neuronophages during delayed neuronal loss in the facial nucleus of the rat. *GLIA* 16:129–139.

Berezovskaya O, Maysinger D, Fedoroff S (1995). The hematopoietic cytokine, colony-stimulating factor 1, is also a growth factor in the CNS: congenital absence of CSF-1 in mice results in abnormal microglial response and increased neuron vulnerability to injury. *Int J Dev Neurosci* 13:285–299.

Bohatschek M, Gschwendtner A, Von Maltzan X, Kloss CUA, Pfeffer K, Labow M, Bluthmann H, Kreutzberg GW, Raivich G (1999). Cytokine-mediated regulation of MHC1, MHC2 and B7-2, in the axotomized mouse facial motor nucleus. *Soc Neurosci Abs* 25:1535

Bohatschek M, Werner A, Raivich G (2001). Systemic lipopolysaccharide injection leads to granulocyte influx into normal and injured brain: Effects of ICAM-1 Deficiency. *Exp Neurol*, (in press).

Bretscher P, Cohn M (1970). A theory of self-nonself discrimination. *Science* 169 1042–1049

Brocker T, Riedinger M, Karjalainen K (1997). Targeted expression of major histocompatibility complex (MHC) class II molecules demonstrates that dendritic cells can induce negative but not positive selection of thymocytes in vivo. *J Exp Med.* 185 541–550.

Bruccoleri A, Harry GJ (2000). Chemical-induced hippocampal neurodegeneration and elevations in TNF alpha, TNF beta, IL-1 alpha, IP-10, and MCP-1 mRNA in osteopetrotic (op/op) mice. *J Neurosci Res* 62:146–155

Bruce AJ, Boling W, Kindy MS, Peschon J, Kraemer PJ, Carpenter MK, Holtsberg FW, Mattson MP (1996). Altered neuronal and microglial responses to excitotoxic and ischemic brain injury in mice lacking TNF receptors. *Nature Med* 2:788–794.

Butter C, O'Neill JK, Baker D, Gschmeissner SE, Turk JL (1991). An immunoelectron microscopical study of the expression of class II major histocompatibility complex during chronic relapsing experimental allergic encephalomyelitis in Biozzi AB/H mice. *J Neuroimmunol* 33 37–42.

Compston A, Zajicek J, Sussman J, Webb A, Hall G, Muir D, Shaw C, Wood A, Scolding N (1997). Glial lineages and myelination in the central nervous system. *J Anat* 190:161–200.

Deckert-Schluter M, Bluethmann H, Rang A, Hof H, Schluter D (1998). Crucial role of TNF receptor type 1 (p55), but not of TNF receptor type 2 (p75), in murine toxoplasmosis. *J Immunol* 160:3427–3436.

De Simone R, Giampaolo A, Giometto B, Gallo P, Levi G, Peschle C, Aloisi F (1995). The costimulatory molecule B7 is expressed on human microglia in culture and in multiple sclerosis acute lesions. *J Neuropathol Exp Neurol* 54 175–187.

Engelhardt B, Martin-Simonet MT, Rott LS, Butcher EC, Michie SA (1998). Adhesion molecule phenotype of T lymphocytes in inflamed CNS. *J Neuroimmunol* 84 92–104.

Fishman PS, Savitt JM (1989). Selective localization by neuroglia of immunoglobulin G in normal mice. *J Neuropathol Exp Neurol* 48 212–220.

Galiano M, Liu ZQ, Roger Kalla, Bohatschek M, Koppius A, Gschwendtner A, Xu SL, Werner A, Kloss C, Bluethmann H, Raivich G (2001). Interleukin-6 (IL6) and the cellular response following facial nerve injury: Effects on lymphocyte recruitment, early microglial activation and axonal outgrowth in IL6-deficient mice. *Eur J Neurosci*, 14:327–341.

Giulian D, Ingeman JE (1988). Colony stimulating factors as promoters of ameboid microglia. *J Neurosci* 8:4707–4717.

Graeber MB, Streit WJ, Kreutzberg GW (1989). Identity of ED2-positive perivascular cells in rat brain. *J Neurosci Res* 22 103–106.

Graeber MB, Streit WJ, Kreutzberg GW (1988a). The microglial cytoskeleton vimentin is localized with activated cells in situ. *J Neurocytol* 17 573–580.

Graeber MB, Tetzlaff W, Streit WJ, Kreutzberg GW (1988b). Microglial cells but not astrocytes undergo mitosis following rat facial nerve axotomy. *Neurosci Lett* 85:317–321.

Hailer NP, Bechmann I, Heizmann S, Nitsch R (1997). Adhesion molecule expression on phagocytic microglial cells following anterograde degeneration of perforant path axons. *Hippocampus* 7:341–349.

Hao C, Guilbert LJ, Fedoroff S (1990). Production of colony-stimulating factor-1 (CSF-1) by mouse astroglia in vitro. *J Neurosci Res* 27:314–323.

Harrison JK, Jiang Y, Chen S, Xia Y, Maciejewski D, McNamara RK, Streit WJ, Salafranca MN, Adhikari S, Thompson DA, Botti P, Bacon KB, Feng L (1998). Role for neuronally derived fractalkine in mediating interactions between neurons and CX3CR1-expressing microglia. *Proc Nat Acad Sci USA* 95:10896–901

Hopkins SJ, Rothwell NJ (1995). Cytokines and the nervous system: expression and recognition. *Trends Neurosci* 18:83–88.

Horvat A, Schwaiger FW, Hager G, Streif R, Probst JC, Ullrich A, Kreutzberg GW (2000). Modified activation of glial cells in protein tyrosine phophatase PTP1C mutant mice during regeneration of the axotomised facial nerve. *J Neurosci* (in press).

Hulkower K, Brosnan CF, Aquino DA, Cammer W, Kulshrestha S, Guida MP, Rapoport DA, Berman JW (1993). Expression of CSF-1, c-fms, and MCP-1 in the central nervous system of rats with experimental allergic encephalomyelitis. *J Immunol* 150:2525–2533.

Hynes RO (1992). Integrins: versatility, modulation, and signaling in cell adhesion. *Cell* 69:11–25.

Ingalls RR, Golenbock DT (1995). CD11c/CD18, a transmembrane signaling receptor for lipopolysaccharide. *J Exp Med* 181 1473–1479.

Jones LL, Kreutzberg GW, Raivich G (1997). Regulation of CD44 in the regenerating mouse facial motor nucleus. *Eur J Neurosci* 9:1854–1863

Jones LL, Doetschman T, Kreutzberg GW, Raivich G (1998a). Neuroglial activation in the injured central nervous system: role of transforming growth factor-β1. *Soc Neurosci Abstr* 24:710.2.

Jones LL, Kreutzberg GW, Raivich G (1998b). Transforming growth factor β's 1, 2 and 3 inhibit proliferation of ramified microglia on an astrocyte monolayer. *Brain Res* 795 301–306.

Jones LL, Liu ZQ, Shen J, Werner A, Kreutzberg GW, Raivich G (2000). Regulation of the cell adhesion molecule CD44 after nerve transection and direct trauma to the mouse brain.*J Comp Neurol,* 426:468–492.

Jung S, Aliberti J, Graemmel P, Sunshine MJ, Kreutzberg GW, Sher A, Littman DR (2000). Analysis of fractalkine receptor CX(3)CR1 function by targeted deletion and green fluorescent protein reporter gene insertion. *Mol Cell Biol* 20:4106–4114.

Jungi TW, Pfister H, Sager H, Fatzer R, Vandevelde M, Zurbriggen A (1997). Comparison of inducible nitric oxide synthase expression in the brains of Listeria monocytogenes-infected cattle, sheep, and goats and in macrophages stimulated in vitro. *Infect Immun* 65:5279–5288.

Kalla R, Liu ZQ, Xu SL, Koppius A, Imai Y, Kloss CUA, Kohsaka S. Gschwendtner G, Möller CJ, Werner A, Raivich G (2001). Microglia and the early phase of immune surveillance in the injured axotomized facial motor nucleus: Impaired microglial activation and lymphocyte recruitment but no effect on neuronal survival or axonal regeneration in the MCSF-deficient mice. *J Comp Neurol* 436:182–201.

Kiefer R, Lindholm D, Kreutzberg GW (1993). Interleukin-6 and transforming growth factor-beta-1 mRNAs are induced in rat facial nucleus following motoneuron axotomy. *Eur J Neurosci* 5:775–781.

Kiefer R, Streit WJ, Toyka KV, Kreutzberg GW, Hartung H (1995). Transforming growth factor-beta-1: A lesion-associated cytokine of the nervous system. *Int J Dev Neurosci* 13:331–339.

Klein MA, Möller JC, Jones LL, Bluethmann H, Kreutzberg GW, Raivich G (1997). Impaired neuroglial activation in Interleukin-6 deficient mice. *GLIA* 19 227–233.

Kloss CUA, Kreutzberg GW, Raivich G (1997). Proliferation of ramified microglia on an astrocyte monolayer: characterization of stimulatory and inhibitory cytokines. *J Neurosci Res* 49:248–254.

Kloss CUA, Werner A, Shen J, Klein MA, Kreutzberg GW, Raivich G (1999). The integrin family of cell adhesion molecules in the injured brain: regulation and cellular localization in the normal and regenerating mouse facial nucleus. *J Comp Neurol* 441:162–178.

Kösel S, Egensperger R, Bise K, Arbogast S, Mehraein P, Graeber MB (1997). Long-lasting perivascular accumulation of major histocompatibility complex class II-positive lipophages in the spinal cord of stroke patients: possible relevance for the immune privilege of the brain. Acta *Neuropathol* 94:532–538.

Krakowski M, Owens T (1996). Interferon-gamma confers resistance to experimental allergic encephalomyelitis. *Eur J Immunol* 26:1641–1646.

Kreutzberg GW (1966). Autoradiographische Untersuchungen über die Beteiligung von Gliazellen an der axonalen Reaktion im Fazialiskern der Ratte. *Acta Neuropathol* 7:149–161.

Liu L, Tornqvist E, Mattsson P, Eriksson NP, Persson JK, Morgan BP, Aldskogius H, Svensson M (1995). Complement and clusterin in the spinal cord dorsal horn and gracile nucleus following sciatic nerve injury in the adult rat. *Neuroscience* 68:167–179.

Lucas R, Juillard P, Decoster E, Redard M, Burger D, Donati Y, Giroud C, Monso-Hinard C, De Kesel T, Buurman WA, Moore MW, Dayer JM, Fiers W, Bluethmann H, Grau GE (1997). Crucial role of tumor necrosis factor (TNF) receptor 2 and membrane-bound TNF in experimental cerebral malaria. *Eur J Immunol* 27:1719–1725.

McGeer PL, Kawamata T, Walker DG, Akiyama H, Tooyama I, McGeer EG (1993). Microglia in degenerative neurological disease. *GLIA* 7:84–92.

Merzbacher L (1909). Untersuchungen über die Morphologie und Biologie der Abräumzellen im Zentralnervensystem. Fischer-Verlag, Stuttgart.

Möller JC, Klein MA, Haas S, Jones L.L, Kreutzberg GW, Raivich G (1996). Regulation of thrombospondin in the regenerating mouse facial nucleus. *Glia* 17:121–132.

Moneta ME, Gehrmann J, Topper R, Banati RB, Kreutzberg GW (1993). Cell adhesion molecule expression in the regenerating rat facial nucleus. *J Neuroimmunol* 45:203–206.

Morganti-Kossmann MC, Kossmann T, Wahl SM (1992). Cytokines and neuropathology. *Trends Pharmacol Sci* 1992, 13 286–291.

Nissl F (1899). Über einige Beziehungen zwischen Nervenzellenerkrankungen und gliösen Erscheinungen bei verschiedenen Psychosen. *Arch Psych* 32:1–21.

Perry VH, Hume DA, Gordon S (1985). Immunohistochemical localization of macrophages and microglia in the adult and developing mouse brain. *Neurosci* 15:313–326.

Raivich G, Gehrmann J, Kreutzberg GW (1991). Increase of macrophage colony-stimulating factor and granulocyte-macrophage colony stimulating factor receptors in the regenerating rat facial nucleus. *J Neurosci Res* 30:682–686.

Raivich G, Gehrmann J, Graeber MB, Kreutzberg GW (1993). Quantitative immunohistochemistry in the rat facial nucleus with iodine-125 iodinated secondary antibodies and in-situ autoradiography non-linear binding characteristics of primary monoclonal and polyclonal antibodies. *J Histochem Cytochem* 41:579–592.

Raivich G, Moreno-Flores MT, Möller JC, Kreutzberg GW (1994). Inhibition of posttraumatic microglial proliferation in a genetic model of macrophage colony-stimulating factor deficiency in the mouse. *Eur J Neurosci* 6:1615–1618.

Raivich G, Jones LL, Kloss CUA, Werner A, Neumann H, Kreutzberg GW (1998a). Immune surveillance in the injured nervous system: T-lymphocytes invade the axotomized mouse facial motor nucleus and aggregate around sites of neuronal degeneration. *J. Neurosci* 18 (1998) 5804–5816.

Raivich G, Haas S, Werner A, Klein MA, Kloss CUA, Kreutzberg GW (1998b). Regulation of MCSF receptors on microglia in the normal and injured mouse central nervous system: A quantitative immunofluorescence study using confocal laser microscopy. *J Comp Neurol* 395:342–358.

Raivich G, Bohatschek M, Kloss CUA, Werner A, Jones LL, Kreutzberg GW (1999a). Neuroglial activation repertoire in the injured brain: Graded response, molecular mechanisms and cues to physiological function. *Brain Res Rev* 30:77–105.

Raivich G, Jones LL, Werner A, Blüthmann H, Doetschmann T, Kreutzberg GW (1999b). Molecular signals for glial activation: Pro- and anti-inflammatory cytokines in the injured brain. *Acta Neurochir* [Suppl] 73:21–30.

Raivich G, Galiano M, Jones LL, Kloss CUA, Werner A, Bluethmann H, Kreutzberg GW (1999c). Lymphocyte infiltration into the injured brain: Role of proinflammatory cytokines IL1, IL6 and TNFa. *Ann Anat* 181 S43–S44.

Río Hortega Pd (1932). Microglia. In: *Cytology and Cellular Pathology of the Nervous System.* (E. Penfield, ed.), Paul B. Hoeber, New York, Vol. II, pp. 481–534.

Robey E, Allison JP (1995). T-cell activation: integration of signals from the antigen receptor and costimulatory molecules. *Immunol Today* 16:306–310.

Roth P, Stanley E (1992). The biology of CSF-1 and its receptor. *Curr Topics Microbiol Immunol* 181:141–167.

Sawada M, Suzumura A, Yamamoto H, Marunouchi T (1990). Activation and proliferation of the isolated microglia by colony stimulating factor-1 and possible involvement of protein kinase C. *Brain Res* 509:119–124.

Schafer MK, Schwaeble WJ, Post C, Salvati P, Calabresi M, Sim RB, Petry F, Loos M, Weihe E (2000). Complement C1q is dramatically up-regulated in brain microglia in response to transient global cerebral ischemia. *J Immunol* 164:5446–5452.

Scharton-Kersten TM, Yap G, Magram J, Sher A (1997). Inducible nitric oxide is essential for host control of persistent but not acute infection with the intracellular pathogen Toxoplasma gondii. *J Exp Med* 185:1261–1273.

Schiefer J, Kampe K, Dodt HU, Zieglgansberger W, Kreutzberg GW (1999). Microglial motility in the rat facial nucleus following peripheral axotomy. *J Neurocytol* 28:439–453.

Schmitt AB, Brook GA, Buss A, Nacimiento W, Noth J, Kreutzberg GW (1998). Dynamics of microglial activation in the spinal cord after cerebral infarction are revealed by expression of MHC class II antigen. *Neuropath Appl Neurobiol* 24 167–176.

Schwaiger FW, Fluegel A, Hager G, Horvat A, Spitzer C, Graeber MB, Kreutzberg GW (1999). Neuronal expression of MCP-1 following a remote lesion of the CNS. *Soc Neurosci Abs* 25:1275.

Seilhean D, Kobayashi K, He Y, Uchihara T, Rosenblum O, Katlama C, Bricaire F, Duyckaerts C, Hauw JJ (1997) Tumor necrosis factor-alpha, microglia and astrocytes in AIDS dementia complex. *Acta Neuropath* 93:508–517.

Sjöstrand J (1966). Studies on glial cells in the hypoglossal nucleus of the rabbit during nerve regeneration. *Acta Physiol Scand* 67/Suppl. 270:1–17.

Streit WJ, Kreutzberg GW (1988). Response of endogenous glial cells to motor neuron degeneration induced by toxic ricin. *J Comp Neurol* 268:248–263.

Streit WJ, Graeber MB, Kreutzberg GW (1989a). Peripheral nerve lesion produces increased levels of major histocompatibility complex antigens in the central nervous system. *J Neuroimmunol* 21:117–123.

Streit WJ, Graeber MB, Kreutzberg GW (1989b). Expression of Ia antigen on perivascular and microglial cells after sublethal and lethal motor neuron injury. *Exp Neurol* 105 115–16.

Streit WJ, Sparks DL (1997). Activation of microglia in the brains of humans with heart disease and hypercholesterolemic rabbits. *J Mol Med* 75 130–138.

Streit WJ, Semple-Rowland SL, Hurley SD, Miller RC, Popovich PG, Stokes BT (1998). Cytokine mRNA profiles in contused spinal cord and axotomized facial nucleus suggest a beneficial role for inflammation and gliosis. *Exp Neurol* 152:74–87.

Svensson M, Aldskogius H (1993). Regeneration of hypoglossal nerve axons following blockade of the axotomy-induced microglial cell reaction in the rat. *Eur J Neurosci* 5:85–94.

Suzumura A, Metzitis SGE, Gonatas NK, Silberberg DH (1987). MHC antigen expression on bulk isolated macrophage-microglia from newborn mouse brain: Induction of Ia antigen expression by γ-interferon. *J Neuroimmunol* 15:263–278.

Suzumura A, Sawada M, Yamomoto H, Marunouchi T (1990). Effects of colony stimulating factors on isolated microglia in vitro. *J Neuroimmunol* 30:111–120.

Torvik A, Skjorten F (1971) Electron microscopic observations on nerve cell regeneration and degeneration after axon lesions. II. Changes in the glial cells. *Acta Neuropathol* 17:265–282.

Uno H, Matsuyama T, Akita H, Nishimura H, Sugita M (1997). Induction of tumor necrosis factor-alpha in the mouse hippocampus following transient forebrain ischemia. *J Cereb Blood Flow Metabol* 17:491–499.

Unsicker K, Meier C, Krieglstein K, Sartor BM, Flanders KC (1996). Expression, localization and function of transforming growth factor-beta s in embryonic chick spinal cord, hindbrain and dorsal root ganglia. *J Neurobiol* 29:262–276.

Werner A, Kloss CUA, Walter J, Kreutzberg GW, Raivich G (1998). Intercellular adhesion molecule-1 (ICAM1) in the regenerating mouse facial motor nucleus. *J Neurocytol* 27 219–232.

Wiktor-Jedrzejczak W, Bartocci A, Ferrante AW, Ahmed-Ansari A, Sell KW, Pollard JW, Stanlay ER (1990). Total absence of colony-stimulating factor 1 in the macrophage-deficient osteopetrotic (op/op) mouse. *Proc Natl Acad Sci USA* 87:4828–4832.

Willenborg DO, Fordham S, Bernard CC, Cowden WB, Ramshaw IA (1996). IFN-gamma plays a critical down-regulatory role in the induction and effector phase of myelin oligodendrocyte glycoprotein-induced autoimmune encephalomyelitis. *J Immunol* 157 3223–3227.

Yang GY, Gong C, Qin Z, Ye W, Mao Y, Betz AL (1998). Inhibition of TNFalpha attenuates infarct volume and ICAM-1 expression in ischemic mouse brain. *NeuroReport* 9:2131–2134.

Yoshida H, Hayashi S-I, Kunisasa Z, Ogaea M, Nishikawa S, Okamura H, Sudo T, Shultz LD, Nishikawa SI (1990). The murine mutation osteopetrosis is in the coding region of the macrophages colony stimulating factor gene. *Nature* 345 442–444.

9

Neuroprotective Roles of Microglia in the Central Nervous System

KAZUYUKI NAKAJIMA AND SHINICHI KOHSAKA

Introduction

The cells known as microglia are widely located throughout the brain, from the early developmental stage into the adult stage (Del Río-Hortega (1932); Jordan and Thomas (1988); Perry and Gordon (1988)). An outstanding feature of these cells is their transformation and proliferation in a variety of brain diseases. Thus, interest has focused on their functions and roles in the central nervous system (CNS) (Kreutzberg (1996)).

Microglia are believed to act primarily as scavengers in the brain (Perry and Gordon (1988)). An immunoeffector role has also been proposed because of their immunological properties (Graeber et al. (1988, 1989); Streit et al. (1988, 1989)). Microglia currently are accepted as "sensors of pathology" in the brain (Kreutzberg (1996)). Accompanying the interest in their cellular properties, much attention has been given to whether microglia are beneficial or harmful *in vivo* There have been increasing numbers of reports that microglia initiate and/or facilitate inflammatory and degenerative processes by producing neurotoxic substances. In contrast, a neurotrophic/neuroprotective role of microglia has also been suggested from *in vitro* and *in vivo* studies showing production of neurotrophic molecules (Nakajima and Kohsaka (1993); Kreutzberg (1996)).

We describe here the ability of microglia to act as neurotrophic/neuroprotective cells in the brain based on our previous biochemical and immunohistochemical studies, and also refer to the cellular mechanisms involved in the regulation of the neuroprotective and neurotoxic functions of microglia.

Interaction Between Neurons and Microglia

In the early stage of brain development, a type of microglia known as ameboid microglia is observed (Ling et al. (1980)) (Fig. 1). Ameboid microglia have a macrophagic shape, with a relatively large cell body and short processes. The key function of these cells is thought to be the elimination of dead cells and cellular debris by phagocytosis. Differentiated neurons and oligodendrocytes undergo programmed cell death during development (Cunningham (1982); Oppenheim (1991); Raff et al. (1993)), and these dead cells are scavenged by ameboid microglia (Ashwell (1990)). Considering that microglia can produce neurotrophic molecules (as described below), ameboid microglia may regulate neuronal cell number by supplying survival factor(s) (Fig. 1). In the developing retina, NGF derived from microglia have been reported to induce apoptosis in neuroepithelial cells (Frade et al. (1998)).

With brain development, ameboid microglia decrease in number and ramified microglia with small cell bodies and long branched processes become more numerous (Fig. 1). It is generally thought that ameboid microglia differentiate into ramified microglia during brain development. However, some researchers have questioned microglial differentiation, based on histological studies in which there is no smooth transition of ameboid microglia to ramified microglia in terms of spatial and temporal distribution, proposing instead that ameboid and ramified microglia are different cell types (Fujita and Kitamura (1975); Miyake et al. (1984)). Thus, the relation between ameboid and ramified microglia remains controversial (Fig. 1).

In the normal adult brain, ramified microglia are uniformly dispersed constituting approximately 5–20% of all glial cells (Lowson et al. (1992)). Ramified microglia are thought to be functionally inactive, or in a resting state, although it is possible that they are involved in brain homeostasis by producing low levels of bi-

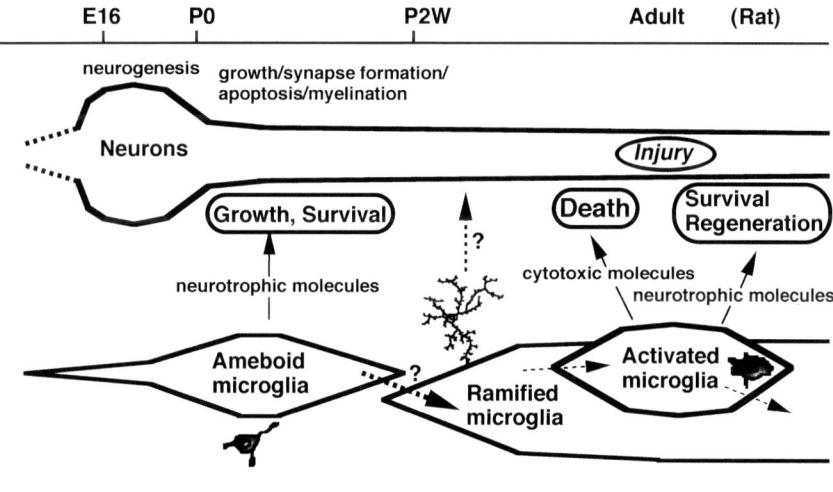

FIGURE 1. Interaction between neurons and microglia during development.

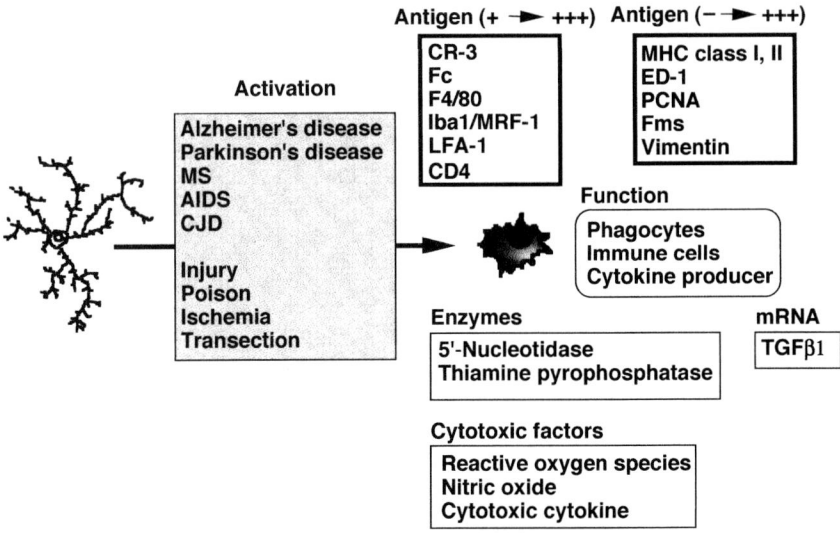

FIGURE 2. Characteristics of activated microglia *in vivo*. MS, multiple sclerosis; AIDS, acquired immunodeficiency syndrome; CJD, Creutzfeldt-Jakob disease; LFA-1, leukocyte function associated-antigen 1.

ologically active molecules (Fig. 1). As long as the brain remains healthy microglial cell density and ramified structure is sustained. However, when the brain is experimentally injured or affected by brain diseases, including degenerative, infectious, or autoimmune diseases, the resident microglia morphologically transform into a type with retracted processes and enlarged cell bodies, and increase in number at the affected site (McGeer et al. (1988); Dickson et al. (1991); Morioka et al. (1992); Benveniste (1997); Brown and Kretzschmar (1997)). This cell type is known as "activated microglia" or "reactive microglia" (Fig. 1). Activated microglia induce or enhance expression of various cellular antigens, receptors and enzyme activities (Fig. 2). At the same time, they are functionally activated, and appear to play an important role in the survival and regeneration of injured neurons (described below). In reversible neuronal injury, such as axotomy of motoneurons, activated microglia gradually return to the morphology of the ramified type as motoneurons regenerate (Fig. 1). The interconversions between ramified microglia and activated microglia seen *in vivo* may be associated with changes in their functional states.

Many pathological studies suggest that activated microglia undermine the survival and/or the regeneration of neurons by producing cytotoxic or harmful molecules (Fig. 1). It is unclear whether microglia actually produce harmful molecules *in vivo*, or kill neurons directly. Therefore, we tried to analyze the relation between microglia and neurons using both *in vitro* and *in vivo* systems.

Role of Microglia in the Survival and Development of Neurons

Neurotrophic Effect of Conditioned Microglial Medium

To analyze the intercellular interaction between microglia and neurons, we first established a survival/growth assay consisting of highly purified microglia and embryonic neurons (Nakajima et al. (1989)).

As a first trial, the effect of microglia on neocortical neurons was examined using cocultures of both cell types. No substantial differences in the number of surviving neurons was observed between the neuronal culture and the neuronal culture mixed with microglia. Although unhealthy or dying neurons were phagocytosed by microglia, healthy neurons were not engulfed by microglia. Microglia in the mixed culture appeared to differentiate between healthy and dying neurons by touching the neuronal surface with their pseudopodia. Thus, microglia in coculture with neurons are phagocytic, but not neurotoxic.

A subsequent trial was carried out to examine the effect of conditioned microglial medium (CMM) on cultured neurons. The addition of CMM significantly enhanced neuronal survival and the neurite outgrowth of neocortical neurons (Nakajima et al. 1989). Similarly, CMM promoted the survival and neurite outgrowth of mesencephalic neurons (Nagata et al. (1993c)). Interestingly, in mesencephalic neurons, dopamine uptake as well as dopamine content and intensity of staining by antityrosine hydroxylase was markedly enhanced, (Nagata et al. (1993c)) (Fig. 3A). These results suggest that certain soluble factor(s) mediating neurotrophic effects are released from cultured microglia.

Moreover, only a weak effect of CMM on other types of neurons, such as cholinergic and glutamatergic neurons, was observed in our system. Following these observations, other researchers reported the effect of CMM on cholinergic neurons (Jonakait et al. (1996)) and glutamatergic neurons (Hegg et al. (1999)). The actions of CMM on different types of neurons are summarized in Figure 3A.

Neurotrophic Substances of CMM

The functional molecules that exert neurotrophic effects on cortical and mesencephalic neurons have been surveyed by comparing the effects of well known authentic factors with those of CMM. Among the several kinds of factors tested, basic fibroblast growth factor (bFGF), interleukin-2 (IL-2), and IL-3 showed activity for promoting the survival and neurite outgrowth of neocortical neurons in our assay system. Supporting these results, bFGF was reported to show neurotrophic effects on various types of neurons (Matsuda et al. (1990)). Basic FGF (Shimojo et al. (1991)) and IL-3 (Ganter et al. (1992)) were found to be produced in microglia (described below). These factors are probably responsible for the

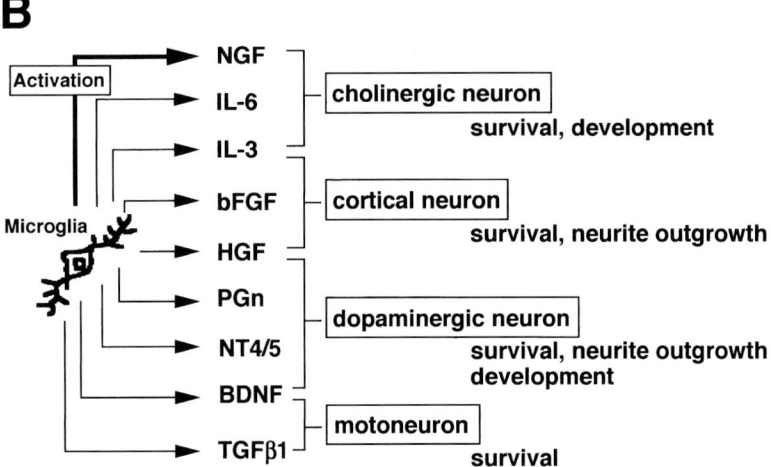

FIGURE 3. Putative neurotrophic molecules in conditioned microglial medium (CMM).

neurotrophic effects of CMM on neocortical neurons (Fig. 3B, Fig. 4). However, it remains uncertain whether IL-2 is produced in microglia.

Similarly, neurotrophic factors for mesencephalic neurons have also been surveyed. However, early studies could not identify functional molecules for enhancing dopamine uptake, suggesting that unknown neurotrophic factors are produced by microglia (Nagata et al. (1993c)). Eventually, the first molecule acting as a neurotrophic factor for dopaminergic neurons emerged in the process of examining the relation between microglia-derived proteases and the neurotrophic effect of CMM. The molecule was plasminogen, a zymogen of active protease plasmin. Plasminogen markedly enhanced neurite outgrowth (Nagata et al. (1993a, 1993b); Nakajima et al. (1993)), dopamine content and dopamine uptake of mesencephalic neurons *in vitro* (Nagata et al. (1993a)). Thus, plasminogen was considered a possible candidate for the neurotrophic factor contained in CMM (Fig. 3B, Fig. 4). However, the molecular mechanism of the neurotrophic effect

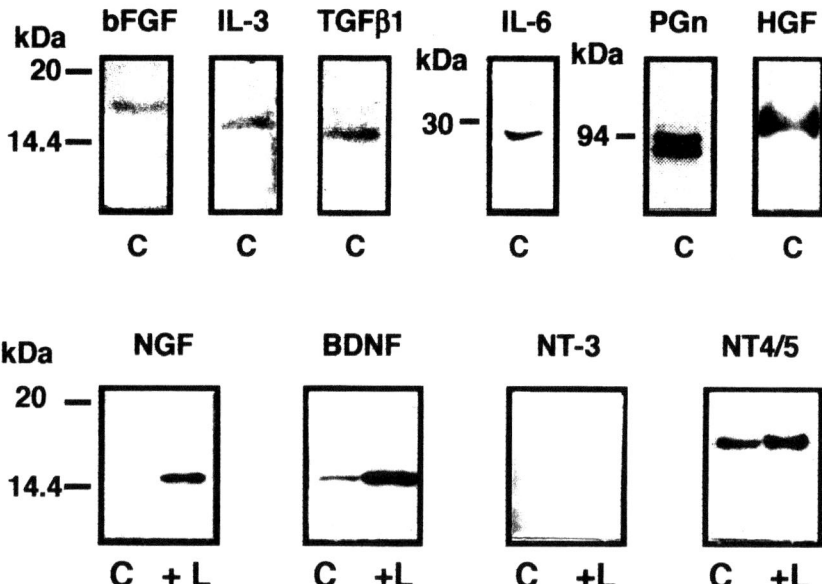

FIGURE 4. Neurotrophic molecules detected in CMM. Microglia were maintained with (+L) or without (C) LPS in the presence of serum-free medium for 6 h, and the CMM was analyzed for each factor by immunoblotting.

exerted by plasminogen remains to be clarified, although a receptor-like protein for plasminogen was identified on the neuronal surface (Nakajima et al. (1994)). Later, hepatocyte growth factor (HGF) was identified as a neurotrophic molecule in CMM. This factor showed activity for promoting the survival and neurite outgrowth of cortical neurons and dopamine uptake by mesencephalic neurons (Hamanoue et al. (1996)) (Fig. 3B, Fig. 4). The neurotrophic effect of HGF is thought to be caused by phosphorylation of its specific receptor, c-met, and the subsequent activation of MAP kinase. HGF has a unique structure, called the kringle domain, which is commonly present on both HGF and on plasminogen. However, the relation between the kringle structures of these molecules and the neurotrophic effects has not been clarified.

Other neurotrophic molecules from microglia have also been demonstrated, or suggested from *in vitro* and *in vivo* studies by other groups. IL-3 exerts a neurotrophic effect on cholinergic neurons (Kamegai et al. (1990)) in addition to its effect on cortical neurons (described above). IL-6 acts as a neurotrophic factor for septal cholinergic neurons (Hama et al. (1991)). These cytokines are released in non-stimulated microglia, as described previously (Woodroofe et al. (1991); Ganter et al. (1992)) (Fig. 3B, Fig. 4).

As suggested by *in situ* hybridization studies performed in the axotomized facial nucleus (Kiefer et al. (1993)), microglia also produce TGFβ1 (Fig. 3B, Fig.

4). TGFβ1 is a known survival factor for motoneurons (Martinou et al. (1990)). A unique property of TGFβ1 is that it requires the action of plasmin for functional activation (Nakajima and Kohsaka (1996)). Thus, a plasminogen-activating system (i.e., or plasmin-generating system) may contribute to the neurotrophic action of TGFβ1.

IL-1β and TNFα, while not released in unstimulated microglia, are induced when microglia are activated by LPS. Although IL-1β is not generally described as a neurotrophic factor, it can enhance the biosynthesis of somatostatin in cultured diencephalic neurons (Scarborough et al. (1989)). TNFα acts to promote the survival of neurons under conditions of glucose-deprivation or in the presence of excitatory amino acids (Cheng et al. (1994)), and TNFα can attenuate N-methyl D-aspartate (NMDA)-dependent neuronal death (Liu et al. (1999)). However, since this cytokine can also damage oligodendrocytes or myelin *in vitro* (Selmaj and Raine (1988)), it is often grouped with the cytotoxic molecules (described below).

Although several factors were identified as neurotrophic factors in CMM (Fig. 3, Fig. 4), the overall neurotrophic activity exerted by CMM can not be explained solely by the effects of these factors. It is likely that other neurotrophic molecule(s) are present in CMM.

Neurotrophin Release from Microglia

The neurotrophins are a family of protein factors composed of at least five structurally related members: nerve-growth factor (NGF) (Thoenen, 1991), brain-derived neurotrophic factor (BDNF) (Leibrock et al. (1989)), neurotrophin-3 (NT-3) (Ernfors et al. (1990b); Jones et al. (1990); Maisonpierre et al. (1990)), neurotrophin-4/5 (NT-4/5) (Berkemeier et al. (1991); Hallbook et al. (1991)), and NT-6 (Gotz et al. (1994)). Among them, NGF, BDNF, NT-3 and NT-4/5 are expressed in mammals.

Neurotrophins such as NGF, BDNF, and NT-3 were originally shown to be produced in target neurons or tissues in both the PNS and CNS (Bandtlow et al. (1990); Ernfors et al. (1990a); Maisonpierre et al. (1990)), but later these factors were found to be produced in glial cells as well. Astrocytes have been found to produce these neurotrophins *in vitro,* including NGF (Furukawa et al. (1986)); Yamakuni et al. (1987)), BDNF (Zafra et al. (1992)), NT-3 (Rudge et al. (1992)), and NT-4/5 (Condorelli et al. (1994)), and to also express NGF *in vivo* (Lu et al. (1991); Altar et al. (1992); Lindholm et al. (1992); Yoshida and Gage (1992)).

Microglia were also studied for the expression of neurotrophins by competitive RT-PCR. This method revealed that cultured microglia express NGF, BDNF and NT-4/5 mRNA under nonstimulated conditions, and, among them, BDNF and NT-4/5 mRNA are increased by stimulation with lipopolysaccharide (LPS) (Miwa et al. (1997)). NGF mRNA was later shown to be increased by stimulation with LPS, although to a lesser degree. BDNF protein was immunocytochemically detected in microglia, and its level was elevated by stimulation with LPS (Miwa et al. (1997)). Other researchers have also reported that NGF (Mallat et al. (1989); Heese et al. (1997)), BDNF, and NT-3 (Elkabes et al. (1996)) are expressed and/or produced in microglia *in vitro* and *in vivo*.

These results prompted us to investigate whether microglia have the ability to release neurotrophins extracellularly by detecting neurotrophin proteins in their CMM. Under normal conditions (without LPS stimulation) microglia were found to constitutively release small amounts of BDNF and NT-4/5 (Fig. 3B, Fig. 4). However, NGF and NT-3 were undetectable. Although it is possible that very small amounts of NGF and NT-3 are released, it is clear that the release of BDNF and NT-4/5 is much more prominent.

How is the release of neurotrophins from microglia regulated? Considering that NT-3 mRNA expression in microglia is region- and development-specific *in vivo* (Elkabes et al. (1996)), the production of neurotrophins appears to be strictly regulated by the surrounding milieu in the brain. However, there is no information concerning the factors responsible for regulating mesencephalic release. Since microglia generally enhance surface antigens and enzyme activities when activated *in vivo* (Fig. 2), the release of neurotrophins was examined during microglial activation. Using LPS stimulation as a model of microglial activation, the amount of released neurotrophins was compared with that from unstimulated microglia. LPS stimulation enhanced the release of BDNF and NT-4/5, and newly induced NGF release. NT-3 however, was undetectable (Fig. 4).

The neurotrophin family includes important proteins that regulate neuronal function and support the survival, and enhance the growth, of various types of neurons in both the PNS and CNS (Barde (1989); Thoenen (1991); Davies (1994); Lindsay et al. (1994)). The finding that the removal of neurotrophins from animals by specific antibody or gene knock-out results in a variety of nervous system abnormalities (Johnson et al. (1980); Ruit et al. (1992); Klein et al. (1993); Ernfors et al. (1994); Farinas et al. (1994); Smeyne et al. (1994); Conover et al. (1995)) also supports the contention that neurotrophins are necessary for a functional nervous system. Therefore, it is easy to envision that neurotrophins released from microglia can influence surrounding neurons in paracrine fashion.

Neuronal survival/death and maturation in the developing CNS and PNS is thought to require regulation by surrounding glial cells. There is persuasive evidence that glial cells have the potency to produce and release a variety of neurotrophic molecules. Microglia were observed in brain parenchyma from the early stage of brain development (E10) (Fig. 1). The finding that microglial cells can produce and release BDNF and NT-4/5 suggests that they play an important role in the regulation of neuronal survival and/or maturation during CNS development. BDNF may be involved in various functions, including synapse modification, neurotransmitter release, long-term potentiation and mechanosensation in addition to its role of enhancing neuronal survival and development (Chao (2000)).

The significance of neurotrophic molecules, including neurotrophins and neurotrophic cytokines, is primarily in their effect on neurons. However, indirect but important neurotrophic actions are also indicated by *in vitro* study regarding the production of cytotoxic molecules in microglia. For example, bFGF and IL-6 as well as BDNF can suppress constitutive NO production in microglia. Neurotrophins, including BDNF, can suppress inducible NO release (Nakajima et al. (1998)). Therefore, these factors may be able to attenuate the release of NO in microglia.

Are microglia neurotoxic?

Neurotoxic molecules

In chronic diseases such as Alzheimer's disease (AD), multiple sclerosis (MS) and acquired immunodeficiency syndrome (AIDS), inflammatory reactions are long-lasting. The accompanying microglial activation is thought to result in the production of cytotoxic molecules and proinflammatory cytokines that may cause indirect (secondary) damage to weak neurons, which may be injured and killed (Fig. 2). Disruption of the blood brain barrier (BBB) permits the infiltration of blood-derived cells, including monocytes/macrophages that also may produce cytotoxic molecules, causing an interaction between resident microglia and infiltrated cells. These events make it difficult to discern the functions of individual cells. Despite this complexity it is speculated that activated microglia in these situations are harmfully activated to produce cytotoxic molecules. The functional change of microglia into cytotoxic cells appears to be under strict control (described below).

An interesting question is whether activated microglia kill neurons. As suggested from their resemblance to tissue macrophages, and from histological observations, activated microglia in culture were found to release several potentially cytotoxic molecules, including superoxide anion (Colton and Gilbert (1987)), nitric oxide (Boje and Arora (1992)), and pro-inflammatory cytokines (Fig. 2). LPS, interferon γ (INFγ) and β-amyloid (McDonald et al. (1997)) are among the stimulators for the production of harmful factors from microglia.

Reactive oxygen species (ROS), including superoxide anion, hydroxy radicals and hydrogen peroxide, are generally hazardous, particularly to myelin and oligodendrocytes, owing to their involvement in lipid peroxidation (Sonderer et al. (1987)). LPS and phorbol-12-myristate-13-acetate (PMA) are stimulators of ROS production from cultured microglia.

Reactive nitrogen oxides such as nitric oxide (NO) are reactive free radicals, with nitrite peroxide as the strongest species. These radicals are believed to inhibit respiratory enzymes, oxidize the SH group of proteins and promote DNA injury, ultimately resulting in neuronal cell death in the brain. LPS and β-amyloid are known as stimulators of NO production. In the presence of INFγ, β-amyloid synergistically stimulates the production of NO and TNF-α (Meda et al. (1995)).

The targets of TNF-α in the CNS have been reported to be oligodendrocytes and myelin because of their susceptibility to TNF-α *in vitro* (Selmaj et al. (1988)). Some neurons have also been reported to suffer apoptosis by treatment with TNF-α (Talley et al. (1995); Downen et al. (1999)). Both microglia and astrocytes produce TNF-α in the CNS (Sawada et al. (1989)). TNFα produced by activated microglia is thought to contribute to inflammatory damage in MS and AIDS dementia and thus TNFα is generally regarded a cytotoxic cytokine. However, its neuroprotective effects following glucose-deprivation and excitatory amino acid-induced neuronal death have been reported (Cheng et al. (1994)). Recent evaluation of TNFα function by gene knockout showed that TNFα is not essential for the initiation and/or progression of inflammation (Liu et al. (1998)). These observations suggest that TNFα is not necessarily neurotoxic.

Eicosanoids including prostaglandins, thromboxanes and leukotriens, are known to be cytotoxic under certain conditions. As microglia express both cycloxygenase-2 and lipoxygenase, they may adversely affect the state of inflammation. Vasoactive amines through production of various eicosanoid such as histamine or an excitotoxic amino acid such as glutamate may play an important role in various degenerative processes.

These potentially cytotoxic products may widely affect the ability of neurons to survive in pathologically damaged brain. However, the release of harmful factors from microglia and the cytotoxicity of these factors have been reported mainly from *in vitro* studies. Microglial cytotoxicity remains to be confirmed *in vivo*.

Function of microglia in eliminating excitatory amino acids

Studies of CNS ischemia suggest that the excitatory neurotransmitter, glutamate is a candidate for triggering neuronal cell death (Rothman (1984)). If levels of extracellular glutamate remain high, they result in neuronal death by activation of the glutamate receptor and subsequent Ca influx. Accordingly, the extracellular concentration of glutamate around synapses must be kept low to prevent neuronal damage because of excessive activation of glutamate receptors. In recent years, glial cells have been found to be important for the elimination of glutamate from the extracellular milieu. To examine whether microglia are associated with neuronal cell death/survival in terms of glutamate toxicity, glutamate transporters were analyzed in the axotomized facial nucleus.

In the facial nerve transection model, microglia in the ipsilateral nucleus are activated. A few days later, they begin to proliferate and wrap the cell body of injured motoneurons, strongly expressing CR-3 and Iba1 (Streit et al. (1988); Graeber et al. (1998)). The proximity of activated microglia to injured motoneuron cell bodies suggests an intimate interaction between the two cell types. Transection of the facial nerve leads to a decrease of GLT-1 (EAAT2), a glial glutamate transporter, in the facial nucleus, but does not affect GLAST (EAAT1), another glial glutamate transporter (Lopez-Redondo et al. (2000)). Selective decreases of GLT-1 are observed also in the motor cortex and spinal cord of amyotrophic lateral sclerosis (ALS) (Rothstein (1995)), in the hippocampus after ischemia (Torp et al. (1995)), and in the brains of epilepsy-prone rats (Akbar et al. (1998)). In the axotomized facial nucleus levels of GLT-1 are significantly decreased 3 to 28 days after surgery, but recover within three months. The largest decrease is observed at three to seven days after axotomy, with an approximately 40% decrease relative to the contralateral side. An immunohistochemical examination was carried out in the axotomized facial nucleus to determine the cell types responsible for GLT-1 expression. Astrocytes were considered likely candidates but, surprisingly, activated microglia around motoneuron cell bodies were seen to be stained strongly with GLT-1 antibody, suggesting that microglia remove glutamate locally by enhancing the expression of GLT-1 (Fig. 5). The activated microglia around motoneurons appear to compensate for the decreased astrocytic function (Fig. 5).

The expression of GLT-1 in microglia was examined immunocytochemically in cultured microglia that were prepared from axotomized facial nucleus, as described by Rieske et al. (1989). The results demonstrated that a certain population of microglia strongly expressed GLT-1 at the protein level, as observed *in vivo*. The regulation of GLT-1 expression in microglia remains to be determined. In addition, the capacity of microglia for glutamate uptake was tested in cultured microglia. Microglia showed specific glutamate uptake activity, and this activity was strongly suppressed in the presence of a competitive inhibitor, L-trans-pyrrolidine-2,4 dicarboxylic acid (PDC), and a specific GLT-1 inhibitor, dihydrokinic acid (DHK). Based on susceptibility to glutamate transporter inhibitors and glutamate transporter profiles obtained by Western blotting, the glutamate transporter functioning in microglia was suggested to be mainly GLT-1.

The survival of axotomized facial motoneurons and subsequent neuronal regeneration may be due, in part, to the increased expression of GLT-1 by activated microglia, suggesting that activated microglia act to eliminate excitotoxin. This phenomenon, in which injured motoneurons remain alive, may be further supported by the properties of microglia, which have an ability to produce and release neurotrophic molecules (Fig. 4, Fig. 5). Thus, it is suggested that activated microglia in the axotomized facial nucleus are serving as neuroprotective cells.

Regulation of microglial function

Activation of microglia

Since microglial activation has been considered closely associated with neuronal death, degeneration and neuroprotection, it has been discussed in detail (Raivich

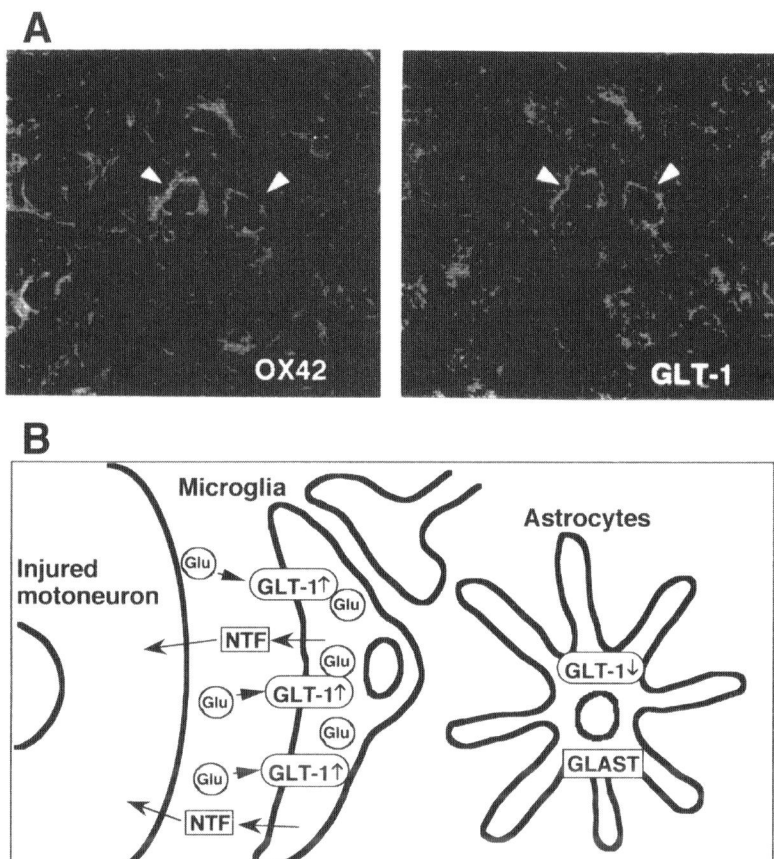

FIGURE 5. Activated microglia as glutamate scavenger. GLT-1, EAAT2; GLAST, EAAT1; NTF, neurotrophic factor.

et al. (1999); Streit et al. (1999)). Our interest is in what molecule(s) activate microglia, and what molecule(s) orient microglia to the neurotrophic type or the cytotoxic type *in vivo* (Fig. 6). Although some information regarding molecules involved in the activation of microglia has been obtained from *in vitro* experiments, we wished to determine physiological substances, not artificial substances. To date, various kinds of stimulators have been considered and predicted (Kreutzberg (1996)). One category is represented by changes in electricity in the ionic milieu around injured neurons. The other category is that of biologically active substances (Fig. 6). Low molecular weight molecules such as peptides and hormones may be included among the candidates. Growth factors or cytokines may also activate microglia. Calcitonin gene-related peptide (CGRP) and ATP (Priller et al. (1995); Honda et al. (2001)) are considered

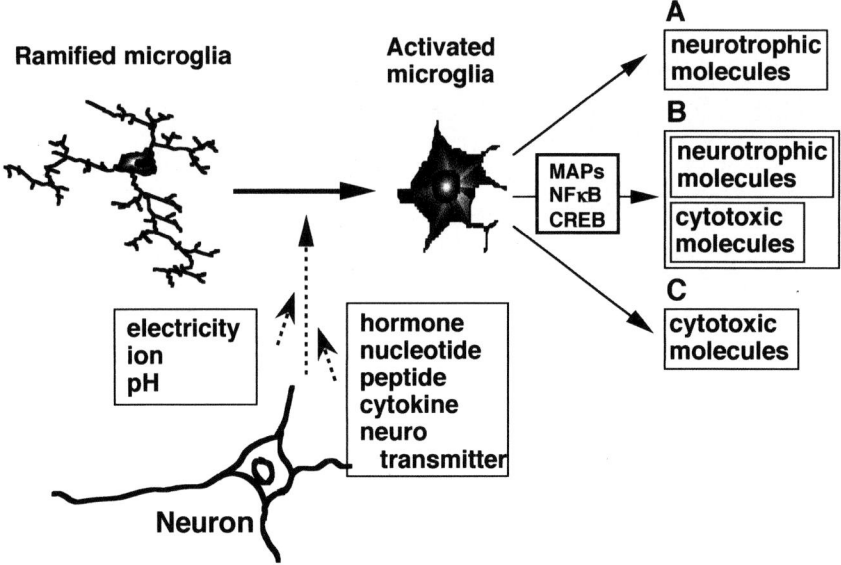

FIGURE 6. Factors responsible for activation of microglia

candidates for the activating factor in the facial nerve transection model. Indeed, ATP could cause biological responses in cultured microglia, such as enhanced release of plasminogen (Inoue et al. (1998)) and IL-1β (Ferrari et al. (1997a)), and activation of NFκB (Ferrari et al. (1997b)). In experimental allergic neuritis (EAN), a cytokine was proposed as the activator of microglia (Gehrman et al. (1992)).

In vitro *analysis of interactions between neurons and microglia*

On the assumption that neurons release diffusible factor(s) for the activation of microglia, the effect of conditioned neuronal medium (CNM) on the function of microglia was examined *in vitro*. Interestingly, CNM promoted the specific activity of acid phosphatase and 5′-nucleotidase and the release of urokinase-type plasminogen activator (uPA) in cultured microglia, but not the release of nitric oxide. 5′-nucleotidase is known to be induced in activated microglia in the axotomized facial nucleus (Kreutzberg (1978)), and uPA is suggested to be induced (Nakajima et al. (1996)). The effect of CNM is in contrast to that of LPS, which markedly enhances the release of nitric oxide from microglia, but not uPA. Although the molecule responsible for the activation of microglia has not yet been identified in the CNM, it is obvious that neurons release soluble factor(s) that activate microglia. It is likely that the factor(s) in CNM are different from LPS. The response of microglia varies depending on the stimuli, and perhaps different kinds of microglial activation result in the release of different kinds of molecules. This model experiment, while preliminary, strongly suggests that neurons release certain soluble molecule(s) for the activation of microglia.

Diversity of microglial function

As mentioned, microglia in culture have the ability to release several neurotrophic molecules in the nonstimulated condition, and the release of some of them is promoted and other molecules are newly induced under the LPS-stimulated condition. On the other hand, cytotoxic molecules, including nitric oxide and TNFα, are also produced under the activated condition. These results indicate that activated microglia simultaneously release both neurotoxic and neurotrophic molecules.

From the neurotrophic/neurotoxic standpoint, microglia can be divided into three states (Fig. 6). The neurotrophic condition (A) is a state where in microglia produce neurotrophic, but not cytotoxic, molecules. This corresponds to nonstimulated microglia in the serum-free condition. Based on expression levels of IL-1β mRNA, this state may be similar to activated microglia *in vivo* (Streit et al. (1999)). The second state (B) is a bifunctional one in which neurotrophic and cytotoxic molecules are produced. This is the state induced with LPS stimulation. The last state (C) is one in which only cytotoxic molecules are produced. This state has not yet been identified in culture. Presumably, this state is under strict control.

It should be emphasized that neurotrophic molecules are released with cytotoxic molecules when microglia are activated by LPS. This means that activated microglia are not necessarily completely cytotoxic.

Determinants that induce microglia to assume neurotrophic or cytotoxic states

It is critical to analyze the intracellular signaling mechanism to distinguish the differences between the neurotrophic and neurotoxic states of microglia. Using the LPS-stimulating model activation of MAP kinases and transcription factors were compared. No activation of extracellular signal regulated kinase (ERK), p38 or c-Jun N-terminal kinase (JNK), nuclear factor kappa beta (NFκB) or cAMP response element binding transcription factor (CREB) was observed in nonstimulated microglia that were neurotrophic, while all these factors were activated in LPS-stimulated microglia that were in the bifunctional state (Fig. 6). This may suggest that the induction of hazardous factors is associated with the activation of MAPs and/or transcription factors, but that the release of neurotrophic molecules is not.

The induction of TNFα in microglia is mediated by the activation of p38 MAP kinase. The LPS-stimulated release of nitric oxide is completely suppressed by pretreatment with specific protein kinase C (PKC) inhibitor, showing that the PKC signaling cascade in LPS-stimulated microglia is associated with the release of nitric oxide. Superoxide anion production may be caused by the PKC activation pathway because it is remarkably induced in the presence of the PKC activator, PMA. Information concerning the signaling cascade by which the neurotrophic and cytotoxic states of microglia are distinguished is still fragmentary, but it will not be long until it is clarified.

Conclusions

Microglia have been suggested to produce neurotrophic molecules based on *in vitro* experiments in which microglia-conditioned media exerted neurotrophic effects on cultured neurons. To date, some molecules have been regarded as effective molecules for neurotrophic substances. Moreover, neurotrophins including BDNF, NT-4/5 and NGF have been found to be produced and released from non-stimulated and activated microglia. The potency of microglia may reflect the survival and/or development of specific populations of neurons during brain development, or the rescue of neurons and neuronal regeneration in the pathological brain. It is surprising that activated microglia promote the release of some neurotrophic molecules because they were considered to act as cytotoxic cells that produced only harmful factors. It is therefore plausible that neurotrophins and neurotrophic cytokines are released along with harmful factors when microglia are activated *in vivo*.

The neuroprotective function of activated microglia was suggested from the immunohistochemical observation of the facial nerve transection model, in which activated microglia around injured motoneurons strongly expressed GLT-1. They appeared to eliminate extracellular glutamate by enhancing the expression of GLT-1, thus acting as neuroprotective cells.

Although microglia are potentially bifunctional from a neurotrophic/neurotoxic standpoint, the molecular mechanism by which the neurotrophic and neurotoxic functions are distinguished is largely unknown. Therefore, it may be necessary to analyze the cellular responces by different kinds of stimuli and intracellular signaling pathways. Such basic studies will help solve the mystery of microglial activation and may provide a better understanding of the regulatory mechanism of microglial functions.

References

Akbar MT, Rattray M, Williams RJ, Chong NW, Meldrum BS (1998). Reduction of GABA and glutamate transporter messenger RNAs in the severe-seizure genetically epilepsy-prone rat. *Neuroscience* 85:1235–1251.

Altar CA, Armanini M, Dugich DM, Bennett GL, Williams R, Feinglass S, Anicetti V, Sinicropi D, Bakhit C (1992). Recovery of cholinergic phenotype in the injured rat neostriatum: roles for endogenous and exogenous nerve growth factor. *J Neurochem* 59:2167–2177.

Ashwell K (1990). Microglia and cell death in the developing mouse cerebellum. *Dev Brain Res* 55:219–230.

Banati RB, Gehrmann J, Schubert P, Kreutzberg GW (1993). Cytotoxicity of microglia. *Glia* 7:111–118.

Bandtlow CE, Meyer M, Lindholm D, Spranger M, Heumann R, Thoenen H (1990). Regional and cellular codistribution of interleukin 1β and nerve growth factor mRNA in the adult rat brain: possible relationship to the regulation of nerve growth factor synthesis. *J Cell Biol* 111:1701–1711.

Barde YA (1989). Trophic factors and neuronal survival. *Neuron* 2:1525–1534.

Benveniste EN (1997). Role of macrophages/microglia in multiple sclerosis and experimental allergic encephalomyelitis. *J Mol Med* 75:165–173.

Berkemeier L, Winslow J, Kaplan D, Nicolics K, Goeddel D, Rosenthal A (1991). Neurotrophin-5: a novel neurotropic factor that activates trk and trkB. *Neuron* 7:857–866.

Boje KM, Arora PK (1992). Microglial-produced nitric oxide and reactive nitrogen oxides mediate neuronal cell death. *Brain Res* 587:250–256.

Brown DR, Kretzschmar HA (1997). Microglia and prion disease: a review. *Histol Histopathol* 12:883–892.

Chao MV (2000). Trophic factors: An evolutionary cul-de-sac or door into higher neuronal function? *J Neurosci Res* 59:353–355.

Cheng B, Christakos S, Mattson MP (1994). Tumor necrosis factors protect neurons against metabolic-excitotoxic insults and promote maintenance of calcium homeostasis. *Neuron* 12:139–153.

Colton CA, Gilbert DL (1987). Production of superoxide anions by a CNS macrophage, the microglia. *FEBS Lett* 223:284–288.

Condorelli DF, Dell'Albani P, Mudo G, Timmusk T, Belluardo N (1994). Expression of neurotrophins and their receptors in primary astroglial cultures: induction by cyclic AMP-elevating agents. *J Neurochem* 63:509–516.

Conover JC, Erickson JT, Katz DM, Blanchi LM, Poueymirou WT, McClain J, Pan L, Helgren M, Ip NY, Baland P, Friedman B, Wilegand S, Vejsada R, Kato AC, Dechiara TM, Yancopoulos GD (1995). Neuronal deficits, not involving motor neurons, in mice lacking BDNF and/or NT-4. *Nature* 375:235–238.

Cunningham TJ (1982). Naturally occurring death and its regulation by developing neural pathways. *Int Rev Cytol* 74:163–186.

Davies AM (1994). The role of neurotrophins in the developing nervous system. *Neurobiol* 25:1334–1348.

Dickson DW, Mattiace LA, Kure K, Huchins K, Lymen WD, Brosnan CF (1991). Biology of disease. Microglia in human disease, with an emphasis on acquired immune deficiency syndrome. *Lab Invest* 64:135–156.

Downen M, Amaral TD, Hua LL, Zhao ML, Lee SC (1999). Neuronal death in cytokine-activated primary human brain cell culture: role of tumor necrosis factor-α. *Glia* 28:114–127.

Elkabes S, DiCicco-Bloom EM, Black I (1996). Brain microglia/macrophages express neurotrophins that selectively regulate microglial proliferation and function. *J Neurosci* 16:2508–2521.

Ernfors P, Ibanez CF, Ebendal T, Olson L, Persson H (1990a). Localization of brain-derived neurotrophic factor mRNA to neurons in the brain by in situ hybridization. *Exp Neurol* 109:141–152.

Ernfors P, Ibanez CF, Ebendal T, Olson L, Persson H (1990b). Molecular cloning and neurotrophic activities of a protein with structural similarities to nerve growth factor: developmental and topographical expression in the brain. *Proc Natl Acad Sci USA* 87:5454–5458.

Ernfors P, Lee K-F, Jaenisch R (1994). Mice lacking brain-derived neurotrophic factor develop with sensory deficits. *Nature* 368:147–150.

Farinas I, Jones KR, Bachus C, Wang X-Y, Reichardt LF (1994). Severe sensory and sympathetic deficits in mice lacking neurotrophin-3. *Nature* 369:658–661.

Ferrari D, Chiozzi P, Falzoni S, Hanau S, Di Virgilio F (1997a). Purinergic modulation of interleukin-1 β release from microglial cells stimulated with bacterial endotoxin. *J Exp Med* 185:579–582.

Ferrari D, Wesselborg S, Bauer MKA, Schulze-Osthoff K (1997b). Extracellular ATP activates transcription factor NF-κB through the P2Z purinoreceptor by selectively targeting NF-κB p65. *J Cell Biol* 139:1635–1643.

Frade JM, Barde YA (1998). Microglia-derived nerve growth factor causes cell death in the developing retina. *Neuron* 20:35–41.

Fujita S, Kitamura T (1975). Origin of brain macrophages and the nature of the so-called microglia. *Acta Neuropathol* VI:291–296.

Furukawa S, Furukawa Y, Satoyoshi E, Hayashi K (1986). Synthesis and secretion of nerve growth factor by mouse astroglial cells in culture. Biochem *Biophys Res Commun* 136:57–63.

Ganter S, Northoff H, Mannel D, Gebicke-Haerter PJ (1992). Growth control of cultured microglia. *J Neurosci Res* 33:218–230.

Gehrmann J, Gold R, Linington C, Lannes-Vieira J, Wekerle H, Kreutzberg GW (1992). Spinal cord microglia in experimental allergic neuritis. Evidence for fast and remote activation. *Lab Invest* 67:100–113.

Gotz R, Koster R, Winkler C, Raulf F, Lottspeich F, Schartl M, Thoenen H (1994). Neurotrophin-6 is a new member of the nerve growth factor family. *Nature* 372:266–269.

Graeber MB, Lopez-Redondo F, Ikoma E, Ishikawa M, Imai Y, Nakajima K, Kreutzberg GW, Kohsaka S. (1998). The microglia/macrophage response in the neonatal rat facial nucleus following axotomy. *Brain Res* 813:241–253.

Graeber MB, Streit WJ, Kreutzberg GW (1988). Axotomy of the rat facial nerve leads to increased CR 3 complement receptor expression by activated microglial cells. *J Neurosci Res* 21:18–24.

Graeber MB, Streit WJ, Kreutzberg GW (1989). Identity of ED2-positive perivascular cells in rat brain. *J Neurosci Res* 22:103–110.

Hallbook F, Ibanez CF, Persson H (1991). Evolutionary studies of the nerve growth factor family reveal a novel member abundantly expressed in Xenopus ovary. *Neuron* 6:1–20.

Hama T, Kushima Y, Miyamoto M, Kubota M, Takei N, Hatanaka H (1991). Interleukin-6 improves the survival of mesencephalic catecholaminergic and septal cholinergic neurons from postnatal, two-week-old rats in cultures. *Neuroscience* 40:445–452.

Hamanoue M, Takemoto N, Matsumoto K, Nakamura T, Nakajima K, Kohsaka S (1996). Neurotrophic effect of hepatocyte growth factor on central nervous system neurons in vitro. *J Neurosci Res* 43:554–564.

Heese K, Fiebich BL, Bauer J, Otten U (1997). Nerve growth factor (NGF) expression in rat microglia is induced by adenosine A_{2a}-receptors. *Neurosci Lett* 231:83–86.

Hegg CC, Thayer SA (1999). Monocytic cells secrete factors that evoke excitatory synaptic activity in rat hippocampal cultures. *Eur J Pharmacol* 385:231–237.

Honda S, Sasaki Y, Ohsawa K, Imai Y, Nakamura Y, Inoue K, Kohsaka S (2001). Exracellular ATP or ADP induce chemotaxis of cultured microglia through $G_{i/o}$-coupled P2Y receptors. *J Neurosci* 21:1975–1982.

Inoue K, Nakajima K, Morimoto T, Kikuchi Y, Koizumi S, Illes P, Kohsaka S (1998). ATP stimulation of Ca2+ -dependent plasminogen release from cultured microglia. *Br J Pharmacol* 123:1304–1310.

Johnson EM Jr, Gorin PD (1980). Dorsal root ganglion neurons are destroyed by exposure in utero to maternal antibody to nerve growth factor. *Science* 210:916–918.

Jonakait GM, Luskin MB, Wei R, Tian XF, Ni L (1996). Conditioned medium from activated microglia promotes cholinergic differentiation in the basal forebrain in vitro. *Dev Biol* 177:85–95.

Jones KR, Reichardt LF (1990). Molecular cloning of a human gene that is a member of the nerve growth factor family. *Proc Natl Acad Sci USA* 87:8060–8064.

Jordan FL, Thomas WE (1988). Brain macrophages; questions of origin and interrelationships. *Brain Res Rev* 13:165–178.

Kamegai M, Niijima K, Kunishita T, Nishizawa M, Ogawa M, Araki M, Ueki A, Konishi Y, Tabira T (1990). Interleukin-3 as a trophic factor for central cholinergic neurons in vitro and in vivo. *Neuron* 2:429–436.

Kiefer R, Lindholm D, Kreutzberg GW (1993). Interleukin-6 and transforming growth factor-β1 mRNAs are induced in rat facial nucleus following motoneuron axotomy. *Eur J Neurosci* 5:775–781.

Klein R, Smeyne RJ, Wurst W, Long LK, Auerbach BA, Joyner AL, Barbacid M (1993). Targeted disruption of the trk B neurotrophin receptor gene results in nervous system lesions and neonatal death. *Cell* 75:113–122.

Kreutzberg GW (1996). Microglia: a sensor for pathological events in the CNS. *Trends Neurosci* 19:312–318.

Kreutzberg GW, Barron KD (1978). 5′-nucleotidase of microglial cell in the facial nucleus during axonal reaction. *J Neurocytol* 7:601–610.

Lawson LJ, Perry VH, Gordon S (1992). Turnover of resident microglia in the normal adult mouse brain. *Neurosci* 48:405–415.

Leibrock J, Lottspeich F, Hohn A, Hofer M, Hengerer B, Masiakowski P, Thoenen H, Barde Y-A (1989). Molecular cloning and expression of brain-derived neurotrophic factor. *Nature* 341:149–152.

Lindsay R, Wiegard S, Alter C, Distefano PS (1994). Neurotrophic factors: from molecule to man. *Trends Neurosci* 17:182–189.

Lindholm D, Castren E, Hengerer B, Zafra F, Berninger B, Thoenen H (1992). Differential regulation of nerve growth factor (NGF) synthesis in neurons and astrocytes by glucocorticoid hormones. *Eur J Neurosci* 4:404–410.

Ling EA, Penney D, Leblond CP (1980). Use of carbon labelling to demonstrate the role of blood monocytes as precursors of the 'ameboid cells' in the corpus callosum of postnatal rats. *J Comp Neurol* 193:631–657.

Liu J, Marino MW, Wong G, Grail D, Dunn A, Bettadapura J, Slavin AJ, Old L, Bernard CC (1998). TNF is a potent anti-inflammatory cytokine in autoimmune-mediated demyelination. *Nat Med* 4, 78–83.

Liu XH, Xu H, Barks JD (1999). Tumor necrosis factor-α attenuates N-methyl-D-aspartate-mediated neurotoxicity in neonatal rat hippocampus. *Brain Res* 851:94–104.

Lopez-Redondo F, Nakajima K, Honda S, Kohsaka S (2000). Glutamate transporter GLT-1 is highly expressed in activated microglia following facial nerve axotomy. *Brain Res Mol Brain Res* 76:429–435.

Lu B, Yokoyama M, Dreyfus CF, Black IB (1991). NGF gene expression in activity growing brain glia. *J Neurosci* 11:318–326.

Maisonpierre PC, Belluscio L, Friedman B, Alderson RF, Wiegard ST, Furth ME, Lindsay RM, Yancopoluos GD (1990). NT-3, BDNF, and NGF in the developing rat nervous system: parallel as well as reciprocal patterns of expression. *Neuron* 5:501–509.

McDonald DR, Brunden KR, Landreth GE (1997). Amyloid fibrils activate tyrosine kinase-dependent signaling and superoxide production in microglia. *J Neurosci* 17:2284–2294.

Mallat M, Houlgatte R, Brachet P, Prochiantz A (1989). Lipopolysaccharide-stimulated rat brain macrophages release NGF in vitro. *Dev Biol* 113:309–311.

Martinou JC, Le Van Thai A, Valette A, Weber MJ (1990). Transforming growth factor b1 is a potent survival factor for rat embryo motoneurons in culture. *Brain Res Dev Brain Res* 52:175–181.

Matsuda S, Saito H, Nishiyama N (1990). Effect of basic fibroblast growth factor on neurons cultured from various regions of postnatal rat brain. *Brain Res* 520:310–316.

McGeer PL, Itagaki S, Boyes BE, McGeer EG (1988). Reactive microglia are positive for HLA-DR in the substantia nigra of Parkinson's and Alzheimer's disease brains. *Neurology* 38:1285–1291.

Meda L, Cassatelia MA, Szendrei GI, Otvos Jr L, Baron P, Villalba M, Ferrari D, Rossi F (1995). Activation of microglial cells by b-amyloid protein and interferon-γ. *Nature* 374:647–650.

Miwa T, Furukawa S, Nakajima K, Furukawa Y, Kohsaka S (1997). Lipopolysaccharide enhances synthesis of brain-derived neurotrophic factor in cultured rat microglia. *J Neurosci Res* 50:1023–1029.

Miyake T, Tsuchihashi Y, Kitamura T, Fujita S (1984). Immunohistochemical studies of blood monocytes infiltrating into the neonatal rat brain. *Acta Neuropathol* (Berl) 62:291–297.

Morioka T, Kalehua AN, Streit WJ (1992). Progressive expression of immunomolecules on microglial cells in rat dorsal hippocampus following transient forebrain ischemia. *Acta Neuropathol* 83:149–157.

Nagata K, Nakajima K, Kohsaka S (1993a). Plasminogen promotes the development of rat mesencephalic dopaminergic neurons in vitro. *Dev Brain Res* 75:31–37.

Nagata K, Nakajima K, Takemoto N, Saito H, Kohsaka S (1993b). Microglia-derived plasminogen enhances neurite outgrowth from explant cultures of rat brain. *Int J Dev Neurosci* 11:227–237.

Nagata K, Takei N, Nakajima K, Saito H, Kohsaka S (1993c). Microglial conditioned medium promotes survival and development of cultured mesencephalic neurons from embryonic rat brain. *J Neurosci Res* 34:357–363.

Nakajima K, Hamanoue M, Takemoto N, Hattori T, Kato K, Kohsaka S (1994). Plasminogen binds specifically to alpha-enolase on rat neuronal plasma membrane. *J Neurochem* 63:2048–2057.

Nakajima K, Hamanoue M, Shimojo M, Takei N, Kohsaka S (1989). Characterization of microglia isolated from a primary culture of embryonic rat brain by a simplified method. *Biomed Res* 10(S3):411–423.

Nakajima K, Kikuchi Y, Ikoma E, Honda S, Ishikawa M, Liu YM, Kohsaka S (1998). Neurotrophins regulate the function of cultured microglia. *Glia* 24:272–289.

Nakajima K, Kohsaka S (1993). Functional roles of microglia in the brain. *Neurosci Res* 17:187–203.

Nakajima K, Kohsaka S (1996). Functional implications of microglia-derived secretory proteases. In: *Topical issues in microglia research,* Singapore (Ling EA, Tan CK, Tan CBC, eds), pp 203–218. Singapore Neuroscience Association.

Nakajima K, Nagata K, Hamanoue M, Takemoto N, Kohsaka S (1993). Microglia-derived elastase produces a low-molecular-weight plasminogen that enhances neurite outgrowth in rat neocortical explant cultures. *J Neurochem* 61:2155–2163.

Nakajima K, Reddington M, Kohsaka S, Kreutzberg GW (1996). Induction of urokinase-type plasminogen activator in rat facial nucleus by axotomy of the facial nerve. *J Neurochem* 66:2500–2505.

Oppenheim RW (1991). Cell death during development of the nervous system. *Annu Rev Neurosci* 14:453–501.

Perry VH, Gordon S (1988). Macrophages and microglia in the nervous system. *Trends Neurosci* 11:273–277.

Priller J, Haas CA, Reddington M, Kreutzberg GW (1995). Calcitonin gene-related peptide and ATP induce immediate early gene expression in cultured rat microglial cells. *Glia* 15:447–457.

Raff MC, Barres BA, Burne JF, Coles HS, Ishizaki Y, Jacobson MD (1993). Programmed cell death and the control of cell survival: lessons from the nervous system. *Science* 262:695–700.

Raivich G, Bohatschek M, Kloss UA, Werner A, Jones LL, Kreutzberg GW (1999). Neuroglial activation repertoire in the injured brain: graded response, molecular mechanisms and cues to physiological function. *Brain Res Rev* 30:77–105.

Rieske E, Graeber MB, Tetzlaff W, Czlonkowska A, Streit WJ, Kreutzberg GW (1989). Microglia and microglia-derived brain macrophages in culture: generation from axotomized rat facial nuclei, identification and characterization in vitro. *Brain Res* 492:1–14.

Río-Hortega D (1932). Microglia. In: *Cytology and Cellular Pathology of the Nervous System,* Vol. 2, New York (Penfield W, ed.), pp 481–534. Hocker.

Rothman S (1984). Synaptic release of excitatory amino acid neurotransmitter mediates anoxic neuronal death. *J Neurosci* 4:1884–1891.

Rothstein JD, Van Kammen M, Levey AI, Martin LJ, Kuncl RW (1995). Selective loss of glial glutamate transporter GLT-1 in amyotrophic lateral sclerosis. *Ann Neurol* 38:73–84.

Rudge JS, Alderson RF, Pasnikowski EM, McClain J, Ip NY, Lindsay RM (1992). Expression of ciliary neurotrophic factor and the neurotrophins-nerve growth factor, brain-derived neurotrophic factor and neurotrophin-3 in cultured rat hippocampal astrocytes. *Eur J Neurosci* 4:459–471.

Ruit KG, Elliott JL, Osborne PA, Yan Q, Snider WD (1992). Selective dependence of mammalian dorsal root ganglion neurons on nerve growth factor during embryonic development. *Neuron* 8:573–587.

Sawada M, Kondo N, Suzumura A, Marunouchi T (1989). Production of tumor necrosis factor-alpha by microglia and astrocytes in culture. *Brain Res* 491:394–397.

Scarborough DE, Lee SL, Dinarello CA, Reichlin S (1989). Interleukin-1β stimulates somatostatin biosynthesis in primary cultures of fetal rat brain. *Endocrinology* 124:549–551.

Selmaj KW, Raine CS (1988). Tumor necrosis factor mediate myelin and oligodendrocyte damage in vitro. *Ann Neurol* 23:339–346.

Shimojo M, Nakajima K, Takei N, Hamanoue M, Kohsaka S (1991). Production of basic fibroblast growth factor in cultured brain microglia. *Neurosci Lett* 123:229–231.

Smeyne RJ, Klein R, Schnapp A, Long LK, Bryant S, Lewin A, Lira SA, Barbacid M (1994). Severe sensory and synpathetic neuropathies in mice carrying a disrupted Trk/NGF receptor gene. *Nature* 368:246–249.

Sonderer B, Wild P, Wyler R, Fontana A, Peterhans E, Schwyzer M (1987). Murine glia cells in culture can be stimulated to generate reactive oxygen. *J Leukoc Biol* 42:463–73.

Streit WJ, Graeber MB, Kreutzberg GW (1988). Functional plasticity of microglia: a review. *Glia* 1:301–307.

Streit WJ, Graeber MB, Kreutzberg GW (1989). Expression of Ia antigen on perivascular and microglial cells after sublethal and lethal motor neuron injury. *Exp Neurol* 105:115–126.

Streit WJ, Walter SA, Pennell NA (1999). Reactive microgliosis. *Prog Neurobiol* 57:563–581.

Talley AK, Dewhurst S, Perry SW, Dollard SC, Gummuluru S, Fine SM, New D, Epstein LG, Gendelman HE, Gelbard HA (1995). Tumor necrosis factor α-induced apoptosis in human neuronal cells: protection by the antioxidant N-acetylcysteine and the genes bcl-2 and crmA. *Mol Cell Biol* 15:2359–2366.

Thoenen H (1991). The changing scene of neurotrophic factors. *Trends Neurosci* 14:165–170.

Torp R, Lekieffre D, Levy LM, Haug FM, Danbolt NC, Meldrum BS, Ottersen OP (1995). Reduced postischemic expression of a glial glutamate transporter, GLT1, in the rat hippocampus. *Exp Brain Res* 103:51–58.

Woodroofe MN, Sarna GS, Wadhwa M, Hayes GM, Loughlin AJ, Tinker A, Cuzner ML (1991). Detection of interleukin-1 and interleukin-6 in adult rat brain, following mechanical injury, by in vivo microdialysis: evidence of a role for microglia in cytokine production. *J Neuroimmunol* 33:227–236.

Yamakuni T, Ozawa F, Hishinuma F, Kuwano R, Takahashi Y, Amano T (1987). Expression of β-nerve growth factor mRNA in rat glioma cells and astrocytes from rat brain. *FEBS Lett* 223:117–121.

Yoshida K, Gage FH (1992). Cooperative regulation of nerve growth factor synthesis and secretion in fibroblasts and astrocytes by fibroblast growth factor and other cytokines. *Brain Res* 569:14–25.

Zafra F, Lindholm D, Castren E, Hartikka J, Thoenen H (1992). Regulation of brain-derived neurotrophic factor and nerve growth factor mRNA in primary cultures of hippocampal neurons and astrocytes. *J Neurosci* 12:4793–4799.

10

Influences of Activated Microglia/Brain Macrophages on Spinal Cord Injury and Regeneration

ALEXANDER G. RABCHEVSKY

Introduction

Damage to the central nervous system (CNS) systematically elicits the activation of both astrocytes and microglia, often termed reactive gliosis. This article is focused on the principal features that characterize cellular events associated with the activation of microglia after spinal cord injury (SCI) that govern the regenerative success or failure of injured neurons. In addition to discussing the role of microglia as immunocompetent cells of the CNS, it addresses the influences of activated microglia/brain macrophages on astrogliosis. The controversial issue of whether reactive microgliosis is a beneficial or harmful process with respect to neuroprotection is addressed, and a resolution of this dilemma is offered by suggesting different interpretations of the term "activated microglia" depending on its usage during experimentation *in vivo* or *in vitro*. Importantly, it provides a critical discussion regarding the distinction and relation between microglia-derived brain macrophages and infiltrating peripheral macrophages, and their conflicting roles in creating a pro-regenerative environment. To this end, evidence is reviewed that suggests that microglia-derived brain macrophages are capable of overriding many of the inhibitory obstacles to regeneration following SCI through their production of growth factors and cytokines, as well as their deposition or modulation of the extacellular matrix in the injured environment.

Phenotypically distinct from astrocytes and oligodendrocytes, ramified microglia are uniformly dispersed through the CNS and are thought to be in a functionally quiescent state involved in ionic homeostasis (Graeber and Streit (1990); Lawson et al. (1990); Barron (1995)). Since the early studies of Río Hortega (Río Hortega, 1932) documenting the phenotypic heterogeneity of microglial cells in the normal and injured CNS, this "third glial element" has more recently become the focus of extensive research. When the CNS is injured or affected by disease, microglial cells become "reactive" and are considered functionally activated. The ensuing morphology is characterized by cellular hypertrophy and retraction of cytoplasmic processes, accompanied by the upregulation and *de novo* expression of surface antigens (Graeber et al. (1988); Graeber et al. (1990); Streit et al. (1999)). "Resting" microglia can be activated by stimuli ranging from peripheral nerve injury to viral infections to direct mechanical CNS trauma. The term "activated microglia" is used to describe proliferating cells that demonstrate changes in their immunophenotype, but have not undergone transformation into brain macrophages. Importantly, these cells are the resident immunocompetent cells (Graeber and Streit (1990)) capable of antigen presentation to infiltrating T-cells that is a requisite to mounting an inflammatory response in the CNS (Hickey and Kimura (1988); Hickey (1991)). In regions of necrosis, activated microglia can transform into brain macrophages that proliferate and phagocytose cellular debris, and eventually return to a resting state (Streit and Kreutzberg (1988b); Graeber et al. (1989)). In light of this functional plasticity (Streit et al. (1988a)), the significance of microglial cell activation in response to CNS insults remains somewhat of an enigma.

Contrasting theories exist regarding the potentially neurotoxic versus neuroprotective roles of microglia in association with disease and injury to the CNS (Streit and Kincaid-Colton (1995)). On the one hand, activated microglia are systematically associated with neurodegenerative pathologies, such as Alzheimer's disease, amyotrophic lateral sclerosis (ALS), multiple sclerosis, and Parkinson's disease, and therefore have been widely reported to initiate or exacerbate neurodegeneration. However, many of these conclusions are based, in part, on experiments *in vitro* that demonstrate neurotoxic properties of cultured microglia that may have little relevance to the situation *in vivo* (see below). In contrast, mounting evidence suggests that microglial cell activation in the injured spinal cord and other regions of the CNS helps to create a pro-regenerative environment through the production of cytokines, growth factors and extracellular matrix (ECM) molecules that may act directly to enhance regeneration, or increase the neurite growth-promoting properties of other cells invading the injured environment (Fagan and Gage (1990); Chamak et al. (1994); Mallat and Chamak (1994); Rabchevsky and Streit (1997); Rabchevsky and Streit (1998); Batchelor et al. (1999); Popovich et al. (1999); Streit et al. (1999); Moon et al. (2000)). One cell type in particular that activated microglia can have profound influences upon is the astrocyte.

Reactive gliosis and the contribution of microglia to post-injury tissue repair

Following spinal cord injury (SCI), peripheral macrophages and activated microglial cells accumulate at lesion sites during the first week, and astrocytic processes delineate the area of necrosis forming a dense astroglial lining. A basal lam-

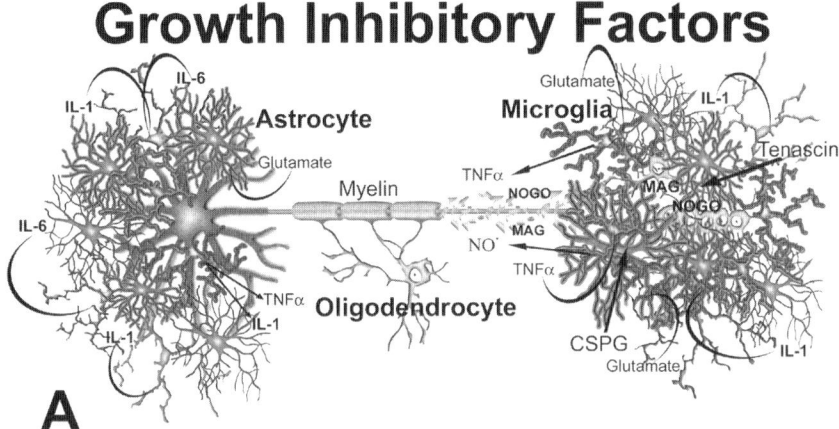

FIGURE 1. Schematic illustrations showing the dichotomous cellular events following spinal cord injury that govern both abortive (**A**) and successful (**B**) regeneration of damaged neurons. Traumatic injury invariably leads to the physiological and morphological activation of microglial cells and astrocytes. Reactive gliosis is important for establishing the regenerative success or failure of injured axons. **A.** The distal axon of an injured neuron encounters activated neuroglial cells that secrete factors that contribute to growth inhibition. The soma is enveloped by activated microglial cells that highly influence the phenotypes of astrocytes and neuronal survival by expressing interleukin-1 (IL-1) and IL-6. The damaged axon undergoes Wallerian degeneration and becomes surrounded by reactive neuroglia, which secrete glutamate that contributes to excitotoxicity. Microglial secretory products are thought to exacerbate demyelination, which releases the potent oligodendrocyte-growth inhibitory molecules myelin-associated glycoprotein (MAG) and Nogo-A (NOGO). Reactive gliosis at the severed nerve endings produces potentially noxious agents (NO·) and harmful cytokines (TNFα) by activated microglia, as well as astrocytic deposition of the inhibitory extracellular matrix (ECM) molecules tenascin-R (tenascin) and chondroitin sulfate proteoglycan (CSPG). (*continued*)

ina (BL) is formed along exposed regions of the brain or spinal cord that essentially mirrors the distribution of astrocytic processes immunoreactive for glial fibrillary acidic protein (GFAP) (Norenberg (1994)). This BL serves an important function in restoring the blood-spinal cord barrier (BSCB) and ionic homeostasis to the injured area, but the accompanying reactive gliosis, characterized by dense astroglial scars that fill the space vacated by dead or dying cells, is thought to inhibit successful regeneration in the injured adult mammalian spinal cord (Figure 1A) (Feringa et al. (1980); Reier et al. (1983); Liuzzi and Lasek (1987)). Despite the common occurrence of reactive gliosis after SCI, limited axonal growth through glial scar tissue does occur, and neurites typically invade traumatized tissue in conjunction with ependymal and mononuclear cells (Guth et al. (1985)). Nevertheless, the putative role that microglia-derived brain macrophages have in modulating astrogliosis or fostering regeneration in the spinal cord is poorly understood.

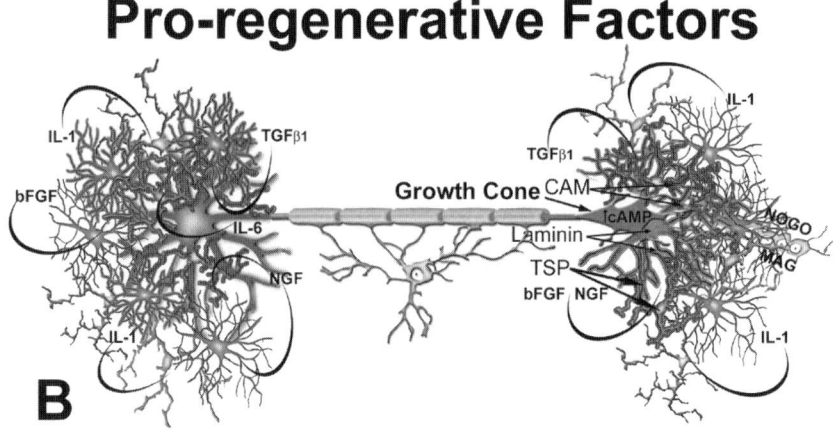

FIGURE 1. (*continued*) **B.** An injured neuron extends the growth cone of its axon through a supporting neuroglial cicatrix expressing pro-regenerative ECM, such as laminin and cell adhesion molecules (CAM). The local production of transforming growth factor-beta (TGF-β1) by reactive microglial cells helps to create a pro-regenerative milieu by modulating astrocytic phenotypes and their inhibitory ECM secretions. Activated microglia also secrete the pro-regenerative ECM molecule thrombospodin (TSP), and, along with astrocytes, their expression of basic fibroblast growth factor (bFGF) and nerve growth factor (NGF) enables growth cones to overcome MAG and NOGO inhibition by increasing cAMP levels.

Reactive microglial cells secrete a variety of cytokines and growth factors both *in vitro* and *in vivo*, such as interleukin-1 (IL-1) and IL-6, tumor-necrosis factor-alpha (TNFα), transforming-growth factor-beta (TGF-β1), basic fibroblast-growth factor (bFGF), nerve-growth factor (NGF), brain-derived neurotrophic factor (BDNF) and neurotrophin 3 (NT-3) (Mallat et al. (1989); Sawada et al. (1989); Fagan and Gage (1990); Shimojo et al. (1991); Giulian et al. (1994); Elkabes et al. (1996); Miwa et al. (1997); Krenz and Weaver (2000)). Additionally, they express receptors for ciliary neurotrophic factor (CNTF), glial cell line-derived neurotrophic factor (GDNF), and low affinity (p75) NGF (Hagg et al. (1993); Junier et al. (1994); Honda et al. (1999)). Collectively, this suggests that the extracellular microenvironment surrounding injured tissue is highly dependent upon the functional states of microglial cells and, accordingly, the repertoire of neuroprotective and growth-promoting factors they secrete and/or bind (Figure 1A and B).

Microglia-derived TGF-β1 and IL-1 have profound effects on astrogliosis

Activated microglial cells produce high amounts of TGF-β1 (Kiefer et al. (1993); Logan et al. (1994); McTigue et al. (2000)), which promotes angiogenesis and enhances cell migration, cell proliferation, production of ECM components, chemotaxis, and neuronal survival (Finch et al. (1993); Kiefer et al. (1995)). The sharp increase in TGF-β1 expression after SCI is tightly correlated with conspicuous activation of microglial cells (Koshinaga and Whittemore (1995); Popovich et al. (1997)) that express TGF-β1 and its receptor βRI (McTigue et al. (2000)). This upregulation in activated cells is thought to help establish a neuroprotective, pro-regenerative environment (Semple-Rowland et al. (1995)). Importantly, the secretion of TGF-β1 and IL-1 by activated microglia regulates astrocyte proliferation, glial scar formation, and vascularization (Giulian et al. (1986); Klagsbrun and D'Amore (1991); Lindholm et al. (1992); Sievers et al. (1993); Pennell and Streit (1997)). With respect to regeneration following injury, however, TGF-β1 significantly increases the production of astrocytic tenascin. The regulation of this growth-inhibiting ECM molecule is mediated by a synergistic action of TGF-β1 and bFGF *in vitro* and after injury *in vivo* (Smith and Hale (1997)). Therefore, the intimate association between microvasculature and astrocytic processes suggests that the activational states of microglia are very important in controlling the development of astroglial scars and governing the success or failure of regeneration.

The infiltration of fibroblasts and Schwann cells invariably occurs if there is a breach in the BSCB with an accompanying loss of resident astrocytes (Blakemore et al. (1986)). These peripheral cells originate from small nerves accompanying blood vessels or nerve roots and have profound influences on the development of astrogliosis (Blakemore (1983)). Long after the formation of a dense astrocytic scar, severed corticospinal axons can maintain massive arborizations in macrophage-filled lesions, and these axons become myelinated, or ensheathed, by endogenous, infiltrating Schwann cells (Li and Raisman (1995)). Importantly, TGF-β1 is a mitogen for Schwann cells (Ridley et al. (1989)), and in microglia/brain macrophage enriched lesions devoid of astrocytes the Schwann cells develop interactions with and myelinate regenerating axons. This may be an important process in sparing and/or rescuing injured axons, as evidenced by significant regeneration of severed spinal axons following grafting of cultured Schwann cells into the transected rat spinal cord (Paino et al. (1994); Xu et al. (1995a); Xu et al. (1995b)). Therefore, because activated microglia are systematically involved in the injury repair process that may, in some instances, contribute to regenerative failure, it is crucial to fully understand the unique biology of microglial cells.

Microglia-derived brain macrophages are functionally distinct from peripheral macrophages

Currently the most plausible theory regarding the debated origin of microglial cells is that they initially derive from a precursor cell related to the monocyte/macrophage lineage associated with embryonic hematopoietic organs

(Theele and Streit (1993)). For that reason, the key challenge that has clouded the precise role of microglia-derived brain macrophages is the lack of cellular markers that distinguish them from their peripheral counterparts. Cell numbers increase following CNS trauma, related to the recruitment and proliferation of microglial cells, as well as blood monocytes that cross the blood-brain barrier (BBB) (Schelper and Adrian (1986); Graeber et al. (1989); Morshead and Van der Kooy (1990); Andersson et al. (1991)). Conversely, the activation and subsequent transformation of activated microglial cells into brain macrophages in the CNS following either peripheral nerve injury or neurodegeneration induced by selective neurotoxins occurs without a compromised BBB or recruitment of peripheral macrophages (Streit and Kreutzberg (1988b); Riva-Depaty et al. (1994)). While bone marrow irradiation chimera studies have reported a small but steady turnover of resident microglial cells derived from blood-borne cells that cross the BBB during development (De Groot et al. (1992); Krall et al. (1994)), there is strong *in vitro* and *in vivo* evidence that cell surface morphology and soma size distinguish activated microglia as a distinct class of mononuclear phagocytes separate from bone marrow precursors, monocytes, or tissue macrophages (Riva-Depaty et al. (1994); Giulian et al. (1995)).

Brain macrophages are intimately associated with the neuropathology in the clinical stages of experimental allergic encephalomyelitis (EAE), a rodent model for multiple sclerosis. However, the main effectors playing a central role in cellular destruction during EAE inflammation are infiltrating blood-borne cells, and not resident microglial cells that are secondarily activated to become brain macrophages when exposed to numerous inflammatory cytokines (Bauer et al. (1995); Rinner et al. (1995)). Phagocytic brain macrophages serve as important effectors for clearing cellular debris and secreting cytokines/growth factors involved in immunological responses to tissue repair at sites of injury (Giulian et al. (1989)), and phagocytosis of myelin debris potentially rids neurite growth-inhibiting molecules expressed on oligodendrocytes, which may indirectly promote regeneration (Schwab et al. (1993)). The question remains, however, as to the function and origin of these immunocompetent cells following physical trauma.

The progressive destruction of axotomized retinal ganglion cells *in vivo* and *in vitro* can be enhanced or suppressed with macrophage-stimulating factors (MSF) or macrophage-inhibiting factors (MIF), respectively (Thanos et al. (1993)). In the absence of a clear understanding of the contributing mechanisms, it has been suggested that suppression of brain macrophages is essential for regeneration to occur. Activated microglia can express molecules that are potentially toxic to neural cells and may, thus, reduce the potential for regeneration (Figure 1A) (Banati et al. (1993)). However, the vast majority of such evidence is derived from manipulating microglial cells in culture. Of critical concern is that advances in the rapid isolation of microglia-derived brain macrophages in mixed glial cultures from the CNS have overlooked the fact that the very act of culturing microglia transforms them into an activated state (Streit (1993)). In this context, it is of little surprise that further stimulation of cultured microglia with lipopolysaccharide (LPS), interferon-gamma (IFN-γ), and/or zymosan A results in the elaboration of various noxious and cytotoxic agents, including reactive oxygen intermediates (Colton and Gilbert (1987)), TNFα (Frei et al. (1987); Remick et al. (1988); Sawada et al. (1989)), glutamate (Piani et al. (1991)), and low molecular weight factors with neurotoxic activities (Giulian et al. (1993a); Giulian et al. (1993b)). While the relevance to the situation *in vivo* remains questionable, there is also evidence that the production of nitric oxide (NO) and TNFα by cultured microglia exposed to such challenges is toxic to oligodendrocytes (Merrill and Zimmerman (1991); Zajicek et al. (1992); Merrill et al. (1993)).

Contrary to the cytotoxic evidence following various challenges *in vitro*, exogenous local administration of the proinflammatory cytokines IL-1, IL-6, and TNFα to rodents several days after SCI results in a reduction in the amount of tissue loss (Klusman and Schwab (1997)). This suggests that the inflammatory response, which invariably leads to microglial activation, is complex and well regulated. Despite the presence of many activated microglia following a spinal cord lesion, the total number of proximal oligodendrocytes is reported to be unchanged (Frei et al. (2000)). This challenges the notion that oligodendrocyte apoptosis is secondary to microglial activation, based solely on apparent colocalization of apoptotic oligodendrocytes and microglial cell processes (Shuman et al. (1997)). Furthermore, when neurons from various CNS regions are cocultured with purified microglia and/or microglia-conditioned medium in the absence of noxious agents, both contact with microglia or growth in microglia-conditioned medium promotes neuronal survival and neuritic extension (Nakajima et al. (1993); Nagata et al. (1993a); Nagata et al. (1993b); Chamak et al. (1994); Kohsaka et al. (1994)). It remains to be determined, however, whether this is attributed to cytokine/growth factor production or deposition of ECM components.

Peripheral macrophage depletion augments the process of self-repair in the injured spinal cord

Functional inhibition of mononuclear phagocytes with chloroquine and colchicine reduces ischemic injury in the spinal cord, which correlates with improved hind-limb motor function (Giulian and Robertson (1990)). One may argue that such suppression of all macrophage populations is critical for the observed recovery, or, conversely, that impeding peripheral macrophage infiltration allows the injury-induced responses of endogenous neuroglial cells to re-establish homeostasis leading to improved functional recovery. The latter hypothesis is supported by depletion of peripheral macrophages following SCI using the selective toxin clodronate. Such treatment results in a conspicuous reduction of infiltrating macrophages in the lesion epicenter that is accompanied by both significant sparing of damaged white matter and improved hind-limb function (Popovich et al. (1999)). Of particular interest at the cellular level is that the significant reduction of peripheral macrophages results in robust axonal sprouting and neovascularization throughout the injured regions that contain numerous morphologically activated microglial cells and unidentified nonneuronal cellular matrices not present in the spinal cords of untreated animals.

Microglia and extracellular matrix

Successful regeneration of severed nerves in the invertebrate CNS is accompanied by increased laminin-immunoreactivity that is coincidental with the appearance of numerous microglial cells, both in time and distribution (Masuda-Nakagawa and Wiedemann (1992); Masuda-Nakagawa et al. (1993); Masuda-Nakagawa et al. (1994)). In mammals, axon growth-inhibiting properties of injured optic nerve white matter is also altered by mononuclear phagocytes that invade the lesions and modify nonpermissive gliotic tissue to promote axonal sprouting (David et al. (1990)), although the mechanisms are undefined. Following peripheral nerve axotomy, regenerating facial motoneurons upregulate the expression of the polymeric ECM protein thrombospondin (TSP) (Möller et al. (1996)), and activated microglial cells surrounding the regenerating facial motoneurons express TSP, suggesting this protein may help to ensure the regeneration of motoneurons. This laminin-like glycoprotein involved in platelet aggregation is thought to function during cell migration in response to injury (Lawler (1986)), and it promotes the adhesion and neuritic outgrowth from both central and peripheral neurons in culture (O'Shea et al. (1991); Osterhout et al. (1992); Chamak et al. (1994)).

Interestingly, TSP is found in the nervous systems of species with regenerative capabilities (Hoffman et al. (1994)) and in ameboid microglia during CNS development (Chamak et al. (1995)). Differentiation of microglia into a ramified phenotype is associated with a reduced or undetectable TSP-immunoreactivity and, conversely, it has been shown that microglia/brain macrophages contribute actively to neurite growth in the injured striatum by the *de novo* expression of TSP (Chamak et al. (1994)). There is also a possible role for TSP in the activation of latent microglial TGF-β1 (Schultz-Cherry et al. (1994)), indicating that synthesis of TSP enhances the established neuroprotective and neurotrophic effects of TGF-β1 (Martinou et al. (1990); Merrill and Zimmerman (1991)). These findings suggest that the expression of TGF-β1 and deposition of TSP by activated microglia favors axonal growth during development and fosters regeneration in the injured mammalian spinal cord (Figure 1B).

The increased appearance of putative inhibitory proteoglycans in spinal cord lesions, most notably chondroitin sulfate proteoglycan (CSPG), has been correlated with the presence of activated macrophages and their secretory products (Fitch and Silver (1997)). However, the appearance of CSPG only occurs in regions with a conspicuous breach in the BSCB characterized by the infiltration of blood-borne cells, extravasation of serum components, and the putative production of numerous pro-inflammatory cytokines. Contrary evidence shows that a novel proteoglycan growth inhibitor, termed injured-membrane proteoglycan (IMP), is primarily found on reactive astrocytes after a lesion, but not on OX-42 positive microglia (Nieto-Sampedro (1999)). But the question still remains whether the reactive microglial cells actively contribute to the spontaneous regeneration reported following neurotrauma.

Influence of microgliosis on regeneration

It has been suggested that activated microglia/brain macrophages permit or support, rather than inhibit, sprouting of injured axons (Fagan and Gage (1990); Blaugrund et al. (1992)), and that regenerative failure in the mammalian CNS may be attributed to the temporal and/or spatial deficiency of brain macrophages in the injured central fiber tracts (Perry et al. (1987); Stichel and Muller (1994); Lazar et al. (1999)). Unlike successful regeneration of peripheral nerves that is accompanied by rapid macrophage recruitment, axotomy of central tracts does not systematically activate microglia in the projection nuclei that undergo progressive retrograde atrophy (Barron et al. (1990)). The absence of proliferative, activated microglia in these nuclei appears to correlate with the inability of the CNS to regenerate central processes. This is exemplified in the lesioned nigrostriatal system, where the ensuing dopaminergic sprouting appears to result from neurotrophic factor production and ECM secretion by activated microglia/brain macrophages at the wound sites (Chamak et al. (1994); Batchelor et al. (1999)). These findings are precisely why care must be taken when interpreting data as reflecting an alteration of microglia-derived brain macrophages if peripheral macrophage involvement may be in question.

Transplantation of microglial cells into the injured spinal cord

Currently, microglial-neuronal interactions can be seen as a double-edged sword, owing in part to two very different experimental approaches that have been employed *in vivo* and *in vitro* (Streit (1993); Gebicke-Haerter et al. (1996); Rabchevsky and Streit (1998)). Few transplantation studies have attempted to combine these approaches to assess whether activated microglia exert beneficial effects in the post-traumatic spinal cord environment. Cultured microglial cells embedded in gelfoam and grafted directly into the injured adult rat spinal cord create a pro-regenerative microenvironment, with prominent neuritic ingrowth, vascularization, and conspicuous laminin and TSP immunoreactivity throughout the grafts (Rabchevsky and Streit (1997)). This study also showed that growth-promoting properties of grafted astrocytes were augmented in regions rich with endogenous infiltrating brain macrophages that possibly modulated their inhibitory effects (McKeon et al. (1995)). The mechanism(s) for the neuritic ingrowth may entail a relation between microglia/brain macrophages and the deposition and/or modulation of laminin and TSP. Endothelial and Schwann cells represent the most likely laminin-producing cell types recruited from the host into the grafted microglial matrices, and they may have contributed to the ingrowth. Altogether, this implies that grafted or endogenous, activated microglia can modify putative inhibitory effects on growing axons in the injured spinal cord, as well as mobilize other host cells with growth-promoting properties (Harvey et al. (1993)).

Another transplantation paradigm has shown nitrocellulose membranes coated with cultured microglial cells and inserted into fetal spinal cord grafts within the injured adult rat spinal cord enhance the regeneration of primary sensory axons following dorsal root injury (Prewitt et al. (1997)). A similar result was observed with membranes treated with TGF-β1 versus those treated with macrophage inhibitory factor (MIF), showing that activated microglia/brain macrophages promote the regeneration of sensory axons, at least in part, through the release of certain secretory products. These findings not only contradict the suggestion that suppression of brain macrophages with MIF treatment is essential for regeneration to occur (Thanos et al. (1993)), but they highlight the fact that different experimental approaches employed in disparate CNS regions may lead to opposite conclusions when the targets of chosen treatments are not clearly defined.

Spinal cord demyelination: regeneration, microgliosis, and effects of Schwann cell grafting

Discrete axotomy of either ascending or descending spinal tracts in the adult rat spinal cord following the disruption of myelin by local intraspinal infusion of serum complement proteins along with the complement-fixing, myelin-specific antibodies to galactocerebroside promotes the regeneration of axons in the dorsal funiculus and those originating from the brainstem (Dyer et al. (1998); Keirstead et al. (1998)). Although this implies that demyelination promotes axonal regeneration in the adult rat spinal cord, it is important to recognize that the regenerative process is characterized by the invasion of numerous brain macrophages that actively engulf disrupted myelin. The successful axonal regeneration may be highly influenced by the demyelinating effects on the microglial and astroglial cell phenotypes, leading to the production of neurotrophic factors and cytokines that promote regeneration (Figure 1B). This is in accordance with the results of combined immunological demyelination plus Schwann cell grafting at the axotomy site, after which enhanced axonal regeneration is observed compared with non-grafted animals (Keirstead et al. (1999)). These results suggest that regenerating axons are able to overcome inhibitors of neurite outgrowth in the presence of axon growth-promoting effects of Schwann cells. What influence(s) endogenous, activated microglia/brain macrophages have on this response is unclear. However, based on their deposition and/or modulation of ECM molecules, and expression of numerous growth factors and cytokines, including TGF-β1, which is a mitogen for Schwann cells (Ridley et al. (1989)), it seems likely that microglia-derived brain macrophages develop interactions with grafted and endogenous Schwann cells to aid in the guidance and myelination of regenerating axons. Future studies using this approach and aimed at depleting the peripheral macrophage population may explain the role that activated microglia have in significant regeneration in the injured mammalian spinal cord. One last point supporting these hypotheses following demyelination is that, shortly following transection of the nigrostriatal tract and subsequent demyelination using ethidium bromide injections, there are large numbers of dopaminergic nigral axons that regenerate through regions that are completely inundated with brain macrophages (Moon et al. (2000)).

Summary

The purpose of this review is to present collective evidence that the activation of microglial cells following injury to the spinal cord and other CNS regions helps to create an environment conducive to tissue sparing and neuroprotection by promoting neovascularization and regeneration of injured neurons through the lesion. It focuses on the principal cellular events associated with the activation of microglia after SCI, and addresses the putative contributions of activated microglia/brain macrophages on astrogliosis and axonal growth inhibition. Whether reactive microgliosis is a beneficial or harmful process is, unfortunately, subject to interpretation, depending upon results from experimentation *in vivo* or *in vitro*. It is evident that isolating a protein expressed uniquely by microglia to differentiate endogenous or grafted microglia/brain macrophages from their peripheral counterparts will be the next big advance in understanding the mechanisms and consequences of microglial activation. Currently, transplantation strategies, demyelination protocols, and selective depletion of peripheral macrophages offer the greatest potential for defining the precise role that microglia-derived brain macrophages have in creating a pro-regenerative environment. In conclusion, the evidence reviewed suggests that microglia-derived brain macrophages are capable of overriding many of the inhibitory obstacles to neuronal regeneration following

SCI through their production of growth factors and cytokines, as well as their deposition and/or modulation of the extracellular matrix in the injured environment.

References

Andersson PB, Perry VH, Gordon S (1991). The kinetics and morphological characteristics of the macrophage-microglial response to kainic acid-induced neuronal degeneration. Neuroscience 42:201–214.

Banati RB, Gehrmann J, Schubert P, Kreutzberg GW (1993). Cytotoxicity of microglia. *Glia* 7:111–118.

Barron KD (1995). The microglial cell. A historical review. *J Neurol Sci* 134 Suppl:57–68.

Barron KD, Marciano FF, Amundson R, Mankes R (1990). Perineuronal glial responses after axotomy of central and peripheral axons. A comparison. *Brain Res* 523:219–229.

Batchelor PE, Liberatore GT, Wong JY, Porritt MJ, Frerichs F, Donnan GA, Howells DW (1999). Activated macrophages and microglia induce dopaminergic sprouting in the injured striatum and express brain-derived neurotrophic factor and glial cell line-derived neurotrophic factor. *J Neurosci* 19:1708–1716.

Bauer J, Huitinga I, Zhao W, Lassmann H, Hickey WF, Dijkstra CD (1995). The role of macrophages, perivascular cells, and microglial cells in the pathogenesis of experimental autoimmune encephalomyelitis. *Glia* 15:437–446.

Blakemore WF (1983). Remyelination of demyelinated spinal cord axons by Schwann cells. In: *Spinal Cord Reconstruction* (Kao CC, Bunge RP, Reier PJ, eds.), pp 281–291. New York: Raven Press.

Blakemore WF, Crang AJ, Curtis R (1986). The interaction of Schwann cells with CNS axons in regions containing normal astrocytes. *Acta Neuropathol* (Berlin) 71:295–300.

Blaugrund E, Duvdevani R, Lavie V, Solomon A, Schwartz M (1992). Disappearance of astrocytes and invasion of macrophages following crush injury of adult rodent optic nerves: implications for regeneration. *Exp Neurol* 118:105–115.

Chamak B, Dobbertin A, Mallat M (1995). Immunohistochemical detection of thrombospondin in microglia in the developing rat brain. *Neuroscience* 69:177–187.

Chamak B, Morandi V, Mallat M (1994). Brain macrophages stimulate neurite growth and regeneration by secreting thrombospondin. *J Neurosci Res* 38:221–233.

Colton CA, Gilbert DL (1987). Production of superoxide anions by a CNS macrophage, the microglia. *FEBS Lett* 223:284–288.

David S, Bouchard C, Tsatas O, Giftochristos N (1990). Macrophages can modify the nonpermissive nature of the adult mammalian central nervous system. *Neuron* 5:463–469.

De Groot CJ, Huppes W, Sminia T, Kraal G, Dijkstra CD (1992). Determination of the origin and nature of brain macrophages and microglial cells in mouse central nervous system, using non-radioactive in situ hybridization and immunoperoxidase techniques. *Glia* 6:301–309.

Dyer JK, Bourque JA, Steeves JD (1998). Regeneration of brainstem-spinal axons after lesion and immunological disruption of myelin in adult rat. *Exp Neurol* 154:12–22.

Elkabes S, DiCicco-Bloom EM, Black IB (1996). Brain microglia/macrophages express neurotrophins that selectively regulate microglial proliferation and function. *J Neurosci* 16:2508–2521.

Fagan AM, Gage FH (1990). Cholinergic sprouting in the hippocampus: a proposed role for IL-1. *Exp Neurol* 110:105–120.

Feringa ER, Kowalski TF, Vahlsing HL (1980). Basal lamina formation at the site of spinal cord transection. *Ann Neurol* 8:148–154.

Finch CE, Laping NJ, Morgan TE, Nichols NR, Pasinetti GM (1993). TGF-beta 1 is an organizer of responses to neurodegeneration. *J Cell Biochem* 53:314–322.

Fitch MT, Silver J (1997). Activated macrophages and the blood-brain barrier: inflammation after CNS injury leads to increases in putative inhibitory molecules. *Exp Neurol* 148:587–603.

Frei E, Klusman I, Schnell L, Schwab ME (2000). Reactions of oligodendrocytes to spinal cord injury: cell survival and myelin repair. *Exp Neurol* 163:373–380.

Frei K, Siepl C, Groscurth P, Bodmer S, Schwerdel C, Fontana A (1987). Antigen presentation and tumor cytotoxicity by interferon-gamma-treated microglial cells. *Eur J Immunol* 17:1271–1278.

Gebicke-Haerter PJ, Van Calker D, Norenberg W, Illes P (1996). Molecular mechanisms of microglial activation. A. Implications for regeneration and neurodegenerative diseases. *Neurochem Int* 29:1–12.

Giulian D, Baker TJ, Shih LC, Lachman LB (1986). Interleukin 1 of the central nervous system is produced by ameboid microglia. *J Exp Med* 164:594–604.

Giulian D, Chen J, Ingeman JE, George JK, Noponen M (1989). The role of mononuclear phagocytes in wound healing after traumatic injury to adult mammalian brain. *J Neurosci* 9:4416–4429.

Giulian D, Corpuz M, Chapman S, Mansouri M, Robertson C (1993a). Reactive mononuclear phagocytes release neurotoxins after ischemic and traumatic injury to the central nervous system. *J Neurosci Res* 36:681–693.

Giulian D, Li J, Bartel S, Broker J, Li X, Kirkpatrick JB (1995). Cell surface morphology identifies microglia as a distinct class of mononuclear phagocyte. *J Neurosci* 15:7712–7726.

Giulian D, Li J, Li X, George J, Rutecki PA (1994). The impact of microglia-derived cytokines upon gliosis in the CNS. *Dev Neurosci* 16:128–136.

Giulian D, Robertson C (1990). Inhibition of mononuclear phagocytes reduces ischemic injury in the spinal cord. *Ann Neurol* 27:33–42.

Giulian D, Vaca K, Corpuz M (1993b). Brain glia release factors with opposing actions upon neuronal survival. *J Neurosci* 13:29–37.

Graeber MB, Streit WJ (1990). Microglia: immune network in the CNS. *Brain Pathol* 1:2–5.

Graeber MB, Streit WJ, Kreutzberg GW (1988). Axotomy of the rat facial nerve leads to increased CR3 complement receptor expression by activated microglial cells. *J Neurosci Res* 21:18–24.

Graeber MB, Streit WJ, Kreutzberg GW (1989). Formation of microglia-derived brain macrophages is blocked by adriamycin. *Acta Neuropathol* 78:348–358.

Graeber MB, Streit WJ, Kreutzberg GW (1990). The third glial cell type, the microglia: cellular markers of activation in situ. *Acta Histochem Suppl* 38:157–160.

Guth L, Barrett CP, Donati EJ, Anderson FD, Smith MV, Lifson M (1985). Essentiality of a specific cellular terrain for growth of axons into a spinal cord lesion. *Exp Neurol* 88:1–12.

Hagg T, Varon S, Louis JC (1993). Ciliary neurotrophic factor (CNTF) promotes low-affinity nerve growth factor receptor and CD4 expression by rat CNS microglia. *J Neuroimmunol* 48:177–187.

Harvey AR, Fan Y, Connor AM, Grounds MD, Beilharz MW (1993). The migration and intermixing of donor and host glia on nitrocellulose polymers implanted into cortical lesion cavities in adult mice and rats. *Int J Dev Neurosci* 11:569–581.

Hickey WF (1991). Migration of hematogenous cells through the blood-brain barrier and the initiation of CNS inflammation. *Brain Pathol* 1:97–105.

Hickey WF, Kimura H (1988). Perivascular microglial cells of the CNS are bone marrow-derived and present antigen in vivo. *Science* 239:290–292.

Hoffman JR, Dixit VM, O'Shea KS (1994). Expression of thrombospondin in the adult nervous system. *J Comp Neurol* 340:126–139.

Honda S, Nakajima K, Nakamura Y, Imai Y, Kohsaka S (1999). Rat primary cultured microglia express glial cell line-derived neurotrophic factor receptors. *Neurosci Lett* 275:203–206.

Junier MP, Suzuki F, Onteniente B, Peschanski M (1994). Target-deprived CNS neurons express the NGF gene while reactive glia around their axonal terminals contain low and high affinity NGF receptors. *Brain Res Mol Brain Res* 24:247–260.

Keirstead HS, Hughes HC, Blakemore WF (1998). A quantifiable model of axonal regeneration in the demyelinated adult rat spinal cord. *Exp Neurol* 151:303–313.

Keirstead HS, Morgan SV, Wilby MJ, Fawcett JW (1999). Enhanced axonal regeneration following combined demyelination plus Schwann cell transplantation therapy in the injured adult spinal cord. *Exp Neurol* 159:225–236.

Kiefer R, Gold R, Gehrmann J, Lindholm D, Wekerle H, Kreutzberg GW (1993). Transforming growth factor beta expression in reactive spinal cord microglia and meningeal inflammatory cells during experimental allergic neuritis. *J Neurosci Res* 36:391–398.

Kiefer R, Streit WJ, Toyka KV, Kreutzberg GW, Hartung HP (1995). Transforming growth factor-beta 1: a lesion-associated cytokine of the nervous system. *Int J Dev Neurosci* 13:331–339.

Klagsbrun M, D'Amore PA (1991). Regulators of angiogenesis. *Annu Rev Physiol* 53:217–239.

Klusman I, Schwab ME (1997). Effects of pro-inflammatory cytokines in experimental spinal cord injury. *Brain Res* 762:173–184.

Kohsaka S, Nakajima K, Hamanoue M, Koizumi S, Inoue K (1994). Microglia-derived plasminogen has neurotrophic effects on the CNS neurons in vitro. *Neuropathol Appl Neurobiol* 20:190.

Koshinaga M, Whittemore SR (1995). The temporal and spatial activation of microglia in fiber tracts undergoing anterograde and retrograde degeneration following spinal cord lesion. *J Neurotrauma* 12:209–222.

Krall WJ, Challita PM, Perlmutter LS, Skelton DC, Kohn DB (1994). Cells expressing human glucocerebrosidase from a retroviral vector repopulate macrophages and central nervous system microglia after murine bone marrow transplantation. *Blood* 83:2737–2748.

Krenz NR, Weaver LC (2000). Nerve growth factor in glia and inflammatory cells of the injured rat spinal cord. *J Neurochem* 74:730–739.

Lawler J (1986). The structural and functional properties of thrombospondin. *Blood* 67:1197–1209.

Lawson LJ, Perry VH, Dri P, Gordon S (1990). Heterogeneity in the distribution and morphology of microglia in the normal adult mouse brain. *Neuroscience* 39:151–170.

Lazar DA, Ellegala DB, Avellino AM, Dailey AT, Andrus K, Kliot M (1999). Modulation of macrophage and microglial responses to axonal injury in the peripheral and central nervous systems. *Neurosurgery* 45:593–600.

Li Y, Raisman G (1995). Sprouts from cut corticospinal axons persist in the presence of astrocytic scarring in long-term lesions of the adult rat spinal cord. *Exp Neurol* 134:102–111.

Lindholm D, Castren E, Kiefer R, Zafra F, Thoenen H (1992). Transforming growth factor-beta 1 in the rat brain: increase after injury and inhibition of astrocyte proliferation. *J Cell Biol* 117:395–400.

Liuzzi FJ, Lasek RJ (1987). Astrocytes block axonal regeneration in mammals by activating the physiological stop pathway. *Science* 237:642–645.

Logan A, Berry M, Gonzalez AM, Frautschy SA, Sporn MB, Baird A (1994). Effects of transforming growth factor beta 1 on scar production in the injured central nervous system of the rat. *Eur J Neurosci* 6:355–363.

Mallat M, Chamak B (1994). Brain macrophages: neurotoxic or neurotrophic effector cells? *J Leukoc Biol* 56:416–422.

Mallat M, Houlgatte R, Brachet P, Prochiantz A (1989). Lipopolysaccharide-stimulated rat brain macrophages release NGF in vitro. *Dev Biol* 133:309–311.

Martinou JC, Le Van Thai A, Valette A, Weber MJ (1990). Transforming growth factor beta 1 is a potent survival factor for rat embryo motoneurons in culture. *Brain Res Dev Brain Res* 52:175–181.

Masuda-Nakagawa LM, Muller KJ, Nicholls JG (1993). Axonal sprouting and laminin appearance after destruction of glial sheaths. *Proc Natl Acad Sci USA* 90:4966–4970.

Masuda-Nakagawa LM, Walz A, Brodbeck D, Neely MD, Grumbacher-Reinert S (1994). Substrate-dependent interactions of leech microglial cells and neurons in culture. *J Neurobiol* 25:83–91.

Masuda-Nakagawa LM, Wiedemann C (1992). The role of matrix molecules in regeneration of leech CNS. *J Neurobiol* 23:551–567.

McKeon RJ, Hoke A, Silver J (1995). Injury-induced proteoglycans inhibit the potential for laminin-mediated axon growth on astrocytic scars. *Exp Neurol* 136:32–43.

McTigue DM, Popovich PG, Morgan TE, Stokes BT (2000). Localization of transforming growth factor-beta1 and receptor mRNA after experimental spinal cord injury. *Exp Neurol* 163:220–230.

Merrill JE, Ignarro LJ, Sherman MP, Melinek J, Lane TE (1993). Microglial cell cytotoxicity of oligodendrocytes is mediated through nitric oxide. *J Immunol* 151:2132–2141.

Merrill JE, Zimmerman RP (1991). Natural and induced cytotoxicity of oligodendrocytes by microglia is inhibitable by TGF beta. *Glia* 4:327–331.

Miwa T, Furukawa S, Nakajima K, Furukawa Y, Kohsaka S (1997). Lipopolysaccharide enhances synthesis of brain-derived neurotrophic factor in cultured rat microglia. *J Neurosci Res* 50:1023–1029.

Möller JC, Klein MA, Haas S, Jones LL, Kreutzberg GW, Raivich G (1996). Regulation of thrombospondin in the regenerating mouse facial motor nucleus. *Glia* 17:121–132.

Moon LD, Brecknell JE, Franklin RJ, Dunnett SB, Fawcett JW (2000). Robust regeneration of CNS axons through a track depleted of CNS glia. *Exp Neurol* 161:49–66.

Morshead CM, van der Kooy D (1990). Separate blood and brain origins of proliferating cells during gliosis in adult brains. *Brain Res* 535:237–244.

Nagata K, Nakajima K, Takemoto N, Saito H, Kohsaka S (1993a). Microglia-derived plasminogen enhances neurite outgrowth from explant cultures of rat brain. *Int J Dev Neurosci* 11:227–237.

Nagata K, Takei N, Nakajima K, Saito H, Kohsaka S (1993b). Microglial conditioned medium promotes survival and development of cultured mesencephalic neurons from embryonic rat brain. *J Neurosci Res* 34:357–363.

Nakajima K, Nagata K, Hamanoue M, Takemoto N, Kohsaka S (1993). Microglia-derived elastase produces a low-molecular-weight plasminogen that enhances neurite outgrowth in rat neocortical explant cultures. *J Neurochem* 61:2155–2163.

Nieto-Sampedro M (1999). Neurite outgrowth inhibitors in gliotic tissue. *Adv Exp Med Biol* 468:207–224.

Norenberg MD (1994). Astrocyte responses to CNS injury. *J Neuropathol Exp Neurol* 53:213–220.

O'Shea KS, Liu LH, Dixit VM (1991). Thrombospondin and a 140 kd fragment promote adhesion and neurite outgrowth from embryonic central and peripheral neurons and from PC12 cells. *Neuron* 7:231–237.

Osterhout DJ, Frazier WA, Higgins D (1992). Thrombospondin promotes process outgrowth in neurons from the peripheral and central nervous systems. *Dev Biol* 150:256–265.

Paino CL, Fernandez-Valle C, Bates ML, Bunge MB (1994). Regrowth of axons in lesioned adult rat spinal cord: promotion by implants of cultured Schwann cells. *J Neurocytol* 23:433–452.

Pennell NA, Streit WJ (1997). Colonization of neural allografts by host microglial cells: relationship to graft neovascularization. *Cell Transplant* 6:221–230.

Perry VH, Brown MC, Gordon S (1987). The macrophage response to central and peripheral nerve injury. A possible role for macrophages in regeneration. *J Exp Med* 165:1218–1223.

Piani D, Frei K, Do KQ, Cuenod M, Fontana A (1991). Murine brain macrophages induced NMDA receptor mediated neurotoxicity in vitro by secreting glutamate. *Neurosci Lett* 133:159–162.

Popovich PG, Guan Z, Wei P, Huitinga I, van Rooijen N, Stokes BT (1999). Depletion of hematogenous macrophages promotes partial hindlimb recovery and neuroanatomical repair after experimental spinal cord injury. *Exp Neurol* 158:351–365.

Popovich PG, Wei P, Stokes BT (1997). Cellular inflammatory response after spinal cord injury in Sprague- Dawley and Lewis rats. *J Comp Neurol* 377:443–464.

Prewitt CM, Niesman IR, Kane CJ, Houle JD (1997). Activated macrophage/microglial cells can promote the regeneration of sensory axons into the injured spinal cord. *Exp Neurol* 148:433–443.

Rabchevsky AG, Streit WJ (1997). Grafting of cultured microglial cells into the lesioned spinal cord of adult rats enhances neurite outgrowth. *J Neurosci Res* 47:34–48.

Rabchevsky AG, Streit WJ (1998). Role of microglia in postinjury repair and regeneration of the CNS. *MRDD Res Rev* 4:187–192.

Reier PJ, Stensaas LJ, Guth L (1983). The astrocytic scar as an impediment to regeneration in the central nervous system. In: *Spinal Cord Reconstruction* (Kao CC, Bunge RP, Reier PJ, eds.), pp 163–195. New York: Raven Press.

Remick DG, Scales WE, May MA, Spengler M, Nguyen D, Kunkel SL (1988). In situ hybridization analysis of macrophage-derived tumor necrosis factor and interleukin-1 mRNA. *Lab Invest* 59:809–816.

Ridley AJ, Davis JB, Stroobant P, Land H (1989). Transforming growth factors-beta 1 and beta 2 are mitogens for rat Schwann cells. *J Cell Biol* 109:3419–3424.

Rinner WA, Bauer J, Schmidts M, Lassmann H, Hickey WF (1995). Resident microglia and hematogenous macrophages as phagocytes in adoptively transferred experimental autoimmune encephalomyelitis: an investigation using rat radiation bone marrow chimeras. *Glia* 14:257–266.

Río Hortega Pd (1932). Microglia. In: *Cytology and cellular pathology of the nervous system* (Penfield W, ed.), pp 481–534. New York: P.B. Hoeber.

Riva-Depaty I, Fardeau C, Mariani J, Bouchaud C, Delhaye-Bouchaud N (1994). Contribution of peripheral macrophages and microglia to the cellular reaction after mechanical or neurotoxin-induced lesions of the rat brain. *Exp Neurol* 128:77–87.

Sawada M, Kondo N, Suzumura A, Marunouchi T (1989). Production of tumor necrosis factor-alpha by microglia and astrocytes in culture. *Brain Res* 491:394–397.

Schelper RL, Adrian EK, Jr. (1986). Monocytes become macrophages; they do not become microglia: a light and electron microscopic autoradiographic study using 125-iododeoxyuridine. *J Neuropathol Exp Neurol* 45:1–19.

Schultz-Cherry S, Ribeiro S, Gentry L, Murphy-Ullrich JE (1994). Thrombospondin binds and activates the small and large forms of latent transforming growth factor-beta in a chemically defined system. *J Biol Chem* 269:26775–26782.

Schwab ME, Kapfhammer JP, Bandtlow CE (1993). Inhibitors of neurite growth. *Annu Rev Neurosci* 16:565–595.

Semple-Rowland SL, Mahatme A, Popovich PG, Green DA, Hassler G, Jr., Stokes BT, Streit WJ (1995). Analysis of TGF-beta 1 gene expression in contused rat spinal cord using quantitative RT-PCR. *J Neurotrauma* 12:1003–1014.

Shimojo M, Nakajima K, Takei N, Hamanoue M, Kohsaka S (1991). Production of basic fibroblast growth factor in cultured rat brain microglia. *Neurosci Lett* 123:229–231.

Shuman SL, Bresnahan JC, Beattie MS (1997). Apoptosis of microglia and oligodendrocytes after spinal cord contusion in rats. *J Neurosci Res* 50:798–808.

Sievers J, Struckhoff G, Puchner M (1993). Interleukin-1 beta does not induce reactive astrogliosis, neovascularization or scar formation in the immature rat brain. *Int J Dev Neurosci* 11:281–293.

Smith GM, Hale JH (1997). Macrophage/Microglia regulation of astrocytic tenascin: synergistic action of transforming growth factor-beta and basic fibroblast growth factor. *J Neurosci* 17:9624–9633.

Stichel CC, Muller HW (1994). Extensive and long-lasting changes of glial cells following transection of the postcommissural fornix in the adult rat. *Glia* 10:89–100.

Streit WJ (1993). Microglial-neuronal interactions. *J Chem Neuroanat* 6:261–266.

Streit WJ, Graeber MB, Kreutzberg GW (1988a). Functional plasticity of microglia: a review. *Glia* 1:301–307.

Streit WJ, Kincaid-Colton CA (1995). The brain's immune system. *Sci Am* 273:38–43.

Streit WJ, Kreutzberg GW (1988b). Response of endogenous glial cells to motor neuron degeneration induced by toxic ricin. *J Comp Neurol* 268:248–263.

Streit WJ, Walter SA, Pennell NA (1999). Reactive microgliosis. *Prog Neurobiol* 57:563–581.

Thanos S, Mey J, Wild M (1993). Treatment of the adult retina with microglia-suppressing factors retards axotomy-induced neuronal degradation and enhances axonal regeneration in vivo and in vitro. *J Neurosci* 13:455–466.

Theele DP, Streit WJ (1993). A chronicle of microglial ontogeny. *Glia* 7:5–8.

Xu XM, Guenard V, Kleitman N, Aebischer P, Bunge MB (1995a). A combination of BDNF and NT-3 promotes supraspinal axonal regeneration into Schwann cell grafts in adult rat thoracic spinal cord. *Exp Neurol* 134:261–272.

Xu XM, Guenard V, Kleitman N, Bunge MB (1995b). Axonal regeneration into Schwann cell-seeded guidance channels grafted into transected adult rat spinal cord. *J Comp Neurol* 351:145–160.

Zajicek JP, Wing M, Scolding NJ, Compston DA (1992). Interactions between oligodendrocytes and microglia. A major role for complement and tumour necrosis factor in oligodendrocyte adherence and killing. *Brain* 115:1611–1631.

11

Opportunities for Axon Repair in the CNS: Use of Microglia and Biopolymer Compositions

JOSHUA B. STOPEK, WOLFGANG J. STREIT, AND EUGENE P. GOLDBERG

Introduction

The failure of severed CNS axons to regenerate and to reconnect after injury is thought to be due to both insufficient expression of pro-regenerative genes by central neurons, and to an inhibitory microenvironment at the lesion site. Regarding the latter, astroglial scarring and presence of inhibitory myelin proteins are believed to represent major obstacles to axon regeneration. In addition, the inflammatory response within the CNS white matter is slow to develop, especially with regard to macrophage recruitment, and it is likely too weak to trigger sufficient local production of required growth-promoting molecules (Perry et al. (1987); Schwartz et al. (1999)). Transplantation of fetal neural tissue has been studied as one method for filling cystic spinal cord lesions, and has shown some promise for promoting functional recovery (Diener and Bregman (1998); Bregman et al. (1997); Anderson et al. (1995), Reier et al. (1994); Bregman et al. (1989)). However, while fetal tissue transplantation may allow filling of a lesion cavity, host fibers in adult animals tend to terminate within fetal transplants, rather than grow across them (Reier et al. (1983)). The same problem may also arise with neural stem cell transplants, which are increasingly being investigated for their benefit in spinal cord repair (Cao et al. (2001); Liu et al. (2000)).

In recent years, it has been reported that transplants consisting of microglia and/or macrophages can enhance the weak CNS inflammatory response and spontaneous regenerative axonal growth in the injured spinal cord (Prewitt et al. (1997); Rabchevsky and Streit (1997)). Stimulation of post-traumatic growth of neural processes into such transplants is most likely mediated by trophic factors and extracellular matrix molecules produced by microglia (see Chapters 9, 10). However, growth of axons into these cellular transplants is haphazard and does not usually result in reconnecting proximal axon stumps with their respective distal segments. Thus, there is an obvious need to provide some sort of scaffolding to assist regenerative axonal growth across a lesion cavity.

This chapter combines perspectives from neurobiology and materials science to consider the development of novel bioerodable, implantable devices that might supply the necessary components to facilitate axonal regrowth across a spinal cord lesion. The prototypical implantable device for discussion here consists of a microporous, channeled polysaccharide foam coated with a polymerizable phospholipid biopolymer, on which microglia may be seeded. The strategy is to provide three essential components to foster axon regeneration: (1) temporary structural support and guidance channels in the form of a polysaccharide scaffold; (2) a favorable surface terrain consisting of a synthetic phospholipid polymer; and (3) trophic and tropic support by microglia.

Current biomedical research efforts have focused on utilizing biomaterial constructs combined with tissue-specific cells, either by combining the two components before implantation, or by encouraging cells to populate the construct upon implantation in a host. Biomaterials vary in composition from biologically derived materials to those that are completely synthetic. Biologically derived materials, for example, may include proteins, polysaccharides and lipids. Synthetic biomaterial scaffolds include the use of synthetic polymers such as polyolefins, polyamides, polyesters and polysiloxanes. Bioerodable polymer scaffolds are becoming more prevalent, and are capable of delivering therapeutic agents, as well as hydrolyzing into byproducts that can be removed by noninflammatory mechanisms. Biopolymer scaffolds may be designed to have surface properties for control over a range of cellular functions, including signaling, adhesion, matrix deposition and cytokine/growth factor secretion. Control may be placed over surface and bulk properties, toxicity and erosion rates. Preliminary research in our laboratories shows promise that these devices could additionally assist microglia in their intrinsic capacity to aid mature CNS neurons to initiate and maintain a regeneration response.

Neurotrophic role of microglia

There is a considerable body of data that supports a neurotrophic, and generally pro-regenerative, role for microglia in the post-traumatic microenvironment. Cell culture findings have shown that cultured microglia can secrete growth factors such as NGF (Mallat et al. (1989)), bFGF (Shimojo et al. (1991)) and NT-3 (Elkabes et al. (1996)). In addition, laminin immunoreactivity was found to be colocalized with OX-42 + brain macrophages in microglial grafts (Rabchevsky and Streit (1997, 1998)), lending credence to earlier *in vitro* studies showing that microglia secrete laminin (Rieske et al. (1989)). Microglia have also been shown to produce and secrete thrombospondin *in vitro* and *in vivo* during motor neuron regeneration Müller et al. (1996); Chamak et al. (1994)). Laminin and thrombospondin, both extracellular matrix components, have been shown to promote neuronal regeneration (Masuda-Nakagawa et al. (1993); O'Shea et al. (1991); Neugbauer et al. (1991)). Nagata et al. (1993) demonstrated that microglia derived IL-1 and TGF-β have been shown to induce NGF production in astrocytes and Schwann cells (Yoshida and Gage (1992); Lindholm et al. (1990); Lindholm et al. (1987)), demonstrating an indirect neurotrophic role for these cytokines. The cell culture studies have also lead to a number of *in vivo* transplantation experiments involving microglia or peripheral macrophages. Microglia transplanted into the injured spinal cord in conjunction with fetal neural transplants have been shown to increase regeneration of DRG sensory fibers (Prewitt et al. (1997)). Regeneration of descending tracts accompanied by improved locomotor function was observed when macrophages alone were implanted into the transected spinal cord (Rapalino et al. (1998)). Other authors have observed similar beneficial effects of macrophage transplants following spinal cord injury, and have attributed these primarily to the enhanced removal of myelin debris, production of extracellular matrix molecules, and the promotion of neovascularization by the grafted cells (Franzen et al. (1998)). In other studies, injections of activated peripheral macrophages were used to overcome the inherent nonpermissive nature of the optic nerve for regeneration (Lazarov-Spiegler et al. (1996, 1998)). The results from these transplantation studies are not entirely surprising, since a critical role of macrophages in peripheral nerve regeneration has been known for some time (for review, see George et al. (1993)). One of the primary pro-regenerative functions attributed to peripheral macrophages in the PNS is the recruitment and activation of Schwann cells (George et al. (1993)), which was also observed after intraspinal grafting of purified microglia (Rabchevsky and Streit (1997)). All in all, studies with grafts consisting of microglia/macrophages strongly support the use of these cells for enhancing the generally poor regenerative potential of CNS axons.

Previous use of biopolymers

Previous studies have examined a variety of biomaterials, but most of the studies have only considered the regeneration in PNS. The materials previously explored have consisted of traditional medical-grade plastics and hydrogels. The use of polylactides (Molander et al. (1982); Seckel et al. (1983)), polyurethanes (Hoppen et al. (1990)), poly-(hydroxyethyl methacrylate) (PHEMA) hydrogels (Woerly et al. (1993); Plant et al. (1995)) and silicone elastomers (Fields et al. (1989)) have demonstrated the ability to enhance regeneration by providing some form of support and guidance. However, the main problem associated with these previous substrates is that the materials almost always induce a foreign-body response characterized by fibrous encapsulation, analogous to glial scarring in the CNS.

Plant and Harvey (2000) implanted a polymer/cell construct, which consisted of polycarbonate tubes filled with lens capsule-derived extracellular matrix coated with cultured neonatal Schwann cells. They implanted these devices into lesion cavities made in the left optic tract, and regenerating axons were traced up to 1.7 mm. There was also immunohistochemical evidence of myelination by Schwann cells and by host oligodendrocytes. Young et al. (1984) implanted expanded microporous Gortex to bridge peripheral transections and observed guided nerve regeneration, but there was no evidence of nerve tissue invading the implant wall. This was likely because Gortex induces the infiltration of fibroblasts and subsequent deposition of a fibrous capsule. Cuadros and Granatir (1987) reported similiar results, and additionally observed reduced nerve conduction in Gortex grafts compared with normal nerve, which was attributed to scar formation. Silicone tubes (Fields and Ellisman (1986)) have shown some success in bridging peripheral nerve gaps (<10 mm), however, such permanent guidance channels, which are impermeable to endogenous neurotrophic factors and ECM molecules, do not readily support axon regeneration. Other synthetic guidance tubes composed of laminin, fibronectin and collagen have shown increased regeneration compared to silicone controls (Archibald et al. (1991); Takahashi et al. (1988); Tong et al. (1994)). Houweling et al. (1998) implanted collagen matrices containing NT-3 and reported partial functional recovery despite the absence of regrowth of corticospinal axons into host tissue caudal to the lesion area. Liu et al. (1997) and Joosten et al. (1995) also reported increased axonal regeneration through collagen tubes and gel matrices, respectively. However, it is important to recognize that, since these synthetic materials are permanent fixtures, they may induce chronic inflammation, secondary injury and subsequent cavitation (Dunnen et al. (1995, 1996)).

Porous hydrophilic sponges made of poly-(2-Hydroxyethyl methacrylate) (HEMA) infiltrated *in vitro* with Schwann cells were implanted into the lesioned rat optic tract to promote axonal regeneration (Plant et al. (1995)). Schwann cells were shown to survive the implantation technique and were immunopositive for the low-affinity nerve-growth factor receptor, S100 and laminin. Immunohistochemical studies also showed that non-neuronal cells, including macrophages, migrated into the implants. Immunopositive axons were seen in approximately two-thirds of the implants, and were closely associated with the transplanted Schwann cells and endogenous glia. However, axons did not appear to be myelinated by either transplanted cells, or migrating cells from host tissue (Plant et al. (1995)).

Biodegradable, synthetic polymers that have also been considered include polylactides (Maquet et al. 2000), poly- (DL-lactide-co-glycolide) (PLGA) (Hadlock et al. (2000)), polyurethanes and poly-(organo)phosphazenes. PLGA, the most common selection, has shown use as a nerve-guidance substrate, although cell morphology varied from that of normal nerve (Molander et al. (1982)). Polyester-urethane tubes contained regenerated tissue, primarily in the center of the lumen, and partly composed of myelinated axons and Schwann cells (Borkenhagen et al. (1998)). Agarose hydrogel scaffolds were engineered to stimulate and guide neuronal process extension in three dimensions and have shown promise *in vitro* (Yu et al. (1999); Bellakonda et al. (1995)). Growth-factor-treated nitrocellulose was implanted in complete transection lesions and quantitative analysis indicated that NGF-treated nitrocellulose supported regrowing host axons for nearly three times the length exhibited by axons associated with non-treated nitrocellulose implants. These results indicated that substrate-bound growth factors have the capacity to enhance the regrowth of ascending sensory axons across a traumatic spinal cord injury site (Houle et al. (1994)). Crosslinked aggregates of degradable polylactide microparticles have been synthesized and successfully seeded with DRG neurons (Schugens et al. (1995)), although it is unknown whether the degradation of such a system would lead to a chronic immune reaction owing to the dispersion of particulate debris.

Rationale for the use of phospholipid modified biopolymer surfaces

Biological significance

It is known that synthesis and insertion of phospholipids is essential in nerve fiber elongation (Posse de Chaves (1995)). The bulk of phospholipids is synthesized in the cell body and has to be transported to growth cones at the proximal axon stump, which may be some distance away. It has been demonstrated that phosphatidylcholine is required for axonal growth (Posse de Chaves (1995)). It is therefore reasonable to believe that a phospholipid-coated bioderodable foam could provide a favorable terrain to support neurite extension and growth by adsorbing molecules necessary for regeneration (lipids, proteins, growth factors and extracellular matrix molecules) from both endogenous glia and implanted microglia. Recent studies in our laboratories have demonstrated that polymer surfaces modified with synthetic phosphatidylcholine biopolymers improve surface properties for microglial/PC-12 cell differentation and proliferation, and attenuate production of proinflammatory cytokines.

Synthetic phospholipids

Natural phospholipid bilayers tend to be mechanically weak and unstable. This makes the synthesis of phospholipid bilayers *in vitro* very difficult. In order to develop stable artifical phospholipid membranes, a large number of synthetic phospholipid analogues, many with polymerizable groups, has been studied in the past

25 years. By acting as a biomimetic membrane, synthetic phospholipid polymers offer the possibility of providing an organized surface that would enhance the tribological properties of a compatible substrate (LaBerge (2000)). Novel methods for surface modification of biopolymers with phospholipids have been developed in our laboratory utilizing gamma radiation initiation polymerization (Hydrograft™), radio frequency (RF) plasma polymerization, radical polymerization, and electropolymerization.

Phospholipid surface modification of biomedical polymers has been studied to reduce the thrombogenicity of blood contact devices. Arterial recanalization for stented balloon angioplasty has demonstrated significant restoration of function to diseased coronary and peripheral arteries using catheter-based techniques. Primary failure of stented angioplasty is attributed to neointimal cell growth (restenosis) and inherently thrombogenic luminal stent surfaces (e.g. stainless steel). In small coronary and peripheral arteries, vessel thrombosis and restenosis following stenting occurs in up to 40% of patients within six months. Many attempts to resolve this problem have been studied and a variety of reports suggests favorable properties of polymer-coated metallic stents, compared with bare metal (Widenhouse (1996); Goldberg (1998); Goldberg (1999); Goldberg (2000)).

Ishihara et al. (1991) (Tokyo Medical and Dental Institute) synthesized phosopholipid polymers that mimic nonadhesive, nonactivating, biosurfaces. Polymers containing 2-Methacryloyloxyethyl phosphorylcholine (MPC), on biopolymer substrates demonstrated increased phospholipid adhesion, decreased protein adsorption and decreased platelet activation (Iwasaki et al. (1999); Iwasaki et al. (1999b); Zhang et al. (1998)). These MPC polymer surfaces are thought to improve hemocompatiblity, since increased phosopholipid adsorption tends to inhibit thrombosis. Ishihara et al. (1991, 1998, 1998b) have broadly studied MPC and its copolymers for applications in biomedical membranes. Initial research concerned the adsorption of blood phospholipids, blood proteins and platelet activation of MPC-*co-n*-butylmethacrylate (BMA)-coated acrylic beads, and then *MPC-co-BMA* membranes. Those studies suggested that the MPC-*co*-BMA surfaces were more effective for adsorption of blood phospholipids, decreased protein adsorption and lower platelet activation compared with PHEMA-coated surfaces. DeFife et al. (1995) reported that biopolymer membranes coated with MPC suppressed IL-1β and IL-6 release up to 95% compared with tissue-culture polystyrene. These data suggest that MPC-implant surfaces might inhibit chronic inflammation and, unlike other permanent implants, could act as stable cell carrier/substrate.

Nakabayashi (2000) has also demonstrated that MPC can inhibit high-molecular-weight protein deposition responsible for thrombosis and immune activation. This may be attributed to the affinity of the MPC moiety for endogenous phospholipids, which can be immediately physisorbed to the polymer substrate surface because of their structure (Kojima et al. (1991)). MPC was also shown to have the ability to self-reorganize to form confluent-surface monolayers on a nanometer scale (Nakabayashi (2000)). The surface modification of Dacron vascular grafts was also demonstrated by using a blend of MPC and segmented poly-(etherurethane) (SPU) to dip-coat a small (2 mm) diameter vascular graft (Yoneyama et al. (1998); Furuzono et al. (2000)). *In vivo* experiments in rabbits showed patency after five days, whereas unmodified SPU coated grafts showed "massive thrombosis" within 90 minutes. Furuzono et al. (2000) reported significantly greater reduction in platelet adhesion on coated grafts than in non-modified controls.

In related research, novel surface modification techniques developed in our laboratory applied synthetic polymer analogues of phosphatidylcholine to stainless steel and biopolymer surfaces including silicone, polymethyl methacrylate (PMMA), polyethylene terephthalate (PET), Dacron™, Gortex™, Mylar™, polyvinyl pyrrolidone (PVP), and various anionic polysaccharide compositions.

Rationale for use of polysaccharide scaffolds

Much research has revolved around patterning ligands on erodable templates, especially using the RGD amino acid sequence, to control cell adhesion and direction of growth via specific receptor interactions (Hubbell (1995); Drumheller and Hubbell (1994); Cook et al. (1997)). Photolithographic techniques that use patterned masks to block light, ions or electrons, and microprinting methods have been used to pattern templates on the micron scale (Lee et al. (1993)). These techniques have been encouraging for control of cells on flat surfaces, but have not addressed the complex architectural design of a clinically practical microenvironment in three dimensions.

The preparation and properties of bioerodable, microporous anionic polysaccharide foams and films and viscous solutions have been studied in our laboratory for more than twenty years with broad application (Seeger et al. (1996); Burns et al. (1995); Goldberg et al. (1993); Yaacobi et al. (1993); Yaacobi et al. (1992); Goldberg et al. (1980)). These studies have focused on carboxymethyl cellulose (CMC), hyaluronic acid (HA) and alginates (ALGIN). Solutions, gels and bioerodable films and foams have been shown to protect tissue from injury and prevent postoperative adhesion formation. These abnormal tissue attachments (postoperative adhesions) are a cause of morbidity following abdominal and gynecological surgery, including intestinal obstruction and infertility, and increase the risk of mortality after cardiac surgery (Burns et al. (1997)). HA, which is a natural component of mammalian ECM, has shown efficacy and safety in providing a lubricous, mechanical barrier for reducing the incidence of adhesions (Seeger et al. (1996); Burns et al. (1994); Goldberg (1997a, b,c)). These anionic polysaccharides do not interfere with the wound healing process in the anastomosis, and are nontoxic, nonmutagenic, nonimmunogenic and nonpyrogenic. Porous gels and foams of CMC, HA and ALGIN can be readily crosslinked to varying degrees, and lyophilized to produce a wide range of pore/channel dimensions (10μm–1mm), as well as a variety of controlled erosion rates (1 day–2 months).

Combining microglia with biopolymers

Previous work has shown that, after spinal cord injury, the implantation of microglia into the lesion site can stimulate endogenous regenerative nerve fiber growth, and the recruitment of other pro-regenerative cells such as Schwann cells (Rabchevsky and Streit (1997)). A primary rationale for our experimental strategy has been to provide structural support and a stable terrain that will fill the lesion site and deliver bioactive moieties to direct tissue repair and a favorable host response. Furthermore, seeding microglia onto scaffold surfaces is intended to provide a continuous source of trophic factors and extracellular matrix molecules, as well as cellular guidance for regenerating axons. We believe such implants can enhance the naturally occurring, spontaneous regenerative response, and will promote robust nerve fiber growth into and across the scaffolding structure, leading to functional recovery.

Studies *in vitro* involving the inclusion of MPC polymers in the culture media, as well as seeding of microglia and PC-12 cells onto small pieces of various biopolymers, have indicated that of the many polymers examined, Poly-MPC-coated surfaces best promoted cell growth. These surfaces stimulated adhesion, differentiation, and proliferation of microglia, as well guided growth through the microstructure. Subsequent seeding of PC-12 cells onto these macrophage cultures resulted in excellent survival and differentiation of PC-12, without addition of exogenous NGF. Thus, it is reasonable to believe that PC-12 cell differentiation is facilitated by NGF produced by microglia (e.g. Mallat et al. (1989); Elkabes et al. (1996); Heese et al. (1997)). We have also obtained encouraging results regarding the noninflammatory properties of the Poly-MPC-coated surfaces. These data have shown that lipopolysaccharide (LPS) mediated induction of IL-1β synthesis by microglia is inhibited by inclusion of Poly-MPC in the culture media.

MPC-surface-modified implants promote microglial adhesion and differentiation.

We have investigated the ability of various polymers to support growth of cultured microglia. Poly-MPC-surface-modified anionic polysaccharide implants (based on carboxymethyl cellulose, alginate and hyaluronic acid) were prepared by gamma irradiation polymerization techniques. Other polymers of interest were also seeded with microglia, including silicone, Gortex™, Dacron™, PET, PHEMA, PMMA, PVP, PMAA and various copolymers of the aforementioned. Most of these polymers were selected because of their past use in nerve guide studies. However, they did not readily support cell growth or differentiation. Microglial proliferation was not apparent, and cell morphology was ameboid and markedly different from the process-bearing morphology in the CNS. Poly-MPC-

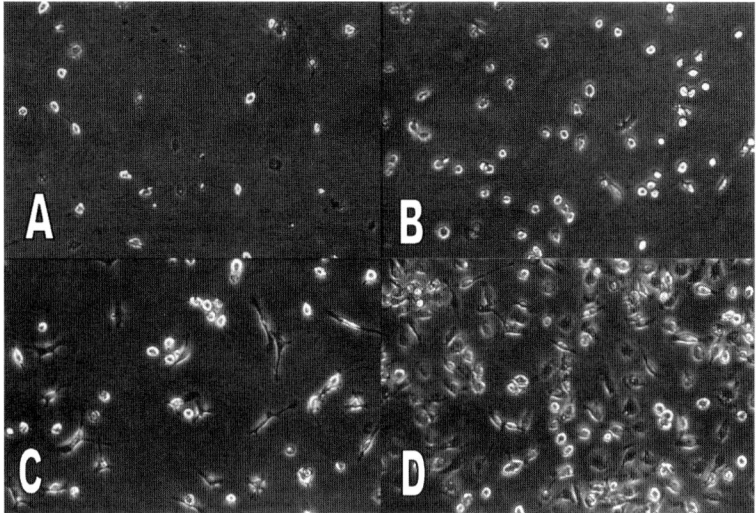

FIGURE 1. Primary rat microglial cells *in vitro* seeded onto polymers (200,000 cells/ml). Panel **A** shows microglia cultured on tissue-culture polystyrene. Panel **B** shows microglia seeded onto anionic polysaccharide foam (no Poly-MPC coating). Panel **C** shows cells seeded onto Poly-MPC surface modified, anionic polysaccharide foam. Notice increased differentiation and process growth, resembling microglial morphology in the CNS environment. Panel **D** shows cells seeded onto an implant modified with a higher surface concentration of Poly-MPC. Notice the robust differentiation and apparent proliferation. Photographs ($\times 200$) were taken 24 hours post-seeding.

coated polysaccharide foams, on the other hand, provided substrates that not only enhanced cell adhesion, but also stimulated cellular differentiation, i.e. microglia showed enhanced process formation and ramification (Figure 1).

PC-12 cells reveal extensive neuritic growth on top of microglia-seeded synthetic phospholipid (MPC) coated foams.

In order to determine whether our composite structures would support neuronal growth and differentiation, PC-12 cells (a common cell line known to develop a neuronal phenotype upon exposure to NGF) were cultured and seeded on scaffolds inundated with microglia. The scaffolds were seeded with PC-12 cells following a one-week incubation period for microglia adhesion and differentiation.

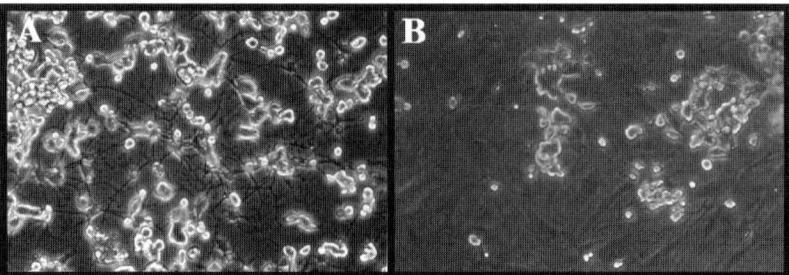

FIGURE 2. **A.** PC-12 differentiation on a Poly-MPC modified anionic polysaccharide foam implant inundated with microglia. **B.** PC-12 cells seeded on top of silicone (PDMS) cultured with microglia.

Materials modified with Poly-MPC clearly showed the most extensive neuritic growth (Figure 2), indicating that the CNS-like, process-bearing morphology of the microglia on the MPC may be accompanied by the upregulation of a number of trophic factors, including NGF, required for growth of neuronal cells.

MPC suppresses the release of pro-inflammatory cytokine IL-1β in LPS-stimulated microglia.

Experiments were conducted to determine whether the presence of MPC would suppress microglial production of the proinflammatory cytokine IL-1β when stimulated by bacterial lipopolysaccharides (LPS). Although stimulation with LPS has long been used to generate activated, IL-1β-producing microglia *in vitro* (Hetier et al. (1988); Yao et al. (1992), it is important to understand that IL-1β production by activated microglia *in vivo* following acute spinal cord injury or stroke is a transient phenomenon (Streit et al. (1998); Pearson et al. (1999)). In contrast, prolonged production of IL-1β occurs during chronic inflammation, and has been associated with neurodegenerative disease states (Mrak and Griffin (2000)). Thus, when considering implantation of biomaterials into the CNS, it is important that the implanted material be lowly inflammatogenic, in order to minimize secondary damage and scarring. Our data shows that MPC attenuates production of IL-1β by LPS-activated microglial cells, and suggest that an MPC-coated scaffold would not elicit a chronic and potentially detrimental immune response upon implantation into the injured spinal cord.

Microporous polysaccharides can provide channeled scaffolding for directed growth.

One of the main challenges associated with our current research is to influence the spatial organization and growth of neurons, not only cell proliferation and differentiation, but facilitation of directed cell growth. This behavior is likely governed by a mix of cell-to-cell interactions, and receptor-mediated interactions between cell-adhesion ligands and substrate surfaces. The unique composite discussed here may yield an implant with a surface designed to enhance natural cell surface interactions. The Poly-MPC phospholipid surface of the implants is intended to facilitate the physiosorption of endogenous lipids and proteins, as well as the alignment of cells necessary to promote axon regeneration, in a manner similar to vascular wound healing (Ishihara et al. (2000). This requires the formation of a sophisticated implant, with a favorable geometry (capable of steering pioneering axons, as well as a more biomimetic terrain) coupled with a substrate (microporous polysaccharide) that can be slowly removed by hydrolysis and dissolution.

The SEM micrographs in Figures 3a and 3b show the microstructure of the lyophilized CMC-foam substrate. These scaffold compositions were cast from 1% CMC solutions. The scaffold pore size varies from 30 to 150 microns. It is important to understand that these scaffold compositions are very hydrophilic, and will fill the lesion site if sized properly. The bioerodability of the scaffold is related to the crosslink density, which may be controlled by crosslinker concentration in the polymerization media. For pratical clinical application, bioerosion rates should afford residence times of 1–10 weeks. Implants with higher crosslink densities will solvate and hydrolyze more slowly.

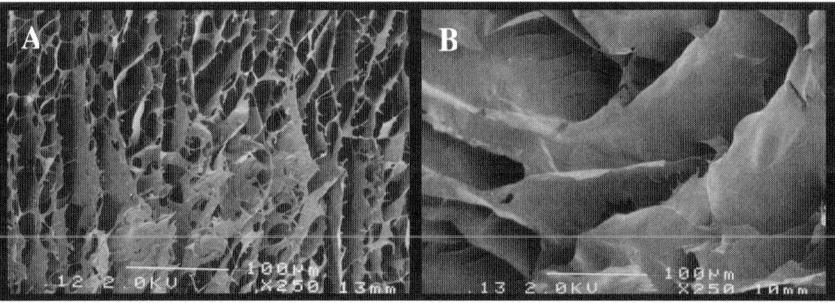

FIGURE 3. **A.** SEM cross-section of lyophilized anionic polysaccharide foam (carboxymethyl cellulose) implant, revealing channel structure. **B.** Detail of cross-section, portraying pore opening and smoothness of channel wall.

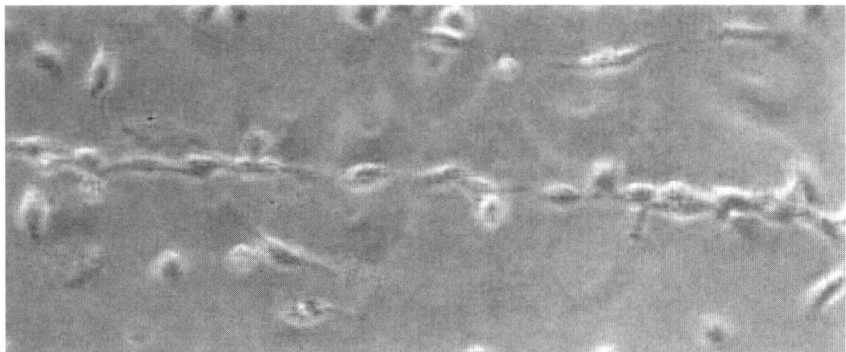

FIGURE 4. Channels in the scaffold provide means of directed growth and three-dimensional spatial organization, which is shown in the alignment of microglia deep within the scaffold channel structure.

Hydrophilicity of the Poly-MPC-polysaccharide composition may also permit the adsorption and subsequent release of trophic factors and ECM molecules. Thus, delivery of neurotrophic factors will not necessarily be limited to the presence of microglia. The coating of the luminal walls of the channels and pores with growth factors, which would be free to diffuse out, also appears feasible in this system. Diffusion of such bioactive molecules is normally based on a concentration gradient, so that areas lacking particular growth factors will readily receive them until equilibrium concentrations are achieved. Overall, these Poly-MPC-modified implants should afford tissue gentle, cell-carrying devices that also should be capable of loading and local release of molecules necessary for regeneration.

To date, it has been established that the architecture of the channeled CMC foam supports the alignment of cells within the implant (Figure 4). It should not be surprising to find regenerating nerve fibers that are guided along paths created by microglia-derived growth factors and extracellular matrix molecules.

Conclusion

Based on studies to date, the continuing investigation of microporous and bioerodable Poly-MPC-modified polysaccharides appears to be a very promising approach for achieving significant advances aimed at the repair of damaged nervous tissue.

Literature cited

Anderson, DK, Howland, DR and Reier, PJ (1995). Fetal neural grafts and repair of the injured spinal cord. *Brain Pathol* 5:451–457.

Archibald, SJ, Krarup, C, Shefner, J, Li, ST, Madison, RD (1991). A collagen-based nerve conduit for peripheral nerve repair, an electrophysiological study of nerve regeneration in rodents and nonhuman primates. *J Comp Neurol* 4:685–696.

Bellakonda, R, Ranieri, JP, Bouche, N, Aebischer (1995). Hydrogel-based three-dimensional matrix for neural cells. *J Biomed Mater Res* 5:663–71.

Borkenhagen, M, Stoll, RC, Neunschwander, P, Suter, UW, Aebischer, P (1998). In vivo performance of a new biodegradable polyester urethane system used as a nerve guidance channel. *Biomaterials* 23:2155–65.

Bregman, BS, Kunkel, B, McAtee, EM and O'Neill, A (1989). Extension of the critical period for developmental plasticity of the corticospinal pathway. *J Comp Neurol* 282:355–370.

Bregman, BS, Diener, PS, McAtee, M, Dai, HN and James, C (1997). Intervention strategies to enhance anatomical plasticity and recovery of function after spinal cord injury. *Adv Neurol* 72:257–275.

Burns, JW, Skinner, K, Colt, J, Sheidlin, A, Bronson, R, Yaacobi, Y, Goldberg, EP (1995). Prevention of Tissue Injury and Postsurgical Adhesions by Precoating Tissues with Hyaluronic Acid *J Surg Res* 59:644–652.

Burns, JW, Colt, MJ, Burgess, LS, Skinner, KC (1997). Preclinical evaluation of seprafilm bioresorbable membrane. *Eur J Surg* 577:40–48.

Cao QI, Zhang YP, Howard RM, Walters WM, Tsoulfas P, Whittemore SR (2001). Pluripotent stem cells engrafted into the normal or lesioned adult rat spinal cord are restricted to a glial lineage. *Exp Neurol* 167:48–58.

Chamak B, Morandi V and Mallat M (1994). Brain macrophages stimulate neurite growth and regeneration by secreting thrombospondin. *J Neurosci Res* 38:221–233.

Cuadros, CL, Grantatir, CE (1987). Nerve regeneration through a synthetic microporous tube (expanded polytetrafluoroethylene): Experimental study in the sciatic nerve of the rat. *Microsurgery* 8:41–46.

Cook, AD, Hrkach, JS, Gao, NN, Johnson, IM, Pajvani, UB, Cannizzaro, SM, Langer, R (1997). Characterization and development of RGD-peptide-modified poly (lactic acid-co-lysine) as an interactive, resorbable biomaterial. *J Biomed Mater Res* 35:513–523.

DeFife, KM, Yun, JK, Azeez, A, Stack, S, Ishihara, K, Nakabayashi, N, Colton, E, Anderson, JM (1995). Adhesion and cytokine production of monocytes on poly(2-methacryloyloxyethyl phosphocholine-co-alkyl methacrylate)-coated polymers. *J Bio Mat Res* 29:431–439.

Diener, PS and Bregman, BS (1998). Fetal spinal cord transplants support growth of supraspinal and segmental projections after cervical spinal cord hemisection in the neonatal rat. *J Neurosci* 18:779–793.

Drumheller, PD, Hubbell, JA (1994). Polymer networks with grafted cell-adhesion peptides for highly biospecific cell adhesive substrates. *Anal Biochem* 222:380–388.

Dunnen, den WFA, Van der Lei, B, Robinson, PH, Holwerda, A, Pennings, AJ, Schakenraad, JM (1995). The Biological performance of a degradable poly(lactic acid-ϵ-caprolactone) nerve guide: influence of tube dimensions. *J Biomed Mater Res* 29:757–766.

Dunnen, den WFA, Stokroos, I, Blaauw, EH, Holwerda, A, Pennings, AJ, Robinson, PH, Schakenraad, JM (1996). Light-microscopic and electron-microscopic evaluation of short-term nerve regeneration using a biodegradable poly(DL-lactid-ϵ-caprolactone) nerve guide. *J Biomed Mater Res* 31: 105–115.

Elkabes, S, DiCicco-Bloom, EM and Black, IB (1996). Brain microglia/macrophages express neurotrophins that selectively regulate microglial proliferation and function. *J Neurosci* 16:2508–2521.

Fields, RD, Le Beau, JM, Longo, FM, Ellisman, MH (1989). Nerve regeneration through artificial tubular implants. *Progr Neurobiol* 33:87–134.

Fields, RD, Ellisman, MH (1986). Axons regenerated through silicone tube splices. I. Conduction properties. *Exp Neurol* 92:48–60.

Furuzono, T, Ishihara, K, Nakabayashi, N, Tamada, Y (2000). Chemical modification of silk fibroin with 2-methacryloyloxyethyl phosphorylcholine. II. Graft-polymerization onto fabric through 2-methacryloyloxyethyl isocyanate and interaction between fabric and platelets. *Biomaterials* 21:327–333.

Franzen, R, Schoenen, J, Leprince, P, Joosten, E, Moonen, G and Martin, D (1998). Effects of macrophage transplantation in the injured adult rat spinal cord: a combined immunocytochemical and biochemical study. *J Neurosci Res* 51:316–327.

Goldberg, EP, Sheets, JW, Habal, M (1980). Peritoneal Adhesions: Prevention with the use of Hydrophilic Polymer Coatings. *Arch Surg* 115:776–780.

Goldberg, EP, Burns, JW, Yaacobi, Y (1993). Prevention of postoperative adhesions by precoating tissues with dilute sodium hyaluronate solutions. *Prog Clin Biol Res* 381:191–204.

Goldberg, EP (1997a). *Pelvic Surgery: Adhesion Formation and Prevention.* New York: Springer-Verlag.

Goldberg, EP (1997b). *Advances in Biomedical Polymers: Problems and Opportunities for Ophthalmic, Cardiovascular and Mammary Implants, and Tissue Protective Surgical Devices.* Soc of Plastics Eng

Goldberg, EP (1997c). *Protection of Tissues during Surgery with Polymer Solution Coatings:* A New Concept. Intnl Congress-Peritoneal Tissue Repair.

Goldberg, EP (1998). *Polydimethylsiloxane (PDMS) coatings for stainless steel endovascular stents: uniform, stable, highly adherent coatings for reduced thrombogenicity and drug delivery.* Soc for Biomaterials.

Goldberg, EP (1999). *Medical device surface modification by pulsed laser ablation deposition (PLAD) of silicone onto stainless steel.* Soc for Biomaterials.

Goldberg, EP (2000). *Phospholipid and silicone modification of metal implant surfaces by electropolymerization.* Soc for Biomaterials, 6th World Congress.

Griffin, JW, George, R and Ho, T (1993). Macrophage systems in peripheral nerves. A review. *J Neuropath Exp Neurol* 52:553–560.

Hadlock, T, Sunback, C, Hunter, D, Cheney, M and Vacanti, JP (2000). A polymer foam conduit seeded with schwann cells promotes guided peripheral nerve regeneration. *Tissue Eng* 6:119–27.

Heese, K, Fiebich, BL, Bauer, J and Otten, U (1998). NF-kappa B modulates lipopolysaccharide-induced microglial nerve growth factor expression. *Glia* 22:401–407.

Hetier, E, Ayala, J, Denèfle, P, Bousseau, A, Rouget, P, Mallat, M and Prochiantz, A (1988). Brain macrophages synthesize interleukin-1 and interleukin-1 mRNAs in vitro. *J Neurosci Res* 21:391–397.

Hoppen, HJ, Leenslag, JW, Pennings, AJ, Van Der Lei, B, Robinson, PH (1990). Two-ply biodegradable nerve guide: Basic aspects of design, construction and biological performance. *Biomaterials* 11:286–290.

Houle, JD and Zeigler, MK (1994). Bridging a complete transection lesion of adult rat spinal cord with growth factor-treated nitrocellulose implants. *J Neural Transplant Plast* 5:115–24.

Houweling, DA, Lankhorst, AJ, Gispen, WH, Bar, PR and Joosten, PA (1998). Collagen containing neurotrophin-3 (NT-3) attracts regrowing injured corticospinal axons in adult rat spinal cord and promotes partial functional recovery. *Exp Neurol* 153:49–59.

Hubbell, JA (1995). Biomaterials in tissue engineering. *Bio-technology* 13:565–576.

Ishihara, K, Ziats, NP, Tierney, BP, Nakabayashi, N, Anderson, JM (1991). *J Bio Mat Res* 25:1397–1407.

Ishihara, K, Iwasaki, Y, Nakabayashi, N (1998). Novel biomedical polymers for regulating serious biological reactions. *Mats Sci & Eng* C6:253–259.

Ishihara, K, Nomura, H, Mihara, T, Kurita, K, Iwasaki, Y, Nakabayashi, N (1998b). Why do phospholipid polymers reduce protein adsorption? *J Biomed Mater Res* 39:323–330.

Ishihara, K (2000). Chemistry of Phospholipid Polymers as Biomaterials. *Phospholipid Biomaterials Soc* for Biomaterials.

Iwasaki, Y, Ijuin, M, Mikami, A, Nakabayashi, N, Ishihara, K (1999). Behavior of blood cells in contact with water soluble phospholipid polymer. *J Biomed Mater Res* 46:360–367.

Iwasaki, Y, Sawada, S, Nakabayashi, N, Khang, G, Lee, HB, Ishihara, K (1999b). The effect of the chemical structure of the phospholipid polymer on fibronectin adsorption and fibroblast adhesion on the gradient phospholipid surface. *Biomaterials* 20:2185–2191.

Joosten, EA, Bar, PR, Gispen, WH (1995). Collagen implants and cortico-spinal axonal growth after mid-thoracic spinal cord lesion in the adult rat. *J Neurosci Res* 41:481–90.

Kojima, M, Ishihara, K, Watanabe, A, Nakabayashi, N (1991). Interaction between biocompatible phospholipid polymer and phospholipids. *Biomaterials.* 12:121–124.

LaBerge, M (2000). Investigation of the tribological properties of phospholipid polymers. *Phospholipid Polymer Biomaterials,* Society of Biomaterials 42–104.

Lazarov-Spiegler, O, Soloman, AS, Zeev-Brann, AB, Hirschberg, DL, Lavie, V and Schwartz, M (1996). Transplantation of activated macrophages overcomes central nervous system regrowth failure. *FASEB J* 10:1296–1302.

Lazarov-Spiegler, O, Solomon, AS and Schwartz, M (1998). Peripheral nerve-stimulated macrophages stimulate a peripheral nerve-like regenerative response in rat transected optic nerve. *Glia* 24:329–337.

Lee, J, Kaibara, M, Iwaki, M, Sasabe, H, Suszuki, Y, Kusakabe, M (1993). Selective adhesion and proliferation of cells on ion-implanted polymer domains. *Biomaterials* 14:958–960.

Lindholm, D, Heumann, R, Meyer, M and Thoenen, H (1987). Interleukin-1 regulates synthesis of nerve growth factor in non-neuronal cells of rat sciatic nerve. *Nature* 330:658–659.

Lindholm, D, Heumann, R, Meyer M and Thoenen, H (1990). Transforming growth factor-β1 stimulates expression of nerve growth factor in the rat CNS. *Neuroreport* 1:9–12.

Liu, S, Peulve, P, Jin, O, Boisset, N, Tollier, J, Said, G and Tadie, M (1997). Axonal regrowth through collagen tubes bridging the spinal cord to nerve roots. *J Neurosci Res* 49:425–432.

Liu, S, Qu, Y, Steward TJ, Howard MJ, Chakrabortty, S, Holekamp, TF, McDonald, JW (2000). Embryonic stem cells differentiate into oligodendrocytes and myelinate in culture and after spinal cord transplantation. *Proc Natl Acad Sci USA* 97:6126–6131.

Mallat, M, Houlgatte, R, Brachet, P, and Prochiantz, A (1989). Lipopolysaccharide-stimulated rat brain macrophages release NGF in vitro. *Dev Biol* 133:309–311.

Maquet, V, Martin, D, Malgrange, B, Franzen, R, Schoenen, J, Moonen, G and Jerome, R (2000). Peripheral nerve regeneration using bioresorbable macroporous polylactide scaffolds. *J Biomed Mater Res* 52:639–651.

Masuda-Nakagawa, LM, Muller, KJ and Nicholls, JG (1993). Axonal sprouting and laminin appearance after destruction of glial sheaths. *Proc Natl Acad Sci USA* 90:4966–4970.

Molander, H, Olsson, Y, Engkvist, O (1982). Regeneration of peripheral nerve through a polylactin tube. *Muscle Nerve* 5:54–57.

Mrak, RE and Griffin, WS (2000). Interleukin-1 and the immunogenetics of Alzheimer disease, *J Neuropathol Exp Neurol*, 59:471–476.

Müller, JC, Klein, MA, Haas, S, Jones, L, Kreutzberg, GW, Raivich G. (1996). Regulation of thrombospondin in the regenerating mouse facial motor nucleus. *Glia* 17:121–132.

Nagata, K, Takei, N, Nakajima K, Saito H and Kohsaka S. (1993). Microglial conditioned medium promotes survival and development of cultured mesencephalic neurons from embryonic rat brain. *J Neurosci Res* 34:357–363.

Nakabayashi, N (2000). Concept and history of phospholipid polymer biomaterials. *Phospholipid Polymer Biomaterials,* Society of Biomaterials 6–14.

Neugebauer, KM, Emmett, CJ, Venstrom, KA and Reichardt, LF (1991). Vitronectin and thrombospondin promote retinal neurite outgrowth: developmental regulation and role of integrins. *Neuron* 6:345–358.

O'Shea, KS, Liu, LH and Dixit, VM (1991). Thrombospondin and a 140 kd fragment promote adhesion and neurite outgrowth from embryonic central and peripheral neurons and from PC12 cells. *Neuron* 7:231–237.

Pearson, VL, Rothwell, NJ and Toulmond, S (1999). Excitotoxic brain damage in the rat induces interleukin-1β protein in microglia and astrocytes: correlation with the progression of cell death. *Glia* 25:311–323.

Perry, VH, Brown, MC, Gordon, S (1987). The macrophage response to central and peripheral nerve injury. *J Exp Med* 165:1218–1223.

Plant, GW and Harvery, AR (2000). A new type of biocompatible bridging structure supports axon regrowth after implantation into lesioned rat optic tract. *Cell Transplant* 9:759–772.

Plant, GW, Harvey, AR, Chirila, TV (1995). Axonal growth within poly (2-hydroxyethyl methacrylate) sponges infiltrated with Schwann cells and implanted into the lesioned rat optic tract. *Brain Res* 671:119–130.

Posse de Chaves, EI, Rusinol, AE, Vance, DE, Campenot, RB, Vance, JE (1997). Role of lipoproteins in the delivery of lipids to axons during axonal regeneration. *J Biol Chem* 49:30766–30773.

Prewitt, CM, Niesman, IR, Kane, CJ. and Houlé, JD (1997). Activated macrophage/microglial cells can promote the regeneration of sensory axons into the injured spinal cord. *Exp Neurol* 148:433–443.

Rabchevsky, AG and Streit, WJ (1997). Grafting of cultured microglial cells into the lesioned spinal cord of adult rats enhances neurite outgrowth. *J Neurosci Res* 47:34–48.

Rabchevsky, AG and Streit, WJ (1998). Role of microglia in post-injury repair and regeneration of the CNS. *Mental Retard Develop Disab Res Rev* 4:187–192.

Rapalino, O, Lazarov-Spiegler, O, Agranov, E, Velan, GJ, Yoles, E, Fraidakis, M, Solomon, A, Gepstein, R, Katz, A, Belkin, M, Hadani, M and Schwartz, M (1998). Implantation of stimulated homologous macrophages results in partial recovery of paraplegic rats. *Nature Med* 4:814–821.

Reier, PJ, Stensaas, LJ and Guth, L (1983). The astrocytic scar as an impediment to regeneration in the central nervous system. In *Spinal Cord Reconstruction,* eds. CC Kao, RP Bunge and PJ Reier, pp. 163–195. New York: Raven Press.

Reier, PJ, Anderson, DK, Schrimsher, GW, Bao, J, Friedman, RM, Ritz, LA and Stokes, BT (1994). Neural cell grafting: Anatomical and functional repair of the spinal cord. In *The neurobiology of central nervous system trauma,* eds. SK Salzman and AI Faden, pp. 288–311. New York: Oxford University Press.

Rieske, E, Graeber, MB, Tetzlaff, W, Czlonkowska, A, Streit, WJ and Kreutzberg GW (1989). Microglia and microglia derived brain macrophages in culture: generation from axotomized facial nuclei, identification and characterization *in vitro. Brain Res* 492:1–14.

Schwartz, M, Moalem, G, Leibowitz-Amit, R, Cohen, IR (1999). Innate and adaptive immune responses can be beneficial for CNS repair. *Trends Neurosci* 22:295–299.

Schugens, CH, Grandfils, CH, Jerome, R, Teyssie, PH, Delree, P, Martin, D, Malgrange, B, Moonen, G (1995). Preparation of a macroporous biodegradable polylactide implant for neuronal transplantation *J Biomed Biomat Res* 29:1349–1362.

Seckel, BR, Chiu, TH, Nylias, E, Sidman, RL (1983). Nerve regeneration through synthetic biodegradable nerve guides: Regulation by the target organ. *Plastic Reconstr Surg* 74:173–181.

Seeger, JM, Kaelin, LD, Staples, EM, Yaacobi, Y, Bailey, JC, Normann, S, Burns, JW, Goldberg, EP (1996). Prevention of postoperative pericardial adhesions using tissue-protective solutions. *J Surg Res* 68:63–66.

Shimojo, M, Nakajima, K, Takei, N, Hamanoue, M, and Kohsaka, S (1991). Production of basic fibroblast growth factor in cultured rat brain microglia. *Neurosci Lett* 123:229–231.

Streit, WJ, Semple-Rowland, SL, Hurley, SD, Miller, RC, Popovich, PG, and Stokes, BT (1998). Cytokine mRNA profiles in contused spinal cord and axotomized facial nucleus suggest a beneficial role for inflammation and gliosis. *Exp Neurol* 152:74–87.

Takahashi, M, Satou, T, Hashimoto, S (1988). In vivo regeneration of peripheral nerve axons and perineurium guided by resorbable collagen film. *Acta Pathol Jpn* 38:1489–1502.

Tong, X, Hirai, K, Shimada, H (1994). Sciatic nerve regeneration navigated by laminin-fibronectin double coated biodegradable collagen grafts in rats. *Brain Res* 663:155–162.

Widenhouse, CW (1996). *Surface Modification of Vascular Prosthesis and Intracorneal Lens Polymers.* Ph.D. dissertation, University of Florida.

Woerly, S, Maghami, G, Duncan, R, Subr, V, Ulbrich, K (1993). Synthetic polymer derivatives as substrata for neuronal adhesion and growth. *Brain Res Bull* 30:423–432.

Yaacobi, Y, Israel, AA, Goldberg, EP (1993). Prevention of postoperative abdominal adhesions by tissue precoating with polymer solutions. *J Surg Res* 55:422.

Yaacobi, Y, Latif, MH, Kaul, K, Maher, MF (1992). Reduction of postoperative adhesions secondary to strabismus surgery in rabbits. *Ophthalmic Surg* 23:123.

Yao, J, Keri, JE, Taffs, RE, Colton, CA (1992). Characterization of interleukin-1 production by microglia in culture. *Brain Research* 591:88–93.

Yoneyama, T, Ishihara, K, Nakabayashi, N, Ito, M, Mishima, Y (1998). Chemical modification of silk fibroin with 2-methacryloyloxyethyl phosphorylcholine. II. Graft-polymerization onto fabric through 2-methacryloyloxyethyl isocyanate and interaction between fabric and platelets. *J Biomed Mater Res* 43:15–20.

Yoshida K and Gage FH (1992). Cooperative regulation of nerve growth factor synthesis and secretion in fibroblasts and astrocytes by fibroblast growth factor and other cytokines. *Brain Res* 569:14–25.

Young, BL, Begovac, DG, Stuart, DG, Goslow, GE (1984). An effective sleeving technique in nerve repair. *J Neurosci Meth* 10:51–58.

Yu, X, Dillon, GP, Bellamkonda, RB (1999). A laminin and nerve growth factor-laden three-dimensional scaffold for enhanced neurite extension. *Tissue Eng* 4:291–304.

Zhang, SF, Rolfe, P, Wright, G, Lian, W, Milling, AJ, Tanaka, S, Ishihara, K (1998). Physical and biological properties of compound membranes incorporating a copolymer with a phosphorycholine head group. *Biomaterials* 19:691–700.

12

Beta Amyloid Protein Clearance and Microglial Activation

SALLY A. FRAUTSCHY, GREG M. COLE, AND MARCH D. ARD

1. Introduction

A. *The problem of amyloid β-peptide (Aβ) accumulation in aging and Alzheimer's disease (AD): Beyond Aβ production*

Progression of AD involves a slow accumulation of Aβ peptide deposited extracellularly in the neuropil and vasculature of the brain. Aβ peptides of MW from 4 to 5 kD (Aβ40, Aβ42 or Aβ43) are normally produced by many cells, but the mutations in early onset familial AD (fAD) cause increased production of the rapidly aggregating Aβ1-42 (Borchelt et al. (1997); Younkin (1995)). However, in approximately 95% of cases of AD, there is Aβ accumulation without genetically increased Aβ production. Thus, in the vast majority of AD cases, which arise out of interaction of aging with genetic risk factors, other aspects of Aβ metabolism appear to be important. For example, the increased risk, and earlier onset, of AD from the apolipoprotein (Apo) E4 allele, which is strongest between 65 and 80 years of age, is not associated with increased Aβ production, but rather with reduced Aβ clearance or enhanced amyloid formation. Other potential genetic risk factors for late-onset AD may also influence AD pathogenesis at levels beyond Aβ production, including alpha-2 macroglobulin (Liao et al. 1998), alpha-1 antichymotrypsin (Thome et al. (1995)), interleukin 1 (Nicoll et al. (2000)), and transforming growth factor beta (Luedecking et al. (2000)). Although factors regulating Aβ degradation and clearance have received little attention compared with factors regulating Aβ production associated with early-onset AD genes, there is now a large enough literature to consider the issues. This review will focus on the removal and degradation of Aβ and Aβ aggregates by microglia, and briefly cover the degradation of soluble Aβ by known proteases.

B. Is Aβ the Culprit Causing AD?

Although extensive evidence suggests that Aβ is probably playing a major causal role in AD, details of its pathogenic role remain elusive. The extensive controversy over which form of Aβ is involved and how it induces neuronal damage falls into four categories. Is neuronal damage in AD caused by: (1) the Aβ deposited as fibrils in plaques; (2) a circulating complex of Aβ oligomers; (3) Aβ on or in neurons; (4) an indirect effect on neurons mediated by chronic inflammation? Although most of the controversy relates to which form of Aβ is problematic, a few researchers believe that Aβ is not so much problematic as it is a response to damage. In truth, the ultimate involvement of Aβ in AD is likely to be a multifactorial pathway involving all the processes described above, as suggested by the following evidence.

Aβ is believed by most to be causal because (a) the familial AD mutations presenilin (PS) 1, PS2 and amyloid precursor protein (APP) are associated with overproduction of Aβ42 (Borchelt et al. (1996); Gomez-Isla et al. (1997); Holcomb et al. (1998); Scheuner et al. (1996); Younkin, (1995)); (b) the major risk factor ApoE4 is associated with over-accumulation of Aβ40 (Ishii et al. (1997); Mann et al. (1997)); (c) Aβ can be toxic in culture (Pike et al. (1997); Pike et al. (1991)) and *in vivo* (Weldon et al. (1998)) (c) overexpression of the APP mutations that increase Aβ lead to neuritic pathology (Games et al. (1995)) and cognitive deficits in transgenic mice (Chen et al. (2000); Hsiao et al. (1996); Janus et al. (2000); Morgan et al. (2000)); and (d) Aβ accumulation correlates with clinical decline (Naslund et al. (2000)). However, controversy as to the magnitude of Aβ's role in AD is created by the following observations: (a) aged individuals with major Aβ deposition and no cognitive deficits can be found; (b) Aβ is not necessarily toxic, and can even be neurotrophic (Whitson et al. (1989); Yankner et al. (1990)); and (c) overexpression of APP in mice does not lead to neurofibrillary tangles unless human tau is overexpressed (Lewis et al., 2001), or unless amyloid fibrils are injected into tau transgenics (Gotz et al., 2001); nor does APP overexpression lead to major neuron loss (Irizarry et al. (1997a); Irizarry et al. (1997b)).

These data do not argue against a role for Aβ in AD, rather they may argue against an exclusive role for the amyloid deposits per se (as opposed to another form of Aβ). This is because: (a) animal models of plaque deposition can show behavioral deficits before extensive deposit formation, or with no deposit formation (Chen et al. (2000); Holcomb et al. (1999); Janus et al. (2000)); (b) there is a subset of non-cognitively-impaired people having high Aβ deposits (Benzing et al. (1993)); and (c) deposits in humans do not always correlate with dementia (Terry et al. (1991)). These data raise doubts about the amyloid plaques, posing a clinical problem, and instead suggest the involvement of another form of Aβ, such as a soluble oligomer playing a causal role. Recent reports from studies on Alzheimer brains (Kuo et al. (1996); Lue et al. (1999)); rodent neurons *in vitro* (Oda et al. (1995)), and rat infusion models *in vivo* (Frautschy, Harris-White,

Finch, Morgan, et al., unpublished) suggest that, unlike amyloid plaques or deposits, soluble Aβ correlates well with neuron damage and memory loss. This may be related to observations that neuronal uptake of soluble Aβ precedes extracellular deposition and toxicity (LaFerla et al. (1997)). Even if soluble Aβ is the major player, it does not mean that amyloid plaques are not players at all. Current data would argue that, even if plaques do not directly damage neurons, they are likely to activate cascades that damage neurons by their induction of a chronic focal inflammatory response (Chao et al. (1996)); (Akiyama et al. (2000)) and aberrant sprouting (Geddes et al. (1986)). A relatively uncharted territory of Aβ action that may prove to be important is whether the observation of Aβ accumulation on neuron plasma membrane in humans with AD (Probst et al. (1991)) and Down's syndrome (Allsop et al. (1989)), in dogs (Torp et al. (2000)) and in Aβ-infused rats (Frautschy et al. (1996)) is provoking neuronal damage. Regardless of how and which Aβ induces damage, a better understanding of how it is cleared is likely to be clinically relevant.

2. *Clearance of soluble Aβ in vitro*

Aβ is produced by cleavage of the β-amyloid precursor protein (APP) with the N-terminal and C-terminal cuts by β- and γ-secretases, respectively. Alternatively, Aβ production can be prevented by another endopeptidase, α-secretase, that cleaves within the Aβ domain. The major pool of soluble Aβ is released from the cells and can be found as a soluble protein in media and biological fluids, including plasma and cerebrospinal fluid. Since this pool of extracellular, soluble Aβ is not accumulating in normal young brains or systemic pools, production must be in equilibrium with clearance. Metabolic labeling studies in APP transgenic mice with a 30 minute labeling pulse suggested a rough estimate of t1/2 for the immunoprecipitable pool of endogenous Aβ between 1 and 2.5 hours based on assumptions of a steady-state equilibrium with the rate of synthesis (Savage et al. (1998)). Injection of small amounts of soluble radio-labeled Aβ into rat hippocampus is followed by a rapid degradation of the Aβ by extracellular proteases with a t1/2 for Aβ of less than one hour (Iwata et al. (2000)). A number of proteases have been shown to degrade Aβ, and are suggested to play a role in its clearance.

A. *Matrix metalloproteases (MMPS)*

Once secreted, Aβ can bind to extracellular matrix (ECM) proteins, including collagen (El Khoury et al. (1996)) and heparan sulfate proteoglycans (Castillo et al. (1997)). Since prolonged residence in the ECM could be an early phase of deposition, MMPs were initially considered strong candidates for degradation of secreted Aβ. For example, MMP-9 can readily cleave Aβ after the leucine at Aβ34

(Backstrom et al. (1996)). MMPs are typically downregulated by transforming-growth factor β (TGFβ) during scar formation after injury, suggesting that TGFβ could promote amyloid deposition by inhibiting ECM degradation. Since CNS Aβ production is primarily neuronal, and MMP-9 is expressed in neurons, one might expect an important role for MMP-9. However, studies of MMP-9 knock-out mice have shown limited and inconclusive evidence of Aβ accumulation (G.P. Lim, personal communication). Crosses with MMP knock-outs with APP transgenics would be required to demonstrate a significant role for MMPs in Aβ degradation *in vivo*.

B. Insulin degrading enzyme (IDE)

A series of studies from D. Selkoe's lab have shown that microglia and neurons secrete a metalloprotease, insulin degrading enzyme or "insulyin" that can degrade soluble Aβ *in vitro*, (Vekrellis et al. (2000b)). This enzyme is expressed in a secretory pathway in CNS neurons and may be capable of degrading Aβ before secretion, or even outside of cells (Vekrellis et al. (2000a)). Whether insulyin makes substantial contributions to Aβ degradation *in vivo* remains uncertain.

C. Neprilysin

Beginning with an *in vivo* approach, T. Saido's lab stereotaxically injected Aβ labeled at multiple sites into rat hippocampus and analyzed the principal fragments produced with, and without, a panel of protease inhibitors. They found rapid loss of intact Aβ (t1/2 < 1 hour) and recovery of proteolytic fragments, demonstrating that proteolysis had occurred (Iwata et al. (2000)). Analysis of fragments and inhibitor studies led them to conclude a major role for neprilysin (also known as neutral endopeptidase, or CD10/EC24.11). Aβ infusions and co-infusions with the neprilysin inhibitor, thiorphan, confirmed that Aβ accumulated when neprilysin was inhibited. Very recent work from this group has demonstrated that crosses between APPsw transgenic mice (Tg2576) and neprilysin knock-out mice result in Aβ deposit formation as early as eight weeks of age (Iwata et al. (2001)). Because neprilysin was found in detergent insoluble glycolipid rich (lipid raft) compartments that are known sites of Aβ accumulation, neprilysin is a strong candidate for a physiologically relevant Aβ-degrading enzyme.

The relative roles of neprilysin and insulyin remain controversial because Selkoe's lab has not been able to confirm reduced Aβ clearance in neprilysin knock-out mice, and a demonstration of an *in vivo* impact of blocking insulyin degradation requires additional studies. Whether an upregulation of either neprilysin or insulyin would be a clinically useful target for slowing Aβ accumulation may depend as much on the impact on the levels of other physiologically important peptide substrates for these enzymes as on the ability to reduce soluble Aβ levels.

D. Carrier-associated Aβ

A large pool of soluble Aβ is normally found bound to high-density lipoprotein, (HDL)-like lipoprotein carriers containing ApoE, ApoJ and possible other apolipoproteins. Because we have recently reviewed the literature on this topic (Cole, Ard (2000)), it will only be briefly summarized here. Radio-iodinated Aβ bound to carriers such as ApoJ may be transcytosed through endothelial cells and cleared to the vasculature (Zlokovic et al. (1994)). Aβ bound to ApoE with ApoE isoform-dependent avidity can be taken up by cells, notably through the low-density lipoprotein receptor-related protein (LRP) on neurons (Holtzman et al. (1995)), and multiple receptors on microglia and related brain macrophages (discussed below). Aβ clearance and degradation, as well as amyloid fibril formation, appear to be ApoE isoform-dependent and related to ApoE effects on AD risk and age of onset.

E. Role of the lysosome

Aβ applied to cultured cells such as fibroblasts or PC12 cells accumulates in the endosomal/lysosomal fraction (Burdick et al. (1997)). A percentage of the Aβ is degraded in lysosomes, but the low pH in lysosomes favors Aβ aggregation and the accumulation of aggregates of both exogenous Aβ and endogenous APP and its C-terminal fragments (Yang et al. (1995)). The major upregulation of neuronal cathepsins in AD may be caused by an accumulation of endosomal/lysosomal Aβ (Cataldo et al. (1995)). However, evidence for intraneuronal Aβ or Aβ fragment accumulation has been difficult to distinguish from the accumulation of APP and non-Aβ fragments. Neuronal receptors for Aβ also remain controversial with putative roles for receptor for advanced glycation end product (RAGE) (Yan et al. (1996, 1997)), alpha-7 nicotinic acetylcholine receptor (Wang et al. (2000)) and lipoprotein receptor related protein (LRP) for lipoprotein or alpha-2 macroglobulin-associated Aβ. This suggests that soluble Aβ arriving in lysosomes would normally be rapidly degraded, but that aggregates may accumulate, or be resecreted.

3. Clearance of Aβ aggregates *in vitro*

Cultured microglia readily accumulate Aβ aggregates in various experimental paradigms. In one of the first *in vitro* papers, rat microglia were plated on Aβ-coated tissue-culture plates (Shaffer et al. (1995)). In this system, microglia readily reduced the total Aβ on the plates, and this apparent degradation was inhibited by factors from astrocyte-conditioned media. In our own studies, soluble Aβ1-42 was applied in the media, but at aggregating concentrations. Ultrastructural studies demonstrated accumulation of Aβ epitopes in lysosomes and phagolysosomes when lysosomal proteolysis was blocked. Western analysis of media and cell pellets indicated both Aβ degradation and cell-dependent accumulation of Aβ multimers or oligomers, even in the absence of lysosomal protease inhibitors (Ard et al. (1996)). Because Aβ degradation was promoted by calf serum, more experiments were conducted to determine the possible role of lipoprotein carriers (Cole et al. (1999)). Additional ultrastructural studies showed accumulations of Aβ in microglial surface connected compartments (SCC), which are known to be caveolin-containing sites enriched in lipid and apoE. We suggested that these SCC may be one microcompartment where Aβ oligomers can accumulate, but Aβ was also seen in secondary lysosomes. In the presence of ApoJ or HDL, net Aβ degradation was enhanced, whereas in the presence of ApoE, Aβ degradation was reduced. Again, lysosomal protease inhibitors blocked this Aβ degradation. In cells exposed to HDL-Aβ, fibrils were essentially absent, and the limited Aβ immunoreactivity present was found in membranous vacuoles probably representing phagolysosomes (Cole et al. (1999))(Fig. 1). These results would be consistent with more or less efficient Aβ degradation dependent on the lipoprotein carriers.

Although microglia readily accumulated Aβ into lysosomal compartments and degraded Aβ applied to cells, surprisingly they failed to either phagocytose or degrade Aβ when plated on unfixed cryostat sections of AD brain (Ard et al. (1996)), as shown in (Fig 2). In a series of papers discussed below, Maxfield's group has demonstrated that preformed iodinated Aβ fibrils are readily recognized by cultured microglia and endocytosed via scavenger-receptor-dependent mechanisms. Cultured microglia appear to be capable of phagocytosing and slowly degrading larger aggregates, and even isolated plaque cores (Dewitt et al. (1998)). In the latter study, phagocytosis of plaque cores was inhibited by astrocyte factors. In summary, microglia *in vitro* are capable of Aβ recognition and uptake, whether presented on the surface of dishes, in the media as preformed fibrils, or isolated plaques. Degradation can occur, but is influenced by aggregation state. Unknown factors from astrocytes or cryostat sections can inhibit this process, raising the question whether phagocytosis and clearance of amyloid occurs normally at any stage of AD pathogenesis, and whether it can be stimulated to slow Aβ accumulation.

4. *Aβ clearance* in vivo

In AD brain microglia are frequently clustered in and around amyloid deposits. Ultrastructural studies have demonstrated that the microglial processes are so closely interdigitated with amyloid fibrils that the amyloid may appear to be within microglia. Nevertheless, the amyloid is typically not found within lysosomes, in other phagocytic structures in microglia, suggesting that phagocytosis of amyloid is blocked (Wisniewski et al. (1989)). In contrast, after stroke or injury, invading macrophages clearly phagocytose, and apparently degrade, amy-

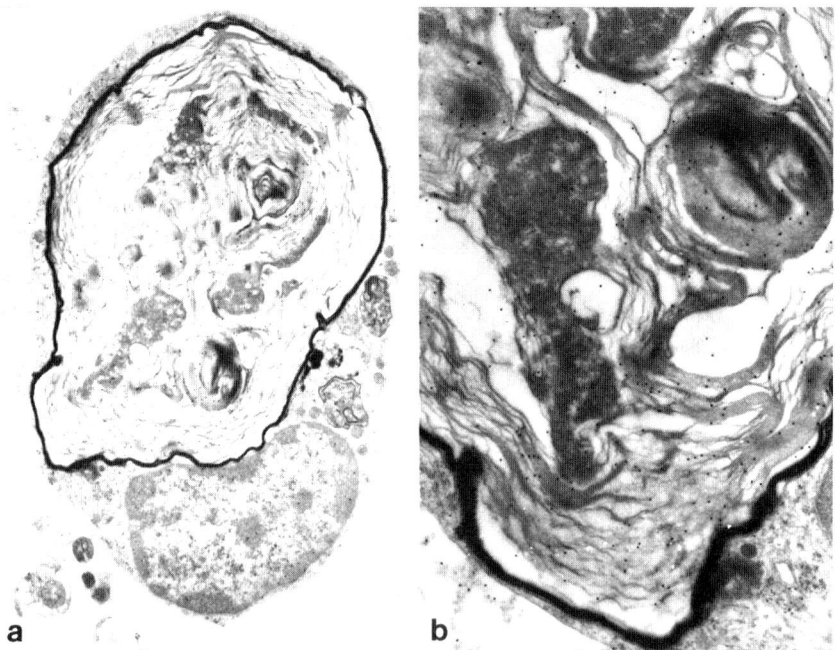

FIGURE 1. Immunogold labeling of Aβ in primary cultures of rat microglia. Cells were incubated overnight (20 hrs) with 7.5µg/ml Aβ1–42 in the presence of 20µg human plasma HDL carrier. Panel A (original magnification, 10,000X.) shows the large membranous inclusions with scattered immunogold labeling better seen at higher magnification (20,000X) in Panel B. (*continued*)

loid fibrils (Wisniewski et al. (1991)). Why phagocytosis of amyloid appears to be blocked in activated microglia surrounding plaques, but permitted in closely-related invading macrophages, can now be addressed in animal models.

A. *Aβ injection and infusion models*

Following cortical or hippocampal injection of isolated plaque cores or amyloid fibrils prepared from cores, we found aggressive phagocytosis of the amyloid and apparent attempted clearance to vessels, ventricles, choroid plexus and meninges (Frautschy et al. (1992)). Phagocytosis of injected synthetic Aβ or amyloid has also been reported by other groups (Games et al. (1992); Weldon et al. (1998)), but few have commented on attempted clearance. In our experience, one month after infusion of Aβ40, Aβ-ir phagocytes are often observed lining the ventricle (Fig. 3A,B). Under these conditions, brain macrophages appear to sequester Aβ and concentrate it, without being completely successful in degrading it. After

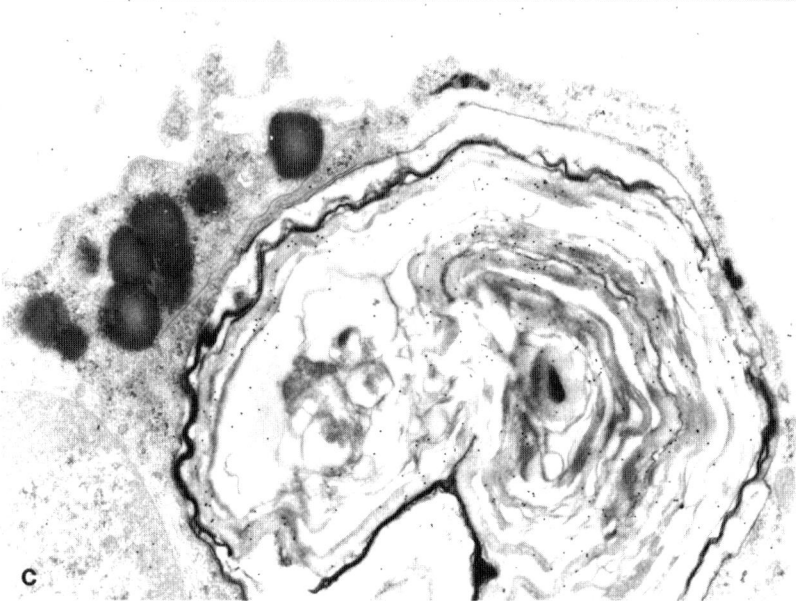

FIGURE 1. (*continued*) In addition to the Aβ immunoreactive membranous vacuoles, cells exposed to HDL also contained electron-dense lipid droplets, as seen to the left of the Aβ labeled vacuole in Panel C. In these experiments, fibrillar Aβ was rarely seen, presumably because the majority of Aβ was bound to the lipoprotein carrier which was supplied in excess (Cole et al. (1999)).

accumulation of indigestible particulates, they attempt to exit the brain into ventricles or vessels. Using other methods, we have also observed evidence of an apparent concentration and accumulation of Aβ by microglia with incomplete degradation. Thus, even sixteen months after cortical injection of 1 μg of soluble Aβ42, Aβ-ir could be seen in microglia at the injection site (Fig. 3C,D). Ultrastructural studies suggested attempted clearance of injected Aβ to vasculature similar to what had previously been observed after injection of plaque cores (Frautschy et al. (1992)). Fig. 4 illustrates the frequent presence of phagocytic microglia adjacent to, or even within, vessels in our rat infusion studies.

Following chronic ventricular infusion of Aβ using Alzet pumps and intraventricular cannulas, we have found aggressive phagocytosis of Aβ (Frautschy et al. (1994)). This appeared to play an important role in preventing plaques because co-injection of a single dose of the anti-inflammatory cytokine TGFβ1 resulted in an inhibition of clearance by phagocytes and the formation of plaque-like deposits (Frautschy et al. (1996)). Although these phagocytes bore monocytic lineage markers, we were not certain whether they were all invading macrophages, or also activated microglia. In both Aβ injection and infusion models, initial injury may

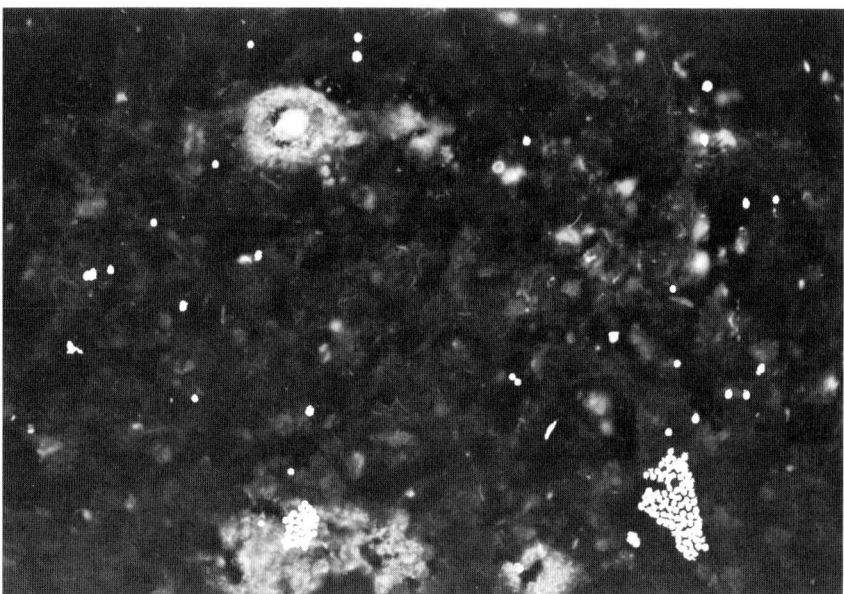

FIGURE 2. Failure of rat microglia to phagocytose plaques from AD brain sections: rat microglia were plated on unfixed cryostat sections from AD brain. Bright fluorescent 3 μM latex beads indicate the location and competence of the phagocytic primary rodent microglia that engulfed them in the final 2 hrs before fixation and immunostaining for Aβ. Intact Aβ deposits are clearly visible; one with a bead-laden phagocytic cell sitting on top of it. While microglia readily phagocytosed aggregated Aβ on culture plates, in these experiments they failed to phagocytose and remove amyloid plaques from sections (Ard et al. (1996)). Original magnification 300X.

allow an influx of macrophages and monocytes to enter the brain, and microglial activation followed by phagocytosis and attempted clearance of amyloid.

B. Transgenic models

As in AD brain, amyloid deposits in aging APP transgenic mice are frequently surrounded by microglia in various activation states (Bornemann, Staufenbiel (2000); Frautschy et al. (1998); Stalder et al. (1999)). However, again as in AD brain, whatever their morphological and immunological indices of activation, the majority of these microglia do not appear to be phagocytosing amyloid. Ultrastructural studies have clearly established this in APPsw mice (M. Staufenbiel, personal communication.) However, several months after entorhinal cortex lesion of amyloid-laden aged PDAPP transgenic mice, microglia in the hippocampal outer molecular layer appear further activated, and amyloid deposits are significantly reduced (Chen

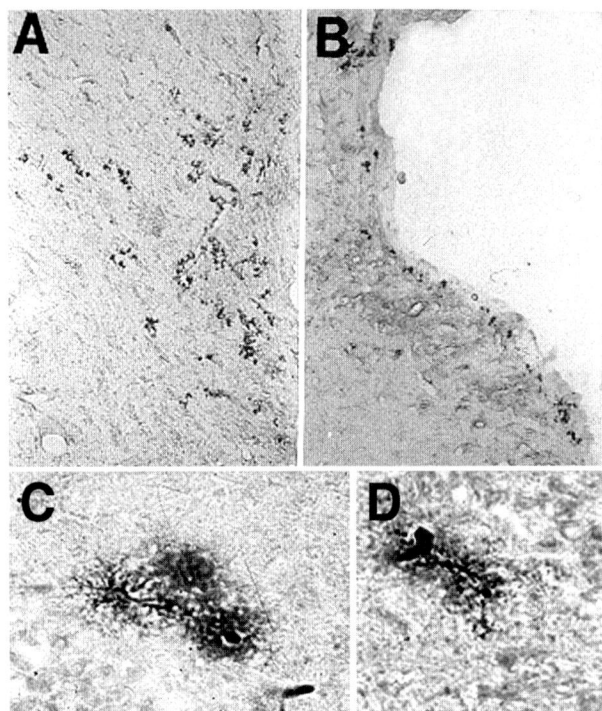

FIGURE 3. Aβ-immunoreactivity in microglial-like cells 1 month and 16 months after Aβ injection or infusion. Aβ immunostaining following intracerebroventricular infusions of Aβ40 into rat ventricles reveals aggressive phagocytosis of Aβ by clusters of amoeboid phagocytes (Panel A). These cells, which can be labeled with OX-42 and other microglia markers (not shown), are frequently seen lining the ventricles (Panel B) and entering, or on vessels, as previously described (Frautschy et al. (1992)). This apparent attempt at Aβ removal was only partly successful, as shown in panels C and D, which illustrate Aβ immunoreactive ramified microglia and deposits 16 months after injection of 1 μg soluble Aβ42 into the hippocampus using methods previously described (Winkler et al. (1994)).

et al. (1998)). Similarly, amyloid deposits in APP transgenics may be reduced by cortical injury, but there is little evidence that microglial clearance of Aβ is a normal physiological process. One piece of evidence that this is so comes from our unpublished work in collaboration with K. Hsiao Ashe on an APP transgenic mouse line with behavioral deficits, but without Aβ deposits. In aging animals from that line, Aβ-ir phagocytes were typically observed streaming through neocortex and lining the third ventricle. Their apparent effective Aβ clearance might explain the lack of deposits in this mouse despite high levels of APP expression (Fig. 5A,B). Microglial streaming or alignment presumably reflects chemotactic responses. *In vitro,* we found rat microglia fed ApoE4 and Aβ1-42 and labeled with fluorescent anti-Aβ antibody were occasionally similarly aligned in their attempt to phagocytose

12. Beta Amyloid Protein Clearance and Microglial Activation   255

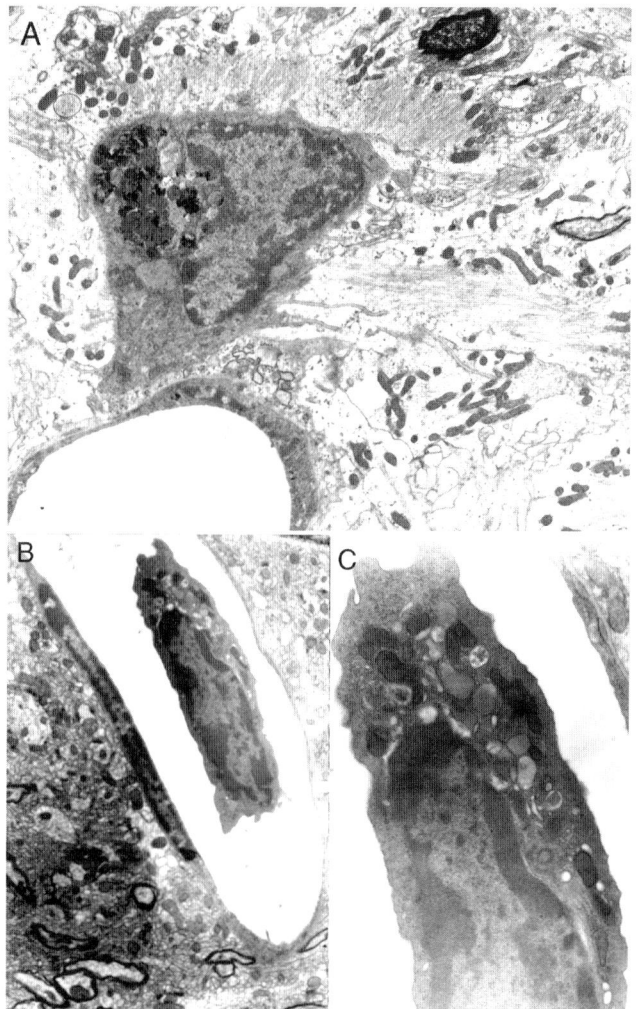

FIGURE 4. Ultrastructure of vessel associated microglia at injection site. In rat infusion studies, phagocytic microglia or macrophages laden with secondary lysosomes and debris can be seen migrating to vessels (Panel A, magnification 3300x) and occasionally entering them (Panel B, magnification 3300x) and at higher magnification (Panel C, magnification 16000x). Presumably, these phagocytic cells correspond to those previously observed carrying injected Aβ and migrating to vessels and ventricles at the light level (Frautschy et al. (1992, 1996)).

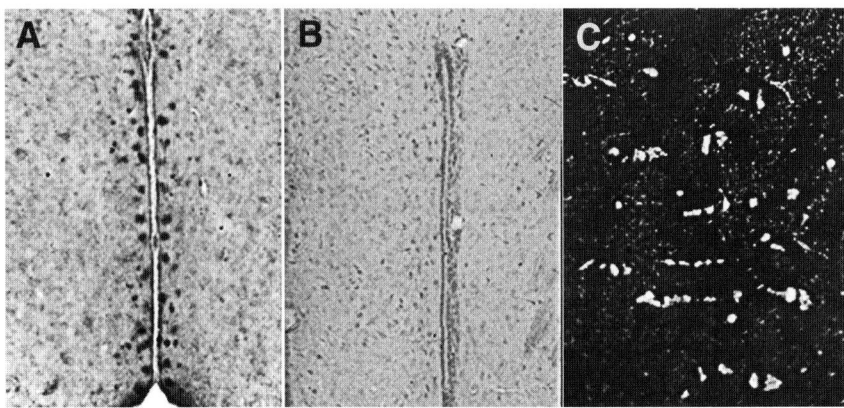

FIGURE 5. Aβ in microglial-like cells of non-plaque-forming APP transgenic mice and in cultured rat microglia: Aβ immunostaining in a low expressing APP transgenic mouse line in FVB background (Hsiao et al. (1995)) with greater than one year survival revealed Aβ-immunoreactive microglia frequently aligned along the ventricles (Panel A). Such staining was not seen with Aβ absorption (Panel B), or in young animals (not shown). Panel C shows fluorescent Aβ immunostaining of arrays of cultured rat microglia on a "lawn" of Aβ aggregates *in vitro*. Similar to Aβ-laden microglia aligning along the ventricles *in vivo*, there is an aligning of motile Aβ-laden rat microglia in cultures prepared as in (Cole et al. (1999)) and exposed to aggregating concentrations of Aβ1–42 in the absence of serum or lipoprotein carriers. These microglia are ameboid, and presumably they are responding to chemotactic signals *in vivo* and *in vitro*. Original Magnification, 100X.

the lawn of deposited Aβ (bright background) (Fig. 5C). Collectively, these results argue that amyloid phagocytosis and clearance (via degradation or transport) by macrophage lineage cells can occur *in vitro* and *in vivo*, but may be normally blocked in AD brain. This raises the question: How do microglia recognize and phagocytose amyloid and other particulates?

5. Aβ receptors on microglia for scavenging and activation

Microglia and related brain macrophages (pericytes, Mato cells) are the major players in removal of particulates. Just as systemic tissue macrophages phagocytize particulates (bacteria, viruses, dust, diesel particles, iron filings, fluorescent beads, etc.), microglia would be expected to attempt to phagocytize and remove Aβ. From this perspective, amyloid accumulation must be accompanied by a failure of this particulate-removal system, which sequentially involves receptor recognition, engulfment, fusion with lysosomes and upregulation of lysosomal hydrolase synthesis. Ideally, pharmaceutical control of Aβ uptake would (1) facilitate clearance of Aβ by microglia, (2) avoid targeting Aβ to neuronal receptors, to limit neurotoxicity, and (3) target microglial receptors that mediate endocytosis without stimulating an inflammatory response. At this point, knowledge of the CNS receptors that bind Aβ is not yet sufficient to craft such an intervention; however, advances are rapidly being made.

A. Receptors that recognize Aβ

Microglia and neurons appear to have several different receptors that recognize Aβ. The predominant cell and Aβ receptor targeted likely depends on many factors, such as a change in the ratio of Aβ42 to Aβ40, or a change from soluble to aggregated form, or changes in lipoprotein carrier content (apoE vs. apoJ). Knowledge of the microglial receptors that normally bind and clear Aβ, and of the form or complex in which they recognize it, may lead to therapies that improve Aβ clearance by taking advantage of receptor properties. For example, cholesterol-lowering drugs or diet might improve the lipoprotein profile in the CNS, affecting clearance of Aβ-lipoprotein complexes.

To date, studies of microglial binding and uptake of Aβ have mostly used free Aβ, not complexed with lipoprotein, which may not be physiologically significant. Aβ recovered from CSF is mostly bound to apoE- and apoJ-containing lipoproteins (Ghiso et al. (1993); Holtzman et al. (1999); Koudinov et al. (1996); LaDu et al. (2000)). On the other hand, in some cases serum was included in the incubation medium, which allows formation of Aβ-lipoprotein complexes during the incubation with cells (Biere et al. (1996)). Further, microglia themselves may synthesize and secrete small amounts of apoE-containing lipoprotein (Stone et al. (1997)).

Receptors for Aβ have generally been demonstrated by competition studies in which appropriate receptor antagonists partly block cellular responses to Aβ. Scavenger receptors of class A are the best-documented microglial receptor for Aβ aggregates. In the human CNS, scavenger receptor is expressed only by microglia and upregulated in senile plaques or brain injury (Bell et al. (1994); Christie et al. (1996b); Honda et al. (1998)). *In vitro,* scavenger receptor mediates Aβ uptake and free-radical production (El Khoury et al. (1996), Paresce et al. (1997, 1996)). Although aggregated Aβ1-42 was endocytosed rapidly and moved to late endosomes/lysosomes, it accumulated in large Aβ-containing granules, instead of being degraded (Paresce et al. (1996, 1997)). Macrophage scavenger receptors comprise class A and class B, with multiple molecularly identified receptors occurring within each class (Yamada et al. (1998)). Scavenger receptors of class B on microglia have greater affinity for methylated bovine serum albumin than fucoidan, and also appear to bind and internalize Aβ aggregates (Paresce et al. (1996)).

The fate of internalized Aβ may differ depending on which scavenger receptor is involved, and on whether the microglial scavenger receptor ferries Aβ into the cell alone, or Aβ is chaperoned by lipoproteins or apolipoproteins. Understanding differential trafficking of lipoproteins and their contents may provide clues to the less thoroughly investigated Aβ trafficking. Scavenger receptor BI mediates selective uptake of HDL lipids while leaving the apolipoproteins free to recirculate extracellularly (Acton et al. (1996)). Other scavenger receptors transport whole lipoproteins into endosomes, where some apolipoproteins, such as apoB, may be degraded, whereas others, such as apoE, may be recirculated to the extracellular fluid. Aβ alone, or bound to lipoprotein carriers, appeared to enter endosomes and accumulate in a late-endosomal compartment (Ard et al. (1996); Chung et al. (1999); Cole et al. (1999)).

However, whereas uptake has been relatively well studied, less is known about degradation. Fibrillar Aβ was not readily degraded and remained in intracellular granules in the microglial cytoplasm for days or weeks (Frackowiak et al. (1992)); from there it was slowly released intact into the extracellular space (Chung et al. (1999)). It is unclear whether amyloid can be completely degraded by cultured microglia, since HPLC analysis of fragments reveals primarily N-terminal cleavage (Chung et al. (2000)). In contrast to amyloid, soluble Aβ was taken up in a nonsaturable manner, presumably by fluid-phase pinocytosis, rather than a receptor-mediated mechanism (Chung et al. (1999)). Chung and colleagues suggested that most of the Aβ taken up by the scavenger receptor or by pinocytosis is re-released in hours, only partially degraded. Others have reported that Aβ taken up is no longer detected by ELISA after overnight incubation, and this apparent degradation is reduced by lysosomal inhibitors (Chu et al. (1998)). Partial degradation may destroy epitopes required for ELISA.

A recent *in vivo* study of APP-transgenic mice found that double-mutant mice combining APP-transgene expression with knock-out of scavenger receptor A showed no measurable difference in the amount of amyloid plaques deposited, compared with mice with intact scavenger receptor (Huang et al. (1999)). It is unclear whether this signifies that scavenger receptor A is not involved in Aβ degradation, or whether multiple receptors can clear Aβ, including scavenger receptor A.

B. Fc-mediated Aβ uptake

Stimulated Aβ uptake can be Fc-mediated using antisera to Aβ (see Section 6 on vaccine).

C. Lipoprotein receptors and Aβ uptake

Various lipoprotein receptors have been identified on microglia, and these may mediate soluble Aβ uptake bound to different classes of lipoproteins. Aβ purified from CSF is bound to apoE- and apoJ-containing lipoproteins (Ghiso et al. (1993); Holtzman et al. (1999); Koudinov et al. (1996); LaDu et al. (2000)). Microglia express the apoE- lipoprotein receptors VLDL receptor (Christie et al. (1996a)), as well as the scavenger receptor for oxidized or modified lipoproteins (Bell et al. (1994); Christie et al. (1996b); Honda et al. (1998)). Rat, but not human, microglia express the low-density lipoprotein receptor-related protein (LRP) (Marzolo et al. (2000); Rebeck et al. (1993)). While there has been substantial investigation of microglial Aβ clearance via scavenger receptors, reports of microglial uptake of Aβ via VLDL receptor are not available. LRP was tested by Paresce and colleagues (Paresce et al. 1996) as a receptor for aggregated Aβ; using the LRP antagonist RAP, these investigators found no inhibition of Aβ uptake in newborn mouse microglia cultures. In this approach, experimental conditions might not have been favorable for LRP-mediated uptake, since the labeling medium did not contain exogenous lipoprotein. Involvement of lipoprotein receptors in Aβ clearance is suggested by the finding that apoE modulates Aβ uptake and accumulation by cultured rat microglia (Cole et al. 1999). Thus, although it is likely that lipoprotein receptors mediate soluble Aβ uptake, more work is needed to identify them and their link to Aβ degradation.

D. Receptors for Aβ-mediated microglial activation

In addition to clearing Aβ from their environment by endocytosing and degrading it, microglia also respond to Aβ by inflammatory activation. Aβ can increase nitric oxide in microglial cell lines primed with interferon (Meda et al. (1995)), but this effect can be prevented by factors such as estrogens, and exacerbated by factors such as glucocorticoids (Harris-White et al. (2001)). It is likely that differential effects of Aβ on activation and toxicity of microglia can occur, depending on prior state of activation and milieu (Korotzer et al. (1993)). In fact, Aβ itself has anti-inflammatory sequences homologous to those in TGFβ, and may under certain environments be anti-inflammatory (Huang et al. (1998)).

At this point, it is unclear whether the same or different receptors mediate inflammation and Aβ degradation. So, manipulating either microglial expression of receptors, or the form of Aβ and Aβ complexes, holds the potential to improve clearance while reducing inflammation, mitigating two sources of potential neurotoxicity in AD. The receptor for advanced glycation end-products (RAGE) is a member of the immunoglobulin superfamily, and is expressed by both microglia and neurons; its expression is increased in AD (Yan et al. (1996); Yan et al. (2000)). Soluble Aβ, as well as Aβ immobilized by adsorption, binds specifically to RAGE (Yan et al. (1996)) and is reported to mediate microglial activation, as shown by haptotactic migration, NF-kappaB activation, and upregulation of TNF-alpha mRNA and protein in the mouse microglial cell line BV-2. Internalization or clearance of the ligand was not assessed. However, it has been suggested that RAGE is not the predominant receptor for Aβ activation (McDonald et al. (1998); McDonald et al. (1997)) in primary microglia from newborn rat brain and the THP-1 human monocyte cell line, which both have low levels of scavenger receptors. RAGE and scavenger ligands failed to mimic the superoxide radical generation elicited by fibrillar Aβ, suggesting other receptors play the principal role in mediating these effects.

The microglial integrin MAC-1 (comprising two subunits, CD11b and CD18) has been found to bind Aβ25-35 (Goodwin et al. (1997)) and mediate nitric oxide release from rat microglia. Binding of Aβ peptide fragment appeared to occur without internalization, since it was unchanged by cytochalasin B. Finally, the chemotactic formyl peptide receptor has been reported to mediate microglial release of IL-1β in response to Aβ aggregates (Lorton et al. (2000)); however, Aβ binding and possible internalization were not specifically studied in these experiments.

6. Regulation of Aβ clearance *in vivo*

Until we know whether it is feasible to stimulate the proteases regulating soluble Aβ levels, therapeutic approaches will be directed at stimulating glial clearance of Aβ. That this approach is feasible has been demonstrated by the developers of the Aβ vaccine.

A. *Vaccine*

Schenk and his collaborators have demonstrated the feasibility of preventing and even reducing pre-existing amyloid deposits in PDAPP transgenic mice by systemic vaccination with preaggregated Aβ1-42 (Schenk et al. (1999)). In a subsequent publication, they demonstrated that the same results can be achieved by using passive Aβ antibody treatment (Bard et al. (2000)). These data on Fc-mediated clearance of Aβ *in vivo* are unprecedented, in that they overturn the dogma that antibodies cannot effectively cross the blood-brain barrier. Further, it has recently been reported that behavioral deficits are alleviated in several APP transgenic lines that were vaccinated before deposition. Involvement of the Fc receptor in microglial clearance of Aβ aggregates has also been characterized in culture. Microglia take up 1.5-fold more fibrillar Aβ microaggregates when they are coated with IgG (Brazil et al. (2000)). An interesting point from this study is that enhanced uptake facilitated by IgG did not improve degradation of the Aβ aggregates, which proceeded only slowly. Recently, it has also been reported that, in the presence of suboptimal levels of Aβ antisera, Fc-mediated phagocytosis is enhanced by C1q, although normally Fc-mediated phagocytosis does not occur *in vitro* without blocking scavenger receptor (Webster and Tenner, unpublished observations). This suggests sensitivity of this pathway is highly variable and dependent on modulating factors. Although Fc stimulation is tightly coupled to NADPH oxidase activation leading to toxic free-radical secretion, to date there has been no evidence of toxicity arising from this or other toxins produced by the Fc-activated microglia. This may be in part because the microglial activation and clearance of amyloid is a surprisingly acute process. Elegant *in vivo* two-photon confocal studies from Hyman's group have recently imaged antibody-stimulated plaque clearance within a matter of a few days (BacSkai et al. (2001)) consistent with the substantial amyloid reductions obtained with only one-month antibody treatment *in vivo* (Bard et al. (2000)). These results suggest that antibody-stimulated microglial amyloid clearance may prove to be both an effective therapeutic and preventive approach.

The efficacy of vaccinations must be tested for both prevention and treatment, and safety issues must be assessed regarding effects of significantly elevated antibody titers in early, and particularly in advanced, stages. This is because clearance of Aβ may accelerate vascular amyloid deposits, damage and hemorrhage or lead to further complement activation. The latter cannot be tested in mouse strains, which may be complement-deficient. Finally, some animal studies suggest that, although vaccination reduces Aβ42, it does not reduce Aβ40, raising the question of whether clinical improvement can occur without reduction of soluble Aβ40. Despite these caveats, the data demonstrating that vaccinations can reduce both pathology and cognitive impairments in mouse models is very encouraging (Chen et al. (2000); Janus et al. (2000); Morgan et al. (2000)).

B. *Nonsteroidal anti-inflammatory drugs (NSAID)*

There is strong epidemiological evidence that chronic NSAID consumption is associated with significant (50% or more) reductions in AD risk (Breitner et al. (1995); Stewart et al. (1997)). Against the background of the vaccine approach, one might suppose that NSAIDs could inhibit microglial phagocytosis of amyloid and increase Aβ loads. In an Aβ infusion model, we found that an anti-inflammatory glucocorticoid reduced plaque size, whereas the specific cyclooxygenase 2 inhibitor; NS-398, increased deposition (Fig. 6), demonstrating potential diversity in the effects of anti-inflammatory agents on Aβ accumulation.

Because the over-the-counter NSAID, ibuprofen, was the most widely used NSAID apparently preventing or delaying AD in several of the epidemiological studies, we decided to test ibuprofen in groups of ten-month-old APPsw (Tg2576) mice. Surprisingly, after six months of ibuprofen treatment, we found 40–50% reductions in amyloid burden compared with controls (Lim et al. (2000)). We have hypothesized that these reductions may be caused by reductions in the pro-amyloidogenic proteins (alpha1 ACT, ApoE, TGFβ), or a stimulation of Aβ clearance. Similarly, D. Morgan's group has very recently found that identical dietary ibuprofen treatment of bigenic APPswX PS1 mutant mice from seven to twelve months have 25% reductions in amyloid burden (Jantzen et al. (2000)). These researchers also found that treatment over the same period with the experimental

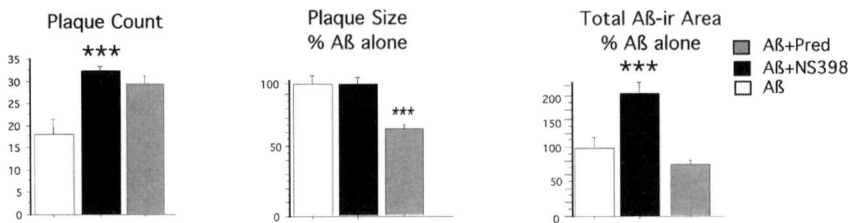

FIGURE 6. Effect of anti-inflammatory agents on Aβ deposition: Aβ deposits in rats infused with Aβ40+42+HDL for one month. Rats were treated for two months before, and one month during, infusion with control chow, or chow containing a COX-2 specific inhibitor or a steroid anti-inflammatory agent prednisolone (50 ppm). Data demonstrated that treatment with prednisolone was associated with a reduction in Aβ-ir deposit size, but did not affect total Aβ-ir area, whereas treatment with the COX-2 inhibitor dramatically increased deposition.

NSAID, nitroflurbiprofen, resulted in larger (50%) reductions in amyloid burden and increased microglial activation as indexed by MHC class II antigen expression. It is important to realize that MHC class II expression is not synonymous with activation of amyloid clearance, and microglia activation is not an expected NSAID effect. Nevertheless, collectively, these results suggest that selected NSAID may be able to reduce amyloid deposition, or even remove the block to amyloid phagocytosis and allow clearance to proceed in the absence of vaccines or added antibody. *In vitro,* nitric oxide production inhibits phagocytosis of Aβ coated beads (Kopec and Carroll (2000)). This suggests that NSAID inhibition of iNOS may actually promote amyloid phagocytosis.

Eventually, an NSAID approach may allow for more refined control of microglial activation state, amyloid phagocytosis and neurotoxin production. Clearly, the ideal approach would result in stimulation of phagocytosis, but inhibition of neurotoxic products.

C. Cytokines

Microglial function is regulated by the cytokine environment and paracrine and autocrine production of cytokines and signal transduction effectors including eicosanoids and nitric oxide. Cytokines can also have a powerful impact on Aβ deposition. For example, IL-1 and IL-6 stimulate synthesis of pro-amyloidogenic alpha1ACT. In another example, very intriguing results have been obtained using GFAP-promoter-driven TGFβ1 transgenic mice by Wyss-Coray, Mucke, Masliah and collaborators (Wyss-Coray et al. (1997)). When crossed with PDAPP transgenics, the bigenic offspring show increased and accelerated vascular amyloid deposition, but very significant reductions in plaque burdens and total amyloid (Akiyama et al. (2000)). Although the mechanism of these effects remains unclear, they clearly show that manipulation of the cytokine environment will profoundly alter amyloid deposition. Further, since TGFβ1 is an anti-inflammatory cytokine, they suggest that small-molecule NSAID can be found that will be able to both control inflammation and reduce amyloid burdens.

D. Steroids

i. Steroid effects on microglial activation and response to Aβ in vitro Aβ induction of nitric oxide is modulated by steroids. Nitric oxide is a reactive oxygen species that can be secreted by microglia during activation and can act as a neurotoxin. Estradiol-17β (E2), the biologically active estrogen, on its own can have a biphasic effect on nitric oxide production and viability, depending on dose and milieu, which may affect the response to Aβ. Nitric oxide and toxicity are stimulated by physiological doses (Harris-White et al. (2001)), but inhibited by high concentrations (Bruce-Keller et al. (2000); Drew and Chavis (2000a)) in an N9 microglial line. This is consistent with E2's known effects in potentiating the immune response. E2 can have both anti-inflammatory and pro-inflammatory effects (Cutolo et al. (1995); Fox et al. (1991)). Despite the fact that E2 can be proinflammatory in the absence of Aβ, E2 can dramatically reduce or eliminate Aβ induction of nitric oxide and associated toxicity (Harris-White et al. (2001)). E2 increases Aβ uptake in N9 cells (Harris-White et al. (2001)), and in human microglia (Li et al. (2000)). This is consistent with a known proinflammatory role of E2 in activating complement and phagocytosis (Pow et al. (1990); Brown et al. (1990); Fan et al. (1996)). Despite increased Aβ uptake, E2 may partly impede intracellular degradation (Harris-White et al. (2001)).

The synthetic glucocorticoid (Gc) dexamethasone (DXM) can reduce (Drew and Chavis (2000b); Tanaka et al. (1997)), or cause a biphasic effect on microgial nitrite (Harris-White et al. (2001)), stimulating nitrite at high doses and potentiating toxicity with IFN. A seemingly paradoxical proinflammatory response to Gcs is a phenomenon common at low doses because, although Gcs inhibit proinflammatory mediators such as interleukins and IFN, they also stimulate other proinflammatory components, including IFN and interleukin receptors (Calandra et al. (1995); Jüttner et al. (1998); Morand, and Leech (1999); Sadeghi et al. (1992); Strickland et al. (1986)). Similar to E2, DXM reduced the nitrite levels induced by Aβ, but in direct contrast to effects with E2 and Aβ, this was associated with significant increases in toxicity (Harris-White et al. (2001)). The combined effects of Gc and Aβ may have been more toxic than E2 because of protective antioxidant effects of E2 (Moosmann and Behl (1999)), or because DXM was enhancing Aβ accumulation. If Aβ accumulates in the endosome/lysosome, this could kill the cell by impairing endosomal/lysosomal permeability (Yang et al. (1998)). Gc-dose dependently enhanced cell pellet Aβ accumulation, also possibly due to increased Aβ uptake. Gcs can differentially alter phagocytosis dependent on milieu, but *in vivo* they do not impair phagocytosis associated with brain injury (Giulian et al. (1989)). Since scavenger receptor-mediated uptake of aggregates is well established (Hyman et al. (2000); Paresce et al. (1996)), Gcs could enhance uptake of Aβ by stimulation of scavenger receptor mRNA and protein as demonstrated in monocytes (Ritter et al. (1999)). Alternatively, because nitric oxide is an important inhibitor of microglial Aβ phagocytosis (Kopec and Carroll (2000)), Gc-induced reductions in NO might also contribute to increased Aβ accumulation by stimulating phagocytosis. In summary, these *in vitro* data are consistent with E2 and DXM's enhancing Aβ uptake and reducing Aβ-induced nitrite through different mechanisms. E2 allowed limited degradation of phagocytosed Aβ to proceed, whereas Gcs significantly impeded degradation.

ii. Steroids in vivo   In rats intracerebroventricularly infused with Aβ, both dietary E2 and Gc treatment appeared to shrink Aβ deposit size with little effect on plaque numbers. Nevertheless, Aβ assayed by ELISA in formic-acid-extracted cortical homogenates was increased with both treatments. This suggested that the reduced size of Aβ deposits resulted from focal concentration of indigestible Aβ, not increased degradation (Harris-White et al. (2001)). Even though microglia can phagocytose and concentrate amyloid fibrils *in vitro*, degradation may be slow and incomplete, so they may eventually regurgitate it undigested (Chung et al. (1999, 2000)). The cellular component of Aβ deposits of E2 treated and Gc-treated rats was different. Aβ-ir deposits of E2-treated rats had large ameboid microglia, whereas the Gc-treated rats had densely stained ramified microglia and vacuolation. These results are similar to data showing that Gc treatment induces shrunken morphology in microglia, and enhances the formation of lysosomal vacuolation (Tanaka et al. (1997)). Gcs also selectively increase the degeneration of ameboid microglia *in vivo*, leaving the remaining microglia ramified (Kaur et al. (1994)). Our *in vivo* data support a role for steroids in promoting increased microglial Aβ uptake leading to plaque compaction without effective degradation. These *in vitro* and *in vivo* data are also consistent with a differential effect of steroids on microglial toxicity associated with Aβ (Gcs cause, and E2 prevents, microglial toxicity associated with Aβ). Elevations in Gc levels with age, stress or illness could contribute to reduced clearance and Aβ deposition. Increased secretion of adrenal steroids is positively correlated with AD severity in cross-sectional studies (Miller et al. (1998); Swanwick et al. (1998); Weiner et al. (1993)), and AD patients have been shown to have increased levels of cortisol secreted per burst (Hartmann et al. (1997); Swanwick et al. (1998)).

Summary

In conclusion, the significance of Aβ binding to microglia is twofold: First, microglia may serve as an important clearance pathway for Aβ in the CNS, preventing neurotoxicity: Second, Aβ binding or uptake may activate microglia to secrete immune mediators, including neurotoxic molecules. Fibrillar Aβ can be taken up by scavenger receptors or Fc receptor. Soluble Aβ may be taken up either via a receptor-independent pathway, or by lipoprotein receptors, which appear to play a role in carrier-mediated Aβ endocytosis. However, uptake does not necessarily lead to degradation, and pathways favoring degradation of soluble or fibrillarAβ are poorly understood. Additional stimulation with vaccine or NSAID therapies may reduce the Aβ burden, but frequency and dosage, as well as time of intervention, need to be worked out to assure safety and efficacy.

References

Acton S, Rigotti A, Landschulz KT, Xu S, Hobbs HH, Krieger M (1996). Identification of scavenger receptor SR-BI as a high density lipoprotein receptor. *Science* 271:518–520.

Akiyama H, Barger S, Barnum S, Bradt B, Bauer J, Cole Gm, Cooper Nr, Eikelenboom P, Emmerling M, Fiebich Bl, Finch Ce, Frautschy S, Griffin Ws, Hampel H, Hull M, Landreth G, Lue L, Mrak R, Mackenzie Ir, Mcgeer Pl, O'Banion Mk, Pachter J, Pasinetti G, Plata-Salaman C, Rogers J, Rydel R, Shen Y, Streit W, Strohmeyer R, Tooyoma I, Van Muiswinkel Fl, Veerhuis R, Walker D, Webster S, Wegrzyniak B, Wenk G, Wyss-Coray T (2000). Inflammation and Alzheimer's disease. *Neurobiol Aging* 21:383–421.

Allsop D, Haga S, Haga C, Ikeda S, Mann DM, Ishii T (1989). Early senile plaques in Down's syndrome brains show a close relationship with cell bodies of neurons. *Neuropathol Appl Neurobiol* 15:531–542.

Ard MD, Cole GM, Wei J, Mehrle AP, Fratkin JD (1996). Scavenging of Alzheimer's amyloid β-protein by microglia in culture. *J Neurosci Res* 43:190–202.

Backstrom JR, Lim GP, Cullen MJ, Tokes ZA (1996). Matrix metalloproteinase-9 (MMP-9) is synthesized in neurons of the human hippocampus and is capable of degrading the amyloid-beta peptide (1–40). *J Neurosci* 16:7910–7919.

BacSkai, BJ, Kajdasz, ST, Christie, RH, Carter, CW, Games, D, Seubert, P, Schenk, D, and Hyman, BT (2001). Imaging of amyloid-deposits in brains of living mice permits direct observation of clearance of plaques with immunotherapy. *Nat Med* 7:369–372.

Bard F, Cannon C, Barbour R, Burke Rl, Games D, Grajeda H, Guido T, Hu K, Huang J, Johnson-Wood K, Khan K, Kholodenko D, Lee M, Lieberburg I, Motter R, Nguyen M, Soriano F, Vasquez N, Weiss K, Welch B, Seubert P, Schenk D and Yednock T (2000). Peripherally administered antibodies against amyloid beta-peptide enter the central nervous system and reduce pathology in a mouse model of Alzheimer disease. *Nat Med* 6(8), 916–9.

Bell MD, López-González R, Lawson L, Hughes D, Fraser I, Gordon S, Perry VH (1994). Upregulation of the macrophage scavenger receptor in response to different forms of injury in the CNS. *J Neurocytol* 23:605–613.

Benzing WC, Mufson EJ, Armstrong DM (1993). Immunocytochemical distribution of peptidergic and cholinergic fibers in the human amygdala: their depletion in Alzheimer's disease and morphologic alteration in non-demented elderly with numerous senile plaques. *Brain Res* 625:125–38.

Biere AL, Ostaszewski B, Stimson ER, Hyman BT, Maggio JE, Selkoe DJ (1996). Amyloid beta-peptide is transported on lipoproteins and albumin in human plasma. *J Biol Chem* 271:32916–32922.

Borchelt DR, Thinakaran G, Eckman CB, Lee MK, Davenport F, Ratovitsky T, Prada CM, Kim G, Seekins S, Yager D, et al. Sisodia S (1996). Familial Alzheimer's disease-linked presenilin 1 variants elevate Abeta1-42/1-40 ratio in vitro and in vivo. *Neuron* 17:1005–1013.

Borchelt DR, Ratovitski T, Van Lare J, Lee MK, Gonzales V, Jenkins NA, Copeland NG, Price DL, Sisodia SS (1997). Accelerated amyloid deposition in the brains of transgenic mice coexpressing mutant presenilin 1 and amyloid precursor proteins. *Neuron* 19:939–945.

Bornemann, KD and Staufenbiel, M (2000). Transgenic mouse models of Alzheimer's disease. *Ann N Y Acad Sci* 908, 260–266.

Brazil MI, Chung H, Maxfield FR (2000). Effects of incorporation of immunoglobulin g and complement component C1q on uptake and degradation of Alzheimer's disease amyloid fibrils by microglia. *J Biol Chem* 275:16941–7.

Breitner JCS, Welsh KA, Helms MJ, Gaskell PC, Gau BA, Roses AD, Vance MAP, Saunders AM, (1995). Delayed onset of Alzheimer's disease with nonsteroidal anti-inflammatory and histamine H2 blocking drugs. *Neurobiol Aging* 16:523–530.

Brown EO, Sundstrom SA, Komm BS, Yi Z, Teuscher C, Lyttle CR (1990). Progesterone regulation of estradiol-induced rat uterine secretory protein, complement C3. *Biol Reprod* 42:713–719.

Bruce-Keller AJ, Keelking JN, Huang FF, Camondola S, Mattson MP (2000). Antiinflammatory effects of estrogen on microglial activation. *Endocrinology* 141:3646–56.

Burdick D, Kosmoski J, Knauer MF, Glabe CG (1997). Preferential adsorption, internalization and resistance to degradation of the major isoform of the Alzheimer's amyloid peptide, A beta 1-42, in differentiated PC12 cells. *Brain Res* 746:275–284.

Calandra T, Bernhagen J, Metz CN, Splegel LA, Bacher M, Donnelly T, Cerami A, Bucala R (1995). MIF as a glucocorticoid-induced modulator of cytokine production. *Nature* 377:68–71.

Castillo GM, Ngo C, Cummings J, Wight TN, Snow AD (1997). Perlecan binds to the β-amyloid proteins (Aβ) of Alzheimer's disease, accelerates Aβ fibril formation, and maintains Aβ fibril stability. *J Neurochem* 69:2452–2465.

Cataldo AM, Barnett JL, Berman SA, Li J, Quarless S, Bursztajn S, Lippa C, Nixon RA (1995). Gene expression and cellular content of cathepsin D in Alzheimer's disease brain: Evidence for early up-regulation of the endosomal-lysosomal system. *Neuron* 14:671–680.

Chao CC, Hu S, Peterson PK (1996). Glia: the not so innocent bystanders. *J Neurovirol* 2:234–239.

Chen K, Soriano F, Lyn W, Grajeda H, Masliah E, Games D (1998). Effects of entorhinal cortex lesions on hippocampal β-amyloid deposition in PDAPP transgenic mice. *Society for Neuroscience* 24 (#592.6):1502.

Chen G, Chen KS, Knox J, Inglis J, Bernard A, Martin SJ, Justice A, McConlogue L, Games D, Freedman SB, Morris RGM (2000). A learning deficit related to age and β-amyloid plaques in a mouse model of Alzheimer's disease. *Nature* 408:975–979.

Christie RH, Chung H, Rebeck GW, Strickland D, Hyman BT (1996a). Expression of the very low-density lipoprotein receptor (VLDL-r), an apolipoprotein-E receptor, in the central nervous system and in Alzheimer's disease. *J Neuropathol Exp Neurol* 55:491–498.

Christie RH, Freeman M, Hyman BT (1996b). Expression of the macrophage scavenger receptor, a multifunctional lipoprotein receptor, in microglia associated with senile plaques in Alzheimer's disease. *Am J Pathol* 148:399–403.

Chu T, Tran T, Yang F, Beech W, Cole GM, Frautschy SA (1998). Effect of chloroquine and leupeptin on intracellular accumulation of amyloid-beta (Aβ) 1-42 peptide in a murine N9 microglial cell line. *FEBS Lett* 436:439–444.

Chung H, Brazil MI, Soe TT, Maxfield FR (1999). Uptake, degradation, and release of fibrillar and soluble forms of Alzheimer's amyloid beta-peptide by microglial cells. *J Biol Chem* 274:32301–32308.

Chung H, Wang R and Maxfield FR (2000). Degradation of β-amyloid peptide by microglia. *Society for Neuroscience* 26, 2286 (#858.10).

Cole GM, Ard MD (2000). Influence of lipoproteins on microglial degradation of alzheimer's amyloid beta–protein. *Micros Res Tech* 50:316–24.

Cole GM, Beech W, Frautschy SA, Sigel JJ, Glasgow C, Ard MD, (1999). Lipoprotein effects on Aβ accumulation and degradation by microglia in vitro. *J Neuroscience Res* 57:504–520.

Cutolo M, Sulli A, Seriolo B, Accardo S, Masi AT (1995). Estrogens, the immune response, and autoimmunity. *Clin Exp Rheumatol* 13:216–226.

Dewitt DA, Perry G, Cohen M, Doller C, Silver J (1998). Astrocytes regulate microglial phagocytosis of senile plaque cores of Alzheimer's disease. *Exper Neurol* 149:329–340.

Drew PD, Chavis JA (2000a). Female sex steroids: effects upon microglial cell activation. *J Neuroimmunol* 111:77–85.

Drew PD, Chavis JA (2000b). Inhibition of microglial cell activation by cortisol. *Brain Res Bull* 52:391–6

El Khoury J, Hickman SE, Thomas CA, Cao L, Silverstein SC, Loike JD (1996). Scavenger receptor-mediated adhesion of microglia to beta-amyloid fibrils. *Nature* 382:716–719.

Fan JD, Wagner BL, McDonnell DP (1996). Identification of the sequences within the human complement 3 promoter required for estrogen responsiveness provides insight into the mechanism of tamoxifen mixed agonist activity. *Mol Endocrinol* 10:1605–1616.

Fox HS, Bond BL, Parsolow TG (1991). Estrogen regulates the IFN-gamma promoter. *J Immunol* 146:

Frackowiak J, Wisniewski HM, Wegiel J, Merz GS, Iqbal K, Wang KC (1992). Ultrastructure of the microglia that phagocytose amyloid and the microglia that produce β-amyloid fibrils. *Acta Neuropathol* 84:225–233.

Frautschy SA, Albright T, Dvorak C, Wolfe DS, Baird A (1994). Transforming growth factor β (TGFβ) modification of β protein immunoreactivity in the hippocampus with and

without β protein infusion and the resulting ultrastructural neuropathology. *Neurobiol Aging* 15(S1):S55–S56.
Frautschy SA, Cole GM, Baird A (1992). Phagocytosis and deposition of vascular β-amyloid in rat brains injected with Alzheimer β-amyloid. *Am J Pathol* 140:1389–1399.
Frautschy SA, Yang F, Calderón L, Cole GM (1996). Rodent models of Alzheimer's disease: rat Aβ infusion approaches to amyloid deposits. *Neurobiol Aging* 17:311–321.
Frautschy SA, Yang F, Irrizarry M, Hyman B, Saido TC, Hsiao K, Cole GM (1998). Microglial response to amyloid plaques in APPsw transgenic mice. *Amer J Pathol* 152:307–317.
Games D, Khan KM, Soriano FG, Keim PS, Davis DL, Bryant K, Lieberburg I (1992). Lack of Alzheimer pathology after β-amyloid protein injections in rat brain. *Neurobiol Aging* 13:569–576.
Games D, Adams D, Alessandrini R, Barbour R, Berthelette P, Blackwell C, Carr T, Clemens J, Donaldson T, Gillespie F, Guido T, Hagoplan S, Johnson-Wood K, Khan K, Lee M, Leibowitz P, Lieberburg I, Little S, Masliah E, McConlogue L, Montoya-Zavala M, Mucke L, Paganini L, Penniman E, Power M, Schenk D, Seubert P, Snyder B, Soriano F, Tan H, Vitale J, Wadsworth S, Wolozin B, Zhao J (1995). Alzheimer-type neuropathology in transgenic mice overexpressing V717F β-amyloid precursor protein. *Nature* 373:523–527.
Geddes JW, Anderson KJ, Cotman CW (1986). Senile plaques as aberrant sprout-stimulating structures. *Exp Neurol* 94:767–776.
Gehrmann J, Banati RB (1995). Microglial turnover in the injured CNS: activated microglia undergo delayed DNA fragmentation following peripheral nerve injury. *J Neuropathol Exp Neurol* 54:680–685.
Ghiso J, Matsubara E, Koudinov A, Choi-Miura NH, Tomita M, Wisniewski T, Frangione B (1993). The cerebrospinal-fluid soluble form of Alzheimer's amyloid beta is complexed to SP-40,40 (apolipoprotein J), an inhibitor of the complement membrane-attack complex. *Biochem J* 293:27–30.
Giulian D, Chen J, Ingeman JE, George JK, Noponen M (1989). The role of mononuclear phagocytes in wound healing after traumatic injury to adult mammalian brain. *J Neurosci* 9:4416–4429.
Gomez-Isla T, Wasco W, Pettingell WP, Gurubhagavatula S, Schmidt SD, Jondro PD, McNamara M, Rodes LA, DiBlasi T, Growdon WB, Seubert P, Schenk D, Growdon JH, Hyman BT, Tanzi RE (1997). A novel presenilin-1 mutation: increased β-amyloid and neurofibrillary changes. *Ann Neurol* 41:809–813.
Goodwin JL, Kehrli MEJr, Uemura E (1997). Integrin Mac-1 and β-amyloid in microglial release of nitric oxide. *Brain Res* 768:279–286.
Götz J, Chen F, van Dorpe J, Nitsch RM (2001). Formation of neurofibrillary tangles in P301L tau transgenic mice induced by Aβ42 fibrils. *Science* 293:1491–1495.
Harris-White ME, Simmons M, Nash D, Miller SA, Chu T, Teter B, Cole GM, Frautschy SA (2001). Estrogen and glucocorticoid effects on microglia and Aβ clearance in vitro and in vivo. *Neurochem Int* 39:435–448.
Hartmann A, Veldhuis JD, Deuschle M, Standhardt H, Heuser I (1997). Twenty-four hour cortisol release profiles in patients with Alzheimer's and Parkinson's disease compared to normal controls: ultradian secretory pulsatility and diurnal variation. *Neurobiol Aging* 18:285–289.
Hayashi T, Yamada K, Esaki T, Muto E, Chaudhuri G, Iquchi Q (1998). Physiological con-

centrations of 17beta-estradiol inhibit the synthesis of nitric oxide synthase in macrophages via a receptor mediated system. *J Cardiovascular Pharmacol* 31:292–8.

Holcomb L, Gordon MN, McGowan E, Yu X, Benkovic S, Jantzen P, Wright K, et al., Duff K (1998). Accelerated Alzheimer-type phenotype in transgenic mice carrying both mutant amyloid precursor protein and presenilin 1 transgenes. *Nat Med* 4:97–100.

Holcomb LA, Gordon MN, Jantzen P, Hsiao K, Duff K, Morgan D (1999). Behavioral changes in transgenic mice expressing both amyloi dprecursor protein and presenilin-1 mutations: lack of association with amyloid deposits. *Behav Genet* 29:177–85.

Holtzman DM, Bales RK, Wu S, Bhat P, Parsadanian M, Fagan AM, Chang LK, Sun Y, Paul SM (1999). Expression of human apolipoprotein E reduces amyloid-B deposition in a mouse model of Alzheimer's disease. *J Clin Invest* 103:R15–R21.

Holtzman DM, Pitas RE, Kilbridge J, Nathan B, Mahley RW, Bu G, Schwartz AL (1995). Low density lipoprotein receptor-related protein mediates apolipoprotein E-dependent neurite outgrowth in a central nervous system-derived neuronal cell line. *Proc Natl Acad Sci USA* 92:9480–9484.

Honda M, Akiyama H, Yamada Y, Kondo H, Kawabe Y, Takeya J, Takahashi K, Suzuki H, Doi T, Sakamoto A, et al. (1998). Immunohistochemical evidence for a macrophage scavenger receptor in Mato cells and reactive microglia of ischemia and Alzheimer's disease. *Biochem Biophys Res Commun* 245:734–740.

Hsiao KK, Borchelt DR, Olson K, Johannsdottir R, Kitt C, Yunis W, Xu S, Eckman C, Younkin S, Price D (1995). Age-related CNS disorder and early death in transgenic FVB/N mice overexpressing Alzheimer amyloid precursor proteins. *Neuron* 15:1203–1218.

Hsiao K, Chapman P, Nilsen S, Eckman C, Harigaya Y, Younkin S, Yang F, Cole G (1996). Correlative memory deficits, Aβ elevation and amyloid plaques in transgenic mice. *Science* 274:99–102.

Huang F, Buttini M, Wyss-Coray T, McConlogue L, Kodama T, Pitas RE, Mucke L (1999). Elimination of the class A scavenger receptor does not affect amyloid plaque formation or neurodegeneration in transgenic mice expressing human amyloid protein precursors. *Am J Pathol* 155:1741–7.

Huang SS, Huang FW, Xu J, Chen S, Hsu CY, and Huang JS (1998). Amyloid beta-peptide possesses a transforming growth factor-beta activity. *J Biol Chem* 273:27640–27644.

Hyman BT, Strickland D, Rebeck GW (2000). Role of the low-density lipoprotein receptor-related protein in beta-amyloid metabolism and Alzheimer disease. *Arch Neurol* 57:646–650.

Irizarry MC, McNamara M, Fedorchak K, Hsiao K, Hyman BT (1997a). APPsw Transgenic Mice develop age-related Aβ Deposits and neuropil abnormalities, but no neuronal loss in CA1. *J Neuropathol Exp Neurol* 56:965–973.

Irizarry MC, Soriano F, McNamara M, Page KJ, Schenk D, Games D, Hyman BT (1997b). Aβ deposition is associated with neuropil changes, but not with overt neuronal loss in the human amyloid precursor protein V717F (PDAPP) transgenic mouse. *J Neurosci* 17:7053–7059.

Ishii K, Tamaoka A, Mizusawa H, Shoji S, Ohtake T, Fraser PE, Takahashi H, Tsuji S, Gearing M, Mizutani T, Yamada S, Kato M, St.George-Hyslop PH, Mirra SS, Mori H (1997). Aβ 1–40 but not Aβ 1–42 levels in cortex correlate with apolipoprotein E E4 allele dosage in sporadic Alzheimer's disease. *Brain Res* 748:250–252.

Iwata N, Tsubuki S, Takaki Y, Watanabe K, Sekiguchi M, Hosoki E, Lee HJ, Hama E, Sekine-Aizawa Y, Saido TC (2000). Identification of the major Aβ1–42-degrading catabolic pathway in brain parenchyma: suppression lead to biochemical and pathological deposition. *Nat Med* 6:143–150.

Iwata N, Tsubuki S, Takaki Y, Shirotani K, Lu B, Gerard NP, Gerard C, Hama E, Lee HJ, Saido TC (2001). Metabolic regulation of brain Abeta by neprilysin. *Science* 292: 1550–2.

Jantzen PT, Gordon MN, Connor KE, DiCarlo G, and Morgan DG (2000). Modification of microglial reactivity in MAPP/MPS1 transgenic mice using NCX-2216 (nitroflurbiprofen). *Society for Neuroscience* 26(1061 (#397.14)).

Janus C, Pearson J, McLaurin J, Mathews PM, Jiang Y, Schmidt SD, Chishti MA, Horne P, Heslin D, French J, Mount HTJ, Nixon RA, Mercken M, Bergeron C, Fraser PE, St.George-Hysolop P, Westaway D (2000). Aβ peptide immunization reduces behavioral impairment and plaques in a model of Alzheimer's disease. *Nature* 408:979–982.

Jüttner S, Bernhagen J, Metz CNRM, Bucala R, Gessner A (1998). Migration inhibitory factor induces killing of leishmania major by macrophages: dependence on reactive nitrogen intermediates and endogenous TNF-alpha. *J Immunol* 161:2383–90.

Kaur C, Wu CH, Wen CY, Ling EA (1994). The effects of subcutaneous injections of glucocorticoids on amoeboid microglia in postnatal rats. *Arch Histol Cytol* 57:449–459.

Kopec KK, Carroll RT (2000). Phagocytosis is regulated by nitric oxide in murine microglia. *Nitric Oxide* 4:103–11.

Korotzer AR, Pike CJ, Cotman CW (1993). β-amyloid peptides induce degeneration of cultured rat microglia. *Brain Res* 624:121–125.

Koudinov AR, Koudinova NV, Kumar A, Beavis RC, Ghiso J (1996). Biochemical characterization of Alzheimer's soluble amyloid beta protein in human cerebrospinal fluid: Association with high density lipoproteins. *Biochem Biophys Res Commun* 223:592–597.

Kuo Y-M, Emmerling MR, Vigo-Pelfrey C, Kasunic TC, Kirkpatrick JB, Murdoch GH, Ball MJ, Roher AE (1996). Water-soluble Aβ (N-40,N-42) oligomers in normal and Alzheimer disease brains. *J Biol Chem* 271:4077–4081.

LaDu MJ, Reardon C, Van Eldik L, Fagan AM, Bu G, Holtzman D, Getz GS (2000). Lipoproteins in the central nervous system. *Ann NY Acad Sci* 903:167–175.

LaFerla FM, Troncoso JC, Strickland DK, Kawas CH, Jay G (1997). Neuronal cell death in Alzheimer's disease correlates with ApoE uptake and intracellular Aβ stabilization. *J Clin Invest* 100:310–320.

Lewis J, Dickson DW, Lin W-L, Chisholm L, Corral A, Jones G, Yen S-H, Sahara N, Skipper L, Yager D, Eckman C, Hardy J, Hutton J, McGowan E (2001). Enhanced neurofibrillary degeneration in transgenic mice expressing mutant Tau and APP. *Science* 293:1487–1491.

Li R, Shen Y, Yang LB, Lue LF, Finch C, Rogers J (2000). Estrogen enhances uptake of amyloid beta-protein by microglia derived from the human cortex. *J Neurochem* 75:1447–54.

Liao A, Nitsch RM, Greenberg SM, Finckh U, Blaccer D, Albert M, Rebeck GW, Gomez-Isla T, Clatworthy A, Binetti Geal (1998). Genetic association of an alpha2-macroglobulin (Val11000l1e) polymorphism and Alzheimer's disease. *Hum Mol Genet* 7:1953–6.

Lim GP, Yang F, Chu T, Chen P, Beech W, Teter B, Tran T, Ubeda O, Ashe KH, Frautschy SA, Cole GM (2000). Ibuprofen suppresses plaque pathology and inflammation in a mouse model for Alzheimer's disease. *J Neurosci* 20(15):5709–14.

Lorton D, Schaller J, Lala A, De Nardin E (2000). Chemotactic-like receptors and abeta peptide induced responses in Alzheimer's disease. *Neurobiol Aging* 21:463–73.

Lue LF, Kuo YM, Roher AE, Brachova L, Shen Y, Sue L, Beach T, Kurth JH, Rydel RE, Rogers J (1999). Soluble amyloid beta peptide concentration as a predictor of synaptic change in Alzheimer's disease. *Am J Pathol* 155:853–862.

Luedecking EK, DeKosky ST, Mehdi H, Ganguli M, Kamboh MI (2000). Analysis of genetic polymorphisms in the transforming growth factor-beta1 gene and the risk of Alzheimer's disease. *Hum Genet* 106:565–9.

Mann DMA, Iwatsubo T, Pickering-Brown SM, Owen F, Saido TC, Perry RH (1997). Preferential deposition of amyloid β protein (Aβ) in the form of Aβ 40 in Alzheimer's disease is associated with a gene dosage effect of the apolipoprotein E E4 allele. *Neurosci Lett* 221:81–84.

Marzolo MP, Von Bernhardi R, Bu G, Inestrosa NC, (2000) Expression of alpha2 macroglobulin receptor/low-density lipoprotein receptor-related protein (LRP) in rat microglial cells. *J Neurosci Res* 60:401–411.

McDonald DR, Bamberger ME, Combs CK, Landreth GE (1998). β-amyloid fibrils activate parallel mitogen-activated protein kinase pathways in microglia and THP1 monocytes. *J Neurosci* 18:4451–60.

McDonald DR, Brunden KR, Landreth GE (1997). Amyloid fibrils activate tyrosine kinase-dependent signaling and superoxide production in microglia. *J Neurosci* 17:2284–94.

Meda L, Cassatella MA, Szendrei GI, Otvos L, Baron P, Villalba M, Ferrari D, Rossi F (1995). Activation of microglial cells by beta-amyloid protein and interferon-gamma. *Nature* 374:647–650.

Miller TP, Taylor J, Rogerson S, Mauricio M, Kennedy Q, Schatzberg A, Tinklenberg J, Yesavage J (1998). Cognitive and noncognitive symptoms in dementia patients: relationship to cortisol and dehydroepiandrosterone. *Int Psychogeriats* 10:85–96.

Moosmann B, Behl C (1999). The antioxidant neuroprotective effects of estrogens and phenolic compounds are independent fom the estrogenic properties. *Proc Natl Acad Sci USA* 96:8867–72.

Morand EF, Leech M (1999). Glucocorticoid regulation of inflammation: the plot thickens. *Inflamm Res* 48:557–60.

Morgan D, Diamond DM, Gottschall PE, Ugen KE, Dickey C, Hardy J, Duff K, Jantzen P, DiCarlo G, Wilcock D, Connor K, Hatcher J, Hope C, Gordon M, Arendash GW (2000). Aβ peptide vaccination prevents memory loss in an animal model of Alzheimer's disease. *Nature* 408:982–985.

Naslund J, Haroutunian V, Mohs R, Davis KL, Davies P, Greengard P, Buxbaum JD (2000). Correlation between elevated levels of amyloid beta-peptide in the brain and cognitive decline. JAMA 283:1571–1577.

Nicoll JA, Mrak RE, Graham DI, Stewart J, Wilcock G, MacGowan S, Esiri MM, Murray LS, Dewar D, Love S, Moss T, Griffin WST (2000). Association of interleukin-1 gene polymorphisms with Alzheimer's disease. *Annals Neurol* 47:365–368.

Oda T, Wals P, Osterburg HH, Johnson SA, Pasinetti G, Morgan TE, Rozovsky I, Blaine

Stine W, Snyder SW, Holzman TF, Krafft GA, Finch CE, (1995) Clusterin (apoJ) alters the aggregation of amyloid β-peptide (Aβ 1-42) and forms slowly sedimenting Aβ complexes that cause oxidative stress. *Exp Neurol* 136:22–31.

Paresce DM, Chung HY, Maxfield FR (1997). Slow degradation of aggregates of the Alzheimer's disease amyloid β-protein by microglial cells. *J Biol Chem* 272:29390–29397.

Paresce DM, Ghosh RN, Maxfield FR (1996). Microglial cells internalize aggregates of the Alzheimer's disease amyloid beta-protein via a scavenger receptor. *Neuron* 17:553–565.

Pike CJ, Ramezan-Arab N, Cotman CW (1997). Beta-amyloid neurotoxicity in vitro: evidence of oxidative stress but not protection by antioxidants. *J Neurochem* 69:1601–1611.

Pike CJ, Walencewicz AJ, Glabe CG, Cotman CW (1991). In vitro aging of β-amyloid protein causes peptide aggregation and neurotoxicity. *Brain Res* 563:311–314.

Pow DV, Perry VH, Morris JF, Gordon S (1989). Microglia in the neurohypophysis associate with and endocytose terminal portions of neurosecretory neurons. *Neuroscience* 33:567–578.

Probst A, Langui D, Ipsen S, Robakis N, Ulrich J (1991). Deposition of β/A4 protein along neuronal plasma membranes in diffuse senile plaques. *Acta Neuropathol* 83:21–29.

Rebeck GW, Reiter JS, Strickland DK, Hyman BT (1993). Apolipoprotein E in sporadic Alzheimer's disease: allelic variation and receptor interactions. *Neuron* 11:575–580.

Ritter M, Buechler C, Langmann T, Orso E, Klucken J, Schmitz G (1999). The scavenger receptor CD163: regulation, promoter structure and genomic organization. *Pathobiology* 67:257–61.

Sadeghi R, Hawrylowicz CM, Chernajovsky Y, Feldmann M (1992). Synergism of glucocorticoids with granulocyte macrophage colony stimulating factor (GM-CSF) but not inteferon gamma (IFN-gamma) or interleukin-4 (IL-4) on induction of HLA class II expression on human monocytes. *Cytokine* 4:287–97.

Savage MJ, Trusko SP, Howland DS, Pinsker LR, Mistretta S, Reaume AG, Greenberg BD, Siman R, Scott RW (1998). Turnover of amyloid β-protein in mouse brain and acute reduction of its level by phorbol ester. *J Neurosci* 18:1743–1752.

Schenk D, Barbour R, Dunn W, Gordon G, Grajeda H, Guido T, Hu K, Huang J, Johnson-Wood K, Khan K, Kholodenko D, Lee M, Liao Z, Lieberburg I, Motter R, Mutter L, Soriano F, Shopp G, Vasquez N, Vandevert C, Walker S, Wogulis M, Yednock T, Games D, Seubert P (1999). Immunization with amyloid-β attenuates Alzheimer-disease-like pathology in the PDAPP mouse. *Nature* 400:173–177.

Scheuner D, Eckman C, Jensen M, Song X, Citron M, Suzuki N, Bird TD, Hardy J, Hutton M, Kukull W, Larson E, Levy-Lahad E, Viitanen M, Peskind E, Poorkaj P, Schellenberg G, Tanzi R, Wasco W, Selkoe D, Younkin S (1996). Secreted amyloid β-protein similar to that in the senile plaques of Alzheimer's disease is increased in vivo by the presenilin 1 and 2 and APP mutations linked to familial Alzheimer's disease. *Nat Med* 2:864–870.

Shaffer LM, Dority MD, Gupta-Bansal R, Frederickson RC, Younkin SG, Brunden KR (1995). Amyloid beta protein removal by neuroglial cells in culture. *Neurobiol Aging* 16:737–745.

Stalder M, Phinney A, Probst A, Sommer B, Staufenbiel M, and Jucker M (1999). Associ-

ation of microglia with amyloid plaques in brains of APP23 transgenic mice [see comments]. *Am J Pathol* 154(6), 1673–84.
Stewart WF, Kawas C, Corrada M, Metter EJ (1997). Risk of Alzheimer's disease and duration of NSAID use. *Neurology* 48:626–631.
Stone DJ, Rozovsky I, Morgan TE, Anderson CP, Hajian H, Finch CE (1997). Astrocytes and microglia respond to estrogen with increased apoE mRNA in vivo and in vitro. *Exp Neurol* 143:313–318.
Strickland RW, Wahl LH, Finbloom DS (1986). Corticosteroids enhance the binding of recombinant interferon to cultured human monocytes. *J Immunol* 137:
Swanwick GR, Kirby M, Bruce I, Buggy F, Coen RF, Coakley D, Lawlor BA (1998). Hypothalamic-pituitary-adrenal axis dysfunction in Alzheimer's disease: lack of association between longitudinal and cross-sectional findings. *Am J Psychiatry* 155:286–289.
Tanaka J, Fujita H, Matsuda S, Toku K, Sakanaka M, Maeda N (1997). Glucocorticoid- and mineralocorticoid receptors in microglial cells: the two receptors mediate differential effects of corticosteroids. *Glia* 20:23–37.
Terry RD, Masliah E, Salmon DP, Butters N, DeTeresa R, Hill R, Hansen LA, Katzman R (1991). Physical basis of cognitive alterations in Alzheimer's disease: synapse loss is the major correlate of cognitive impairment. *Ann Neurol* 30:572–580.
Thome J, Baumer A, Kornhuber J, Rosler M, Riederer P (1995). Alpha-1-antichymotrypsin bi-allele polymorphism apolipoprotein-E tri-allele polymorphism and genetic risk of Alzheimer syndrome. *J Neural Transm* 10:207–212.
Torp R, Head E, Milgram NW, Hahn F, Ottersen OP, Cotman CW (2000). Ultrastructural evidence of fibrillar beta-amyloid associated with neuronal membranes in behaviorally characterized aged dog brains. *Neuroscience* 96:495–506.
Vekrellis K, Chiu S, Mansourian S, Selkoe D (2000a). Insulin-degrading enzyme is the major Aβ-degrading protease in human control and Alzheimer's disease brains. *Society for Neuroscience* 26:1573 (#588.5).
Vekrellis K, Ye Z, Qiu WQ, Walsh D, Hartley D, Chesneau V, Rosner MR, Selkoe DJ (2000b). Neurons regulate extracellular levels of amyloid β-protein via proteolysis by insulin-degrading enzyme. *J Neurosci* 20:1657–1665.
Wang HY, Lee DH, Davis CB, and Shank RP (2000). Amyloid peptide abeta(1-42) binds selectively and with picomolar affinity to alpha7 nicotinic acetylcholine receptors. *J Neurochem* 75(3), 1155–61.
Weiner MF, Vobach S, Svetlik D, Risser RC (1993). Cortisol secretion and Alzheimer's disease progression: a preliminary report. *Biol Psychiatry* 34:158–161.
Weldon DT, Rogers SD, Ghilardi JR, Finke MP, Cleary JP, O'Hare E, Esler WP, Maggio JE, Mantyh PW (1998). Fibrillar β-amyloid induces microglial phagocytosis, expression of inducible nitric oxide synthase, and loss of a select population of neurons in the rat CNS in vivo. *J Neurosci* 18:2161–2173.
Whitson JS, Selkoe DJ, Cotman CW (1989). Amyloid β protein enhances the survival of hippocampal neurons in vitro. *Science* 243:1488–1490.
Winkler J, Connor DJ, Frautschy SA, Behl C, Waite JJ, Cole GM, Thal LJ (1994). Lack of long-term effects after β-Amyloid protein injections in rat brain. *Neurobiol Aging* 15:601.
Wisniewski HM, Barcikowska M, Kida E (1991). Phagocytosis of β/A4 amyloid fibrils of the neuritic neocortical plaques. *Acta Neuropathol* 81:588–590.
Wisniewski HM, Weigel J, Wang KC, Kujawa M, Lach B (1989). Ultrastructural studies of the cells forming amyloid fibers in classical plaques. *Can J Neurol Sci* 16:535–542.

Wyss-Coray T, Masliah E, Mallory M, McConlogue L, Johnson-Wood K, Lin C, Mucke L (1997). Amyloidogenic role of cytokine TGF-beta-1 in transgenic mice and in Alzheimer's disease. *Nature* 389:603–605.

Wyss-Coray T, Lin C, Yan F, Yu GQ, Rohde M, McConlogue L, Masliah E, Mucke L (2001). TGF beta 1 promotes microglial amyloid-beta clearance and reduces plaque burden in transgenic mice. *Nat Med* 7:612–8.

Yamada Y, Doi T, Hamakubo T, Kodama T (1998). Scavenger receptor family proteins: roles for atherosclerosis, host defence and disorders of the central nervous system. *Cell Mol Life Sci* 54:628–640.

Yan SD, Chen X, Fu J, Chen M, Zhu H, Roher A, Slattery T, Zhao L, Nagashima M, Morser J, Migheli A, Nawroth P, Stern D, Schmidt AM (1996). RAGE and amyloid-beta peptide neurotoxicity in Alzheimer's disease [see comments]. *Nature* 382:685–91.

Yan SD, Roher A, Chaney M, Zlokovic B, Schmidt AM, Stern D (2000). Cellular cofactors potentiating induction of stress and the cytotoxicity by amyloid beta-peptide. *Biochim Biophys Acta* 1502:145–157.

Yan SD, Zhu H, Fu J, Yan SF, Roher A, Tourtellotte WW, Rajavashisth T, Chen X, Godman GC, Stern D, Schmidt AM (1997). Amyloid-β peptide-receptor for advanced glycation endproduct interaction elicits neuronal expression of macrophage-colony stimulating factor: a proinflammatory pathway in Alzheimer's disease. *Proc Natl Acad Sci USA* 94:5296–5301.

Yang AJ, Chandswangbhuvana D, Margol L, Glabe CG (1998). Loss of endosomal/lysosomal membrane impermeability is an early event in amyloid Aβ1–42 pathogenesis. *J Neurosci* 52:691–8.

Yang AJ, Knauer M, Burdick DA, Glabe C (1995). Intracellular Aβ1–42 aggregates stimulate the accumulation of stable, insoluble amyloidogenic fragments of the amyloid precursor protein in transfected cells. *J Biol Chem* 270:14786–14792.

Yankner BA, Duffy LK, Kirschner DA (1990). Neurotrophic and neurotoxic effects of amyloid beta protein: reversal by tachykinin neuropeptides. *Science* 250:279–282.

Younkin SG (1995). Evidence that Aβ42 is the real culprit in Alzheimer's disease. *Ann Neurol* 37:287–288.

Zlokovic BV, Martel CL, Mackic JB, Matsubara E, Wisniewski T, McComb JG, Frangione B, Ghiso J (1994). Brain uptake of circulating apolipoproteins J and E complexed to Alzheimer's amyloid beta. *Biochem Biophys Res Commun* 205:1431–1437.

13

Microglia and Aging in the Brain

CALEB E. FINCH, TODD E. MORGAN, IRINA ROZOVSKY, ZHONG XIE, RICHARD WEINDRUCH, AND TOMAS PROLLA

1. Introduction

Microglial activation during normal aging in the CNS is puzzling. In general, microglial activation is associated with neuron death, blood-brain barrier disruption, or invading lymphocytes, whereas, as we shall discuss, there is little evidence for such changes during normal aging. Microglial age changes may also interact with Alzheimer's disease, which increases markedly during aging, as well as with inflammatory processes of aging in peripheral tissues.

This is an emerging subject and readers will encounter many frustrating gaps. Few studies have used the same rodent models and histological and immunocytochemical criteria. By the attention we give to several descriptive reports, some readers may think they are reading a pathology journal from the nineteenth century. It seems fair to say that the molecular biology of microglia is not as developed as that of macrophage-monocytes in other organ systems, an immense field that may have many times more investigators. For general perspectives, see other chapters in this volume, and valuable reviews of Raivich et al. (1999); Streit et al. (1999); Akiyama et al. (2000); Kato and Walz (2000); and Thomas (1992, 1999).

The understanding of brain microglia may rapidly evolve through the new technologies of genomics and proteomics. For example, we shall describe a recent microarray survey of mRNA from the aging mouse brain that shows selective changes in genes that are associated with inflammation and oxidative stress, some of which are attenuated by caloric restriction. From this evidence, we argue that inflammatory processes are a prominent feature in normal brain aging, which may be intensified in Alzheimer's disease, and in some circumstances may be attenuated by metabolic interventions. We outline a unifying hypothesis on aging, inflammation, and oxidative stress, in which monocyte-macrophage cells may have a key role: macrophages and vascular endothelial cells, which occur in all organs, may share subsets of inflammatory mRNAs induced during aging throughout the body.

2. Perspectives on microglia

In general, two classes of macrophage-monocytes are described in the brain: microglia, which are in the brain parenchyma and behind the blood-brain barrier, and perivascular macrophages. Microglia are considered to be the facultative (professional) macrophages of the brain with implied roles of immune surveillance. Activated microglia express the MHC class II antigens, although there is little contact with T-cells for antigen presentation. Microglia can be stained for immunoglobulins, which may be bound by the Fcγ receptor, and which are thought to be derived from the trace quantities of immunoglobulins found in the cerebrospinal fluid (Raivich et al. (1999)). The few peripheral lymphocytes behind the blood-brain barrier in the normal brain, e.g. ~1 cell/mm^2 of 20 micron section in C57BL/6J mice (Raivich et al. (1998)), are hypothesized to have a surveillance function (Wekerle et al. (1986)). The brain has no lymphatic drainage, or the equivalent of peripheral lymph nodes where antigens can be stored in specialized macrophages, such as in Langerhans cells (Barker and Billingham (1977)). However, macrophages in the choroid plexus have some characteristics of Langerhans cells, through which they may participate in immune-mediated functions (Matyszak et al. (1992)).

Microglia in adult brains may be mostly maintained by slow influx of peripheral monocytes, as indicated by bone-marrow transplantation experiments in adult mice (e.g. Eglitis and Mezey (1997)). The turnover rate and regional movements of microglia are not precisely known in the healthy adult brain. Endogenous generation is probably less important because DNA synthesis and mitosis is relatively rare in adult brain microglia (Lawson et al. (1993)), unless stimulated by injury (Raivich et al. (1999)). Monocyte entry into the brain in response appears to be dependent on the complement receptor-type 3 (CR3), which mediates the adhesion of leukocytes to vascular endothelia in other tissues (Andersson et al. (1992)).

In the healthy adult brain, most microglia are considered to be in a resting state, with negligible endocytic or phagocytic activities and low "activation". In Table 1, we summarize the general characteristics and antigen presentations of several microglial phenotypes. Like peripheral macrophage-monocytes, brain microglia display remarkable phenotypic plasticity. The wide range of shapes and biochemical activities is sensitive to the activities of neighboring neurons and other glial cell types, as well as to systemic physiological conditions. The morphotype with fine, spidery extensions is characterized as "resting ramified". In white matter, microglial processes are oriented parallel to axons, whereas grey matter microglia are stellate. The term activation, however, does not imply production of activities that necessarily endanger neurons because microglia have a wide range of biochemical and cellular activities that can be induced during activation, which in some examples are neuroprotective (Bruce-Keller et al. (1999); Streit et al. (1999)).

TABLE 1: Characteristic phenotypes and markers of microglia

Antigen	Resting	Activated	Mouse	Rat	Human
CD11a/LFA-1 alpha	+/−	++	6	5, 9	1, 3
CD11b/CR3	+	+++	13, 19	8, 17, 18	1
CD11c	+	++	ND	ND	1
CD18/LFA-1β	+	++	ND	16	1
MHC class I	+/−	++	ND	11	ND
MHC class II	+/−	+++	22	17, 18	10, 15
F4,80	+	+++	12, 19	ND	ND
LCA	+	++	ND	18	70
ED-1	+/−	+++	ND	11, 18	ND
Lectin (RCA, GSL)	+	++	2	4, 20, 21	7, 21, 14

LFA: lymphocyte function-associated antigen-1
CR3: complement type 3 receptor
MhC: major histocompatibility complex
F4,80: macrophage-specific plasma-membrane glycoprotein
RCA: Ricinus communis agglutinin-1
GSL: Griffonia simplicifolia B4 isolectin
ED-1: macrophage-specific membrane antigen
LCA: leucocyte common antigen (tyrosine phosphatase receptor)
ND: Not Documented

1. Akiyama H, McGeer PL (1990). *J Neuroimmunol* 30:81–93.
2. Ashwell K (1990). *Brain Res Dev Brain Res* 55:219–230.
3. Bo L, Peterson JW, Mork S, Hoffman PA, Gallatin WM, Ransohoff RM, Trapp BD (1996). *J Neuropathol Exp Neurol* 55:1060–1072.
4. Cammer W, Zhang H (1993). *Glycobiology* 3:627–631.
5. Dalmau I, Vela JM, Gonzalez B, Castellano B (1997). *Brain Res Dev Brain Res* 103:163–170.
6. Deckert-Schulter M, Schulter D, Hof H, Wiestler OD, Lassmann H (1994). *J Neuropathol Exp Neurol* 53:457–468.
7. Engel S, Wehner HD, Meyermann R (1996). *Acta Neurochir Suppl* (Vienna)* 66:89–95.
8. Graeber MB, Streit WJ, Kreutzberg GW (1988). *J Neurosci Res* 21:18–24.
9. Grau V, Herbst B, van der Meide PH, Steiniger B (1997). *Glia* 19:181–189.
10. Hayes GM, Woodroofe MN, Cuzner ML (1987). *J Neurol Sci* 80:25–37.
11. Kato H, Kogure K, Liu XH, Araki T, Itoyama Y (1996). *Brain Res* 734:203–212.
12. Lawson LJ, Perry VH, Dri P, Gordon S (1990). *Neuroscience* 39:151–170.
13. Long JM, Kalehua AN, Muth NJ, Hengemihle JM, Jucker M, Calhoun ME, Ingram DK, Mouton PR (1998). *J Neurosci Methods* 84:101–108.
14. Mannoji H, Yeger H, Becker LE (1986). *Acta Neuropathol* (Berl) 71:341–343.
15. McGeer PL, Itagaki S, Tago H, McGeer EG (1987). *Neurosci Lett* 79:195–200.
16. Moneta ME, Gehrmann J, Topper R, Banati RB, Kreutzberg GW (1993). *J Neuroimmunol* 45:203–206.
17. K, Ogawa M, Yoshida M (1994). *Neuroreport* 5:1224–1226.
18. Perry VH, Matyszak MK, Fearn S (1993). *Glia* 7:60–67.
19. Perry VH, Hume DA, Gordon S (1985). *Neuroscience* 15:313–326.
20. Streit WJ, Kreutzberg GW (1987). *J Neurocytol* 16:249–260.
21. Suzuki H, Franz H, Yamamoto T, Iwasaki Y, Konno H (1988). *Neuropathol Appl Neurobiol* 14:221–227.
22. Williams AE, Lawson LJ, Perry VH, Fraser M (1994). *Neuropathol Appl Neurobiol* 20:47–55.

Many neurodegenerative conditions activate microglia, e.g. stroke and trauma. Raivich et al. (1999) distinguish several stages of microglial activation by the induction of molecular markers (integrins, iNOS, MHC class I and II, thrombospondin) and cellular activities (phagocytosis, proliferation). For example, in mice, facial motor nerve axotomy causes retrograde neuron death in the facial motor nucleus, with these different stages of microglial activation (Raivich et al. (1999)). In this model, axotomy also causes transient infiltration of lymphocytes (CDA11a/alphaL), which surround the activated microglia, but without breakdown of the blood-brain barrier (Raivich et al. (1998)). Activation can also persist during prolonged disease processes, e.g. in microglia around senile plaques in Alzheimer's disease that are found throughout clinical stages lasting five to ten years (see Frautschy, this volume). After a stroke causing demyelination, microglia may retain lipid droplets (presumably myelin debris) for at least ten years; at longer times after stroke, debris-containing microglia may migrate to the perivascular space, where they are recognized as perivascular macrophages (Kosel et al. (1997)).

In general, perivascular macrophages express high levels of MHC class II antigens characteristic of activation, which in the absence of an earlier lesion, is thought to be induced by their contact with blood proteins (e.g. Thomas (1992, 1999)). Perivascular macrophages in the posterior pituitary (neural lobe, or neurohypophysis) appear to have a physiological role in clearing neuropeptides. In adult female rats, these cells envelop "apparently healthy" neurosecretory fibers and contain phagosomes with neurosecretory granules. As further evidence for a physiological role, increased hypothalamic neurosecretion in response to osmotic stress also increases the microglial enclosure of neurosecretory terminals (Mander and Morris (1994)). These microglia are thought to degrade vasopressin and other neurohypophyseal peptides (Pow et al. (1989); Lawson et al. (1993)). Future molecular studies may show which aspects of these activation phenotypes are shared with age changes in microglia.

3. Age-related changes in microglia

a. Hippocampus, cerebral cortex, and white matter tracts

We review the literature on aging in microglia by species because laboratory rodents do not acquire senile plaques and extracellular deposits of amyloid unless made transgenic with human Alzheimer mutations. Amyloid β-peptide deposits in Alzheimer disease, which are also common in aging human brains even if not diagnosed as Alzheimer's disease, can cause local activation of microglia (see Frautschy, this volume). Amyloid β-peptide deposits may also cause increased flux of blood-born monocytes, as observed in a blood-brain barrier model (Fiala et al. (1998)). Similarly, peripheral amyloid peptides are associated with the local accumulation of tissue macrophages and oxidized proteins, e.g. glycoxidated alpha 2-microglobulin amyloid in hemodialysis patients (Miyata et al. (1994)). Thus, the aging rodent brain is an important model that may help resolve age changes that are independent of cerebral amyloid deposits.

TABLE 13.2. Age-related microglial changes in rodents

marker	hippocampus	cortical grey matter (1)	cortical white matter (2)
morphology		a, 3–12 mo.: 0 12–27 mo.: +	
ED-1 lysosomal marker of most macrophages		b, 6–25 mo.: ++	b, 6–25 ma.: ++
OX-6 MHc class II antigen	d, hilus 3–24 mo.: ++ e, dentate all layers 9–23; +++ b, d, dentate molec layer 3–24 mo.: 0	c, 2–12 mo.: +++	c, 2–12 mo.: +++ d, 3–24 mo.: ++
OX42 complement receptor, CR3bi	d, hilus 3–24 mo.: + d, dentate molec layer, 3–24 mo.: 0	c, 2–12 mo.: 0	c, 2–12 mo.: 0 d, 3–24 mo.: ++
Mac-1 CR3 receptor	f, hilus 5-13-28 mo.: 0 f, dentate all layers 5-13-28 mo.: 0		
TGF-β1 mRNA	d, hilus 3–24 mo: + d, dentate molec layer 3–24 mo.: 0		d, 3–24 mo.: 0

(1) neocortex layers
(2) neocortex myelinated tracts and corpus callosum
−, decrease of >25–50%; +, increase of 25–99%; ++, 100–250%; +++, >250%
a. Vaughan and Peters (1974); rat, Sprague-Dawley.
b. Perry et al. (1993); rat, Lister hooded.
c. Ogura et al. (1994); rat, male Wistar.
d. Morgan et al. (1999); rat, male F344 xBN hybrid.
e. Hauss-Wegrzniak et al. (1999); rat, male, F344.
f. Long et al. (1999); mouse, male C57BL/6J.

(1). Rodents

The few studies to examine the normal aging brain accord with the regional selectivity of changes in number and activation (Table 2). Other solitary reports on glial aging in mice are reviewed by Scott and Mandybur (1996). However, these studies differed in the anatomical sections examined, in the cell markers used, and in genotype (see Note A p. 296). Another concern is species differences among rodents, which may alter outcomes or impressions of aging. According to Perry et al. (1993), "... microglia of the rat are more branched than those of the mouse," whereas axotomy of the facial nerve causes greater cytokine responses and T-cell infiltration in mouse than rat (Raivich et al. (1998)). These caveats aside, there is good agreement on regions that do and do not show microglial activation.

White matter tracts show consistent activation, starting with a very low baseline of activated microglia in juvenile rodents (one to two months, postweaning through sexual maturation). In the corpus callosum, the numbers of activated microglia increased by >50% between three, six and twenty-four months in two

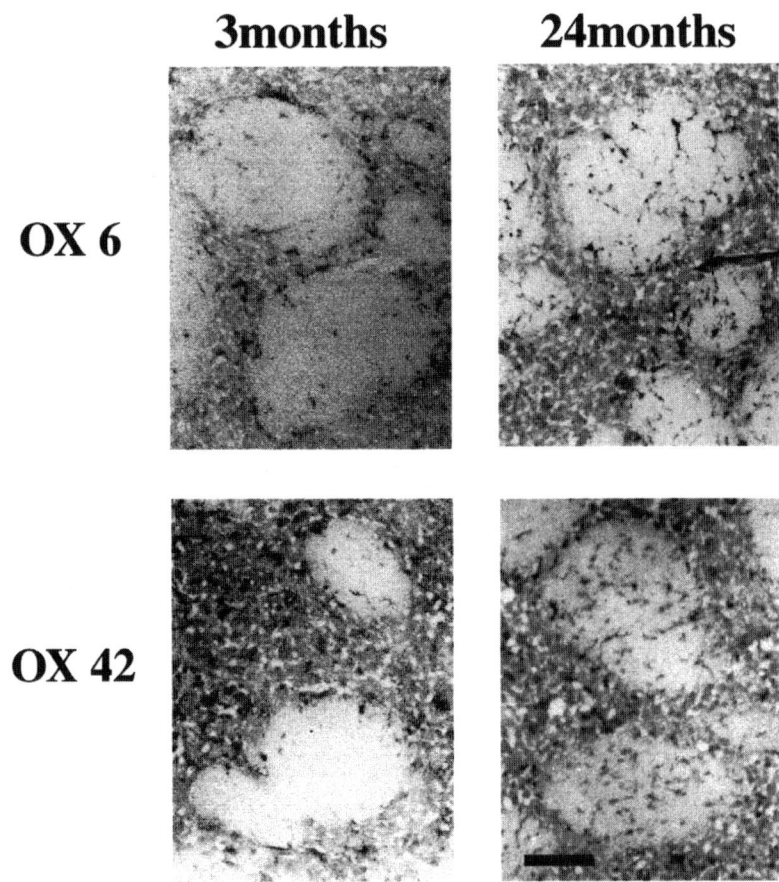

FIGURE 1. Photomicrograph of corticostriatal bundles within the caudate-putamen of 3-month and 24-month rat brains processed for OX6 (MhC class II) and OX42 (CR3) immunoreactivity. Scale bar = 50 μm. Taken from Morgan et al. (1999).

genotypes of male rats (Morgan et al. (1999); Perry et al. (1993)) (Table 2). Intermediate ages show emergent trends: in a third genotype, three- and twelve-month-old rats showed microglial activation by OX6, but not by OX42 immunohistochemistry. Other white matter locations of microglial activation during aging are globus pallidus, particularly around the myelinated corticostriatal tracts (Fig. 1) (Morgan et al. (1999)), and in the dorsal horn of the spinal cord in both white and grey matter (Stuesse et al. (2000)).

In contrast, the hippocampus shows little if any microglial activation. A definitive recent study used rigorous stereological methods ("optical dissector" West (1999)). In C57BL/6J male mice, a strain used in several studies discussed here, age did not alter the number of microglia across the lifespan (number of cell bodies per area) in hippocampal regions (Long et al. (1999)). Microglia were immunostained for the complement-3 (CR-3) receptor; the regions included the hilus of the hippocampus and the neuropil (molecular layer) and neuron layers of the dentate gyrus and the CA1 pyramidal neuron layer of the hippocampus. Similarly, conventional histometry did not detect changes in the density of OX-42 immunopositive microglia in the outer molecular layer of the dentate gyrus of male rats (Morgan et al. (1999), Perry et al. (1993)) These negative results do not exclude subtler changes that might be detected by other markers. For example, an osmotic stress that increased engulfment of neurosecretory terminals activities in posterior pituitary microglia (Section 2) also increased ICAM-1 and CD45 (LCA) expression; these markers are absent in resting parenchymal microglia and are induced by neurodegeneration (Lawson et al. (1993)).

Because of the close interaction of microglia with astrocytes, it is pertinent to review the pioneering papers of Landfield, Lynch, and colleagues that described regionally selective hypertrophy of astrocytes during aging (Landfield et al. (1977, 1978, 1981); Lindsey et al. (1979)). The hippocampus of aging F344 rats showed prominent hypertrophy of astrocytes, but without change in total number (Lindsey et al. (1979)). This study emphasized the neuropil in a region of the dentate gyrus, the outer molecular layer mentioned above, which receives the terminals of the perforant path. This major pathway from the entorhinal cortex to the hippocampus shows a trend for loss of synapses during aging (Geinisman et al. (1977, 1995, 1999)). The astrocytic hypertrophy began in middle-age, but was *less* extensive within the myelinated fibers of the perforant path (Landfield et al. (1977); Lindsey et al. (1979)). "Dark" cells, provisionally identified as a microglia, were found with astrocytes in an increased number of glial clusters (Landfield et al. (1981)). If such glial clusters can be reconfirmed in current rodents, it would be interesting to examine them for T-cells, which are found near glial clusters after facial nerve axotomy (Raivich et al. (1998)).

Lastly, we mention the first quantitative ultrastructural analysis of glial aging in the cerebral cortex (Vaughan and Peters (1974)). In Sprague-Dawley rats, aged 3, 12, 24, and 27 months, the number of microglia increased by 65%, whereas the numbers of astrocytes and oligodendroglia were unchanged. Together, these reports on various aging models suggest that microglial activation during aging in the absence of brain lesions has a high degree of regional selectivity. Additional molecular approaches may show whether microglia or astrocytes are activated independently of each other in the same microregion.

(2). Human and monkey

In apparent contrast to rodents, human and primate hippocampal microglia become activated during aging, according to three reports from 1997. In the most detailed study, DiPatre and Gelman (1997) characterized normal young and nondemented elderly without "acute neuronal necrosis" or neuropathology in the hippocampus. Activated microglia, identified by ferritin immunohistochemistry, were 50–130% more numerous in eight hippocampal-entorhinal cortex regions of normal elderly. The ferritin-positive cells had thickened processes, which is a characteristic of microglial activation in rodents. The greatest increases were in the CA2 and dentate zones of the hippocampus. Age-matched Alzheimer brains showed more activated microglia. Moreover, the density of microglia in the dentate gyrus, which receives perforant path axons that degenerate in Alzheimer disease, was correlated with the density of senile plaques at the perforant path origin in the entorhinal cortex layers 2 and 3. This correlation is intriguing: as discussed above, Landfield and colleagues (1977) found the greatest hypertrophy of astrocytes in the same zone in aging rats, which do not develop senile plaques or amyloid deposits during aging. Overall, the increases of activated microglia during normal aging are 25–75% less than in Alzheimer brains. DiPatre and Gelman (1997) concluded that many activated microglia in Alzheimer brains are "... in vast areas of neuropil between senile plaques." Two other studies using different histochemical markers also showed increased microglial activation during normal aging, in the absence of senile plaques and Alzheime's disease. Microglia expressing the MHc class II antigen (HLA-DR) in normal elderly had the resting-ramified structure in cortical grey (Perlmutter et al. (1992)), which differed from the thickened processes of ferritin-immunopositive microglia (DiPatre and Gelman (1997)). However, both studies agreed that cortical microglia in Alzheimer brain are mainly reactive with thickened processes. Furthermore, Streit and Sparks (1997), using LN-3 for HLA-DR, also found increased numbers of activated microglia in clusters in the hippocampus, comparing three groups: <50, 50–70, and >70, including one 90-year-old without senile plaques. A history of heart disease appeared to increase the activation, particularly in the younger groups. Future work may show whether aging changes in microglia contribute to Alzheimer neurodegeneration independently of the senile plaque.

Female macaques show activated microglia in white matter, as observed in aging rodents. The area of immunostaining for the MHC class II increased >2-fold between young adult (5–9 y) and "mature" animals (10–18 y) (Sheffield and Berman (1998)). This mature group is equivalent to middle-aged, in view of the 25 year life expectancy and 40 year maximum lifespan of this species (Finch and Sapolsky (1999)). The white matter region at the base of the precentral gyrus included the hippocampus and parasubiculum and in the older group had 25-fold microglial immunostaining compared with the young. In normal, aging humans white matter, atrophy is a general trend that progresses from middle age without loss of grey matter or cognition, e.g. the imaging study of Guttmann et al. (1998).

b. Hypothalamus

The hypothalamus contains subpopulations of microglia with specialized functions, some of which interact with neuroendocrine aging. A collaboration of the

Finch lab with James Brawer and Hyman Schipper identified a subpopulation of microglia in the arcuate nucleus of middle-aged female mice and rats with increased phagocytic activity. Moreover, this activity was attenuated by long-term ovariectomy (Schipper et al. (1981)), whereas similar age-like changes were induced by intense exposure of young male or female rats to estradiol (Brawer et al. (1980)). Astrocytes with dense inclusions increased during aging, and again the age changes were attenuated by long-term ovariectomy. Neuroendocrine dysfunctions in the preovulatory surge of gonadatropins show parallel changes to the hypothalamic microglia, with attenuation by long-term ovariectomy and acceleration of age-like changes in young rodents exposed to sustained physiological levels of estradiol (Finch et al. (1984); Kohama et al. (1989)). The primary cell targets of estradiol in these "neurotoxic effects" of estradiol are not known. Miller et al. (1989) re-examined the effects of long-term ovariectomy in C57BL/6J mice, the same strain used by Schipper et al. (1981). Using cell structure and toluidine-blue-staining, the density of all glia decreased with aging in the arcuate nucleus, with larger decreases of glial density in long-term ovariectomized mice; however, the numerically large differences in the arcuate nucleus did not show significant effects of age or treatment by ANOVA conducted on three hypothalamic regions. Because this experimental design did not resolve cell types among astrocytes or microglia, or the specialized subpopulations with inclusion bodies, C.E.F. disputes the statement of Scott and Mandybur (1996) that ". . . The data of Schipper et al. (1981) are contradicted by Miller et al. (1989). . . ." Schipper has since shown that the astrocyte inclusion bodies contain peroxidase activity and redox-active iron, which are probably derived from degenerating mitochondria (Schipper (1996)).

c. Microglial inclusion bodies

Inclusion bodies in microglia from aging rats suggest increased phagocytic or endocytic activities, which may be associated with increased lysosomal mRNA detected by microarray (see below). As we mentioned, a subpopulation of hypothalamic microglia in aging female rats had inclusion bodies (Schipper et al. (1981)). Similarly in cerebral cortex, microglial processes from aging rats had ". . . heterogeneous membrane-bound dense material; . . . inclusions are particularly prominent in older animals . . . of dense material . . ." (Vaughan and Peters (1974), pp. 418–422). Aging pigment (lipofuscin) is prominent in some microglia, e.g. around the choroid plexus of aging mice (Sturrock (1988)). Perivascular macrophages also show vacuolation and inclusion bodies in aging mice: large, flocculant inclusion bodies with degenerating myelin in the senescence-accelerated mouse (SAM-P/10) (Lee EY et al. 2000) and PAS-positive granules in C57BL/6 mice (Lamar et al. (1976)). In a different example, Landfield et al. (Fig. 4 of 1981) described a microglial cell with numerous heterogeneous inclusions apposed to a hippocampal pyramidal neuron. This cell interaction was not considered to be neuronophagia, a microglial attack on degenerating neurons, which can include synaptic stripping (e.g. Graeber et al. (1993); Raivich et al. (1999)). Neuronophagia by glia was considered important in early studies of aging (e.g. Andrew and Cardwell (1940)) and merits revisiting in relation to the reports indicating age-related decreases of synaptic density (Geinisman (1995)) and receptors (Severson and Finch (1980)) in various brain regions.

d. At the blood-brain barrier

The cerebrovasculature shows age changes that are common throughout the body, including thickening of the basal lamina, e.g. in the hippocampus of F344 rats (Topple et al. (1991)) and classic histochemical aging changes of hyalinization and PAS staining, which are associated with oxidation (Sobin et al. (1992); Lamar et al. (1976)). Age changes are also reported in vascular pericytes, which are a specialized macrophages of the vasculature that can be activated by injury; whether pericytes convert to microglia is uncertain (Thomas (1999)), since data are limited and inconclusive.

Two studies of aging rats used the ED-2 membrane antigen, which detects macrophages, but not monocytes. Liu et al. (1996) reported >2-fold increases of ED2 immunopositive perivascular cells by two years, whereas a survey of Perry et al. (1993) did not observe age changes. Finally, a sample of three old mice of a little-studied ASH/TO strain showed increased numbers of phagocytic cells in the choroid plexus (Sturrock (1988)). The increase of perivascular macrophages during aging does not imply a breakdown of the blood-brain barrier transport system, which appears to maintain its integrity during aging, despite reduced transport of some compounds (Mooradian (1994)).

e. In vitro *models*

Primary glial cultures can be established from adult rodent brains and show promise as models for *in vivo* glial age changes that are distinct from the clonal senescence ("Hayflick phenomenon") of serially cultured cells, in which the end-point is loss of cell replication. A suspension of glial cells is plated once and grown to confluence, when microglia are separated by shaking from the more strongly adherent astrocytes. Several of us showed that primary microglia from neocortex of 24-month-old rats had a higher proportion of cells with activated phenotypes of ameboid shape and OX6 immunopositivity (Rozovsky et al. (1998)). Moreover, microglia from old F344 rats are less sensitive to the antiproliferative actions of TGF-β1, and to the down regulation of the NO production induced by LPS (Rozovsky et al. (1998)).

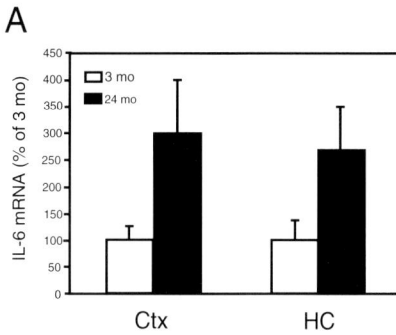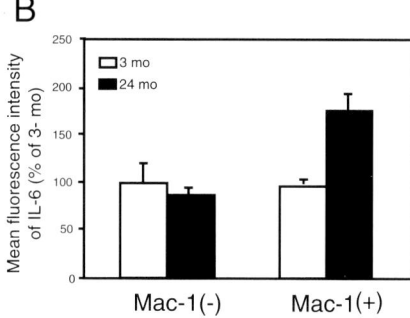

FIGURE 2. A: IL-6 mRNA levels are upregulated in cerebral cortical or hippocampal glial cultures originated from old (24 mo.), compared to that from young (3 mo.) F344 rat brains. B: Elevated IL-6 expression in brain glia cultures originated from 24 mo. vs. 3 mo. old mice. IL-6 increase is predominantly attributed to an increased subpopulation of MAC-1 (C3R, a murine microglial marker) positive microglial cells and with elevated expression by individual microglia (B adopted from Ye and Johnson (1999)).

Microglia cultured from 24-month-old F344 rats had increased basal and LPS-induced levels of IL-6 mRNA (Fig. 2A) which corroborates the increased IL-6 peptide in FACS resolved microglia from aging BALB/c mice (Fig. 2B) (Ye and Johnson (1999)). Similar increases of IL-6 peptide were found in cerebral cortex, hippocampus and cerebellum of aging mice (Ye and Johnson (1999)). Of great interest is the cytotoxicity of activated microglia from aging brains. Pilot data indicate an age-related decrease in neurotoxic factors released by LPS- stimulated microglia from aging rat cerebral cortex (Xie et al. (2000)). These data appear to be consistent with age-related reduction of microglial responses to LPS *in vivo* (see below).

4. Aging and response of microglia to injury

Because microglia are prominent in Alzheimer's disease (see Frautschy, this volume), a major question is how age effects microglial responses to brain lesions. Answers to this question could have a major impact on animal models for Alzheimer's disease, which have largely used young rodents. Quite divergent effects of age on microglial responses are shown in different lesion models. First, we discuss three studies showing *smaller* microglial responses to lesions in the cerebral cortex, hippocampus, and spinal cord, followed by two reports of *greater* response to lesions in cerebral cortex and striatum.

Infusion of LPS into the lateral ventricle for 33 days induces a strong inflammatory response in the hippocampus of young rats (three and nine months), with major increases in the numbers of activated microglia in the hippocampus, increased cytokines, and neuron structural abnormalities (Hauss-Wegrzyniak et al. (1999, 2000)). However, in 24-month-old rats, LPS did not further increase the numbers of activated microglia above the vehicle controls, which remained at about 50% of the levels induced by LPS in younger rats. Whatever the mechanisms, the microglia of aging brains clearly responded differently to this standard inflammatory stimulus.

Deafferenting lesions of the hippocampus (perforant path model), which induce compensatory sprouting in the deafferented zone, showed marked impairments by twelve months in the induction of insulin-like growth factor-1 (IGF-1), as detected by *in situ* hybridization for IGF-1 mRNA (Woods et al. (1998)). IGF-1 supports axonal growth and is induced in microglia just before the onset of compensatory sprouting (Woods et al. (1998)). In contrast, no age differences were found in the IGF-1 mRNA around the lesion cavity, which contains peripheral macrophages that enter the brain after disruption of the blood-brain barrier by the surgical procedure. The microglia affected by aging were not characterized in detail in this lesion model.

The delayed induction of IGF-1 mRNA in aging rats is consistent with delayed clearance of neuronal debris and delayed onset of sprouting in aging rats (Hoff et al. (1982); Anderson et al. (1986)). Lysosomal enzymes also showed delayed and smaller induction during aging in this model (Vijayan and Cotman (1983)). The lysosomal enzymes acid phosphatase and β-glucuronidase are found in activated microglia and astrocytes. In the third example, the activation of microglia in spinal cord after sciatic nerve constriction was 50% smaller in 24-month-old rats than at five and fifteen months. (Stuesse et al. (2000)).

In contrast, two studies with different approaches show increased microglial activation in response to lesions in aging models. To explore the age factor in Alzheimer's disease, the cerebral cortex of three species was injected with fibrillar amyloid beta-peptide (200 pg in 1 microliter; human sequence, residues 1–40) (Geula et al. (1998)). The microglial activation and neuronal loss was much greater in old rhesus monkeys and marmosets, whereas two-year-old rats had much less microglial response and neurotoxicity. Further studies may consider species differences in the Aβ peptide, which is identical in primates and humans, but differs from rodents (Johnstone et al. (1991)). We also mention a preliminary report of nigrostriatal lesions of the basal ganglia as a model for Parkinson disease, in which microglial activation was greater in 24-month-old rats in the striatum ipsilateral to the lesion (OX6, MHC class II immunoreactivity) (Morgan and Gordon (1996)). In parallel were greater responses of astrocytic GFAP (Gordon et al. (1997); full report). The time course of microglial responses was not altered by aging, unlike the hippocampal responses described above.

How can we reconcile these opposing outcomes of aging on microglial responses to lesions? The simplest explanation is that brain regions differ markedly in responses to aging, which has been described as "mosaic" aging (Soong et al. (1992);

Morgan et al. (1999)). An extreme example of mosaic aging is the 1000-fold regional difference in the accumulation of mutant mitochondrial genomes (Kearn-Sayers deletion of mtDNA) in normal aging human brains: striatum>>>cerebral cortex (Soong et al. (1992)). Although there is no information on whether the aging rat brain shows the same mtDNA mosaicism, many glial changes of aging are also quite regional, as discussed above.

Lastly, we discuss an important pharmacological finding in the study of LPS infusions (Hauss-Wegrzyniak et al. (1999)). One group was treated with a novel NSAID (NO-Flurbiprofen), which releases nitric oxide (NO), which greatly attenuated the microglial activation in the younger rats, but not in the older. The release of NO by this drug in the gastric mucosa is thought to minimize the pathogenic changes, such as reduced blood flow, that ensue from suppression of prostaglandin synthesis. This is the first indication that age alters responses of microglia to anti-inflammatory drugs, which could influence therapeutic interventions and drug design. A future concern is that the increasing use of anti-inflammatory drugs could confound studies of microglia in postmortem human brains, e.g. MacKenzie and Munoz (1998) showed a three-fold reduction in the numbers of activated microglia in nondemented elderly who were long-term users of NSAIDs (non-steroidal anti-inflammatory drugs), *versus* non-users.

5. Messenger RNA and aging

Of the many mRNAs induced during microglial activation in young rodents, few have been examined in aging rodent brains. *In situ* hybridization is generally required for cell type localization because many inflammatory factor mRNAs are expressed in several cell types. We give two examples of mRNAs induced by lesions in both neurons and microglia: TGF-β1 mRNA in neurons (Zhu et al. (2000)) and microglia (Morgan et al. (1993); Pasinetti et al. (1993)) and C1qB mRNA in neurons (Rozovsky et al. (1994)) and microglia (Morgan et al. (2000)). Both show modest age-related increases on northern blots (Pasinetti et al. (1999)), which raise the further question of which cell types are contributory. We do not yet have an answer. By *in situ* hybridization of two-year-old rat brains, TGF-β1 mRNA was not increased in glial-rich neuropil zones of the corpus callosum or hippocampus (Morgan et al. (1999)), nor was C1qB mRNA increased in the hippocampal hilus of aging rat brains previously characterized for GFAP (Morgan, Patel and Finch, unpublished). As a positive example of this approach, the increase of GFAP mRNA during aging as detected by Northern blot hybridization in specific brain regions (Nichols et al. (1993); Goss et al. (1990)) was associated with increased mRNAs per astrocyte in the outer molecular layer of the dentate gyrus (Yoshida et al. (1996); Morgan et al. (1999)).

Many of the mRNAs studied for aging in the lab of C.E.F. were originally selected from two cDNA libraries (May et al. (1990); Poirier et al. (1991); Day et al. (1992); Nichols et al. (1994)). We sought shared subsets of mRNAs that are increased in Alzheimer brains; that are increased to a lesser extent during normal aging; and that are regulated by gonadal and adrenal steroids. Clones were selected by differential plaque hybridization, which is dominated by high prevalence mRNAs that only represent a minor fraction of the brain mRNA inventory. Among the cDNAs identified by these screens were GFAP (increased during aging, Alzheimer's disease and in perforant path models; regulated by sex steroids and glucocorticoids and TGF-β1) and complement C1qB (increased during aging, Alzheimer's disease, and in lesion models), which is present in microglia and neurons (Rozovsky et al. (1994); Lampert-Etchells et al. (1993); Pasinetti et al. (1999)).

Further discovery of mRNA changes during aging was enabled by microarray technology. Using the Affymetrix system, the labs of R.W. and T.P. have identified a new group of mRNA candidates for increases during aging that could prove to be associated with microglia, or other monocyte lineage cells (Lee CK et al. (1999, 2000)). A comparison of neocortex and cerebellum from 5- and 30-month-old mice (C57BL/6 males)(see Note A p. 296) showed that aging is associated with specific mRNA changes (see also http://www.genetics.wisc.edu/prolla/set2/aging.html). Of the 6347 genes surveyed on the Affymetrix microarray, about 1% increased by >1.7-fold increases in the neocortex and cerebellum. A subset of these mRNA age changes was validated by quantitative PCR (RTQ-PCR) (Lee CK et al. (2000)). GFAP mRNA and C1q mRNA increases were among several previously identified by the lab of C.E.F.

That relatively few mRNAs show pronounced age changes is consistent with analysis of mRNA inventory using solution hybridization to single copy DNA in the lab of C.E.F. (Colman et al. (1980)). Rats of two strains showed no age changes in the yield of polysomal poly(A)mRNA, or in its sequence complexity. The statistics set an upper limit of 5% to the number of different mRNA changes in rat brain, which approximates the 1-2% of mRNAs that change using oligonucleotide arrays. At that time, many believed that aging was caused by gross impairments of gene expression (Cutler (1975)). This view was not supported by analysis of brain RNA kinetic complexity, the yield of RNA in most brain regions of C57BL/6J mice (Chaconas and Finch (1973)), or the histologically normal cell phenotypes of neurons and glia in aging rodents, which depend on precisely maintained, cell-specific patterns of gene expression.

For the microarray subset of age changes, we have provisionally assigned functional classes based on observations largely on cells in non-neural tissues, from which regulatory mechanisms can be inferred. Of the 60 mRNAs that increased during aging, about 20% can be associated with immune or inflammatory responses. Of immediate interest to the present topic of microglia are increased mRNA for the microglial and macrophage migration factors, ICAM-2, Exodus-2, and MPS1, and CD40L receptor, which is involved in lymphocyte activation. The increase of lysozyme C could be associated with microglial phagocytosis and inclusion bodies, which increase with age, as we discussed. These and other monocyte-macrophage mRNAs could also be expressed in brain microglia and perivascular macrophages. In view of the presence of T-cells in microglial clusters after certain neuronal lesions, as described, it will be important to look for T-cell mRNAs in microglial clusters of the aging brain.

The complement (C) system is represented by increased mRNAs for the initiator of the classical cascade C1 (subunits C1qA, C1qB and C1qC) and C4 in aging mice. Studies from the lab of C.E.F showed that C1qB is induced in neurons and microglia by experimental lesions (Johnson et al. (1992); Pasinetti et al. (1992); Rozovsky et al. (1994)), that the brain can make enzymatically active C1q (Goldsmith et al. (1997)), and that C1qB and C4 mRNAs increase during Alzheimer disease (Lampert-Etchells et al. (1993)). These findings indicate that the C-proteins found in senile plaques of Alzheimer's disease are made by local inflammatory responses (reviewed by Akiyama et al. (2000)). A modest increase of C1qB mRNA was found in aging rat striatum (Pasinetti et al. (1999)). Contrary to expectations that the brain as an immune-privileged site should not normally produce factors that mediate the B- and T-cell-based immune responses, recent work shows that C-system mRNAs can be made by all brain cell types. Moreover, the classical C-system cascade can be activated independently of antigen-antibody complexes, by debris from cell degeneration, including DNA (single or double stranded) and myelin (reviewed in Goldsmith et al. (1997)). We caution that the induction of complement mRNAs during aging does not constitute evidence that the C-system is enzymatically activated.

How general are these age changes between brain regions? GFAP increases appear to be general age changes in most CNS regions. Moreover, a subset of the increases associated with inflammatory genes was found in both cerebral cortex and cerebellum (Table 3). Taken together, our observations support the hypothesis that many aspects of brain aging are consequences of heightened inflammatory processes and oxidative stress.

6. Mechanisms in microglial activation

Why do microglia become activated during aging? We discuss four possibilities: neuron death; changes in neuron activity; changes in response to other glia; and response to oxidation.

A great many histopathological observations show increased expression of MHC class II antigens in association with dying neurons. Thus, it seems plausible that glial changes are secondary to neuron degeneration. Here we find ourselves on one of gerontology's most treacherous paths, that of variable neuron loss and

TABLE 13.3. Shared mRNA increases during aging in cerebellum (cb) and neocortex (ct) (selected examples)

gene name	increase during aging	functions
CD40L receptor	2x(ct); NC(cb)	lymphocyte activation
C1qA, C1qB and C1qC	1.7–4.4x	initiator of the classical complement cascade; induced in neurons and microglia (see Section 5)
COX-1	2x(ct); NC(cb)	prostaglandin H synthetase
Exodus-2	1.8x(ct); 4.3x(cb)	Exodus-2/SLC/6Ckine/TCA4; chemokine which is a T-cell attractant and inhibits proliferation of marrow progenitor cells
GFAP	2.3x(ct); 3.8x(cb)	intermediate filament of astrocyte; induced by oxidative stress and neuronal dysfunction
ICAM-2	2x(ct); NC(cb)	intracellular adhesion molecule expressed during brain development in astrocytes and microglia
lysozyme C	2.2x(ct); 6.4x(cb)	digestive enzyme secreted by stomach and also secreted by leptomeningeal cells (Ohe et al., 1996)
MPS1	1.9–3.9x	cytokine; modulates hemopoeisis

NC: no change

individual variations in aging (Note B). We argue that microglial activation is too general, and arises too early during aging, to be secondary to neuron death in all examples. At least four independent samples of two-year-old rats show significant neuron loss, in the range of 10–30% in the pyramidal neuron layers of the hippocampus (Landfield et al. (1978, 1981); Lindsey et al. (1979); Meaney et al. (1988)). In marked contrast, recent studies using stereological techniques have failed to find pyramidal neuron loss with age (Rasmussen et al. (1996)). These different outcomes of highly competent investigators are not likely to result from technical problems. An alternative hypothesis is that the variation in between studies is traceable to sporadic neuron death (Note B.)

Microglial activation could be involve age-related changes in electrical activities. The microglial MHC class II marker is activated by spreading depression (Gehrmann et al. (1993)) and epileptiform seizures (Shaw et al. (1994)), whereas blocking of conduction induces GFAP (Canady et al. (1994)). In each case, glial responses are transient and the neurons recover. The mechanisms of microglial activation could include ion channels that are sensitive to extracellular potassium, or to increased cytokines, particularly TNF (Gehrmann et al. (1993)).

Because astrocyte activation appears to begin at middle age (12–18 months), which is at least six months before neuron loss was detected (Landfield et al. (1978, 1981); Lindsey et al. (1979); Meaney et al. (1988)), this schedule appears to preclude astrocytic activation being secondary to neuron death. Recall here that two studies of aging rodents showed white matter microglial activation in animals that did not show microglial activation in the hippocampus. One may also conclude that astrocyte activation is not sufficient for microglial activation, at least in the hippocampus. In white matter, microglial activation during aging could be either a cause of, or consequent to, alterations in oligodendroglia. As described, white matter shows marked atrophy during normal aging in humans, whereas in rats oligodendroglia become vacuolated (Vaughan and Peters (1974)). Both observations suggest chronic progressive demyelination. More studies are needed to show whether the activated microglia in white matter are phagocytosing myelin (Aldskogius and Kozlova (1998); Bruce-Keller et al. (1999)).

Oxidized lipids in myelin sheaths could be a factor in microglial activation. Because protein oxidation increases during aging in rodent brain (Butterfield et al. (1997); Dubey et al. (1996)), it is plausible that lipid oxidation also increases during aging, particularly in the outer sheaths of myelin that are remote from cytoplasmic repair processes. Oxidized lipids can activate microglia and peripheral macrophages, and are recognized by scavenger receptors (Bruce-Keller et al. (1999)). Here we can learn from studies of oxidative stress in diabetes, in which the pathological sequelae of diabetes are being debated as a cause or a consequence. Glycated adducts, or advanced glycation and products (AGE) form in association with hyperglycemia of diabetes in many tissues with long-lived proteins and lipids. Macrophages recognize AGEs through scavenger receptors that, in turn, activate NF-kappaB and oxidative stress responses (Yan et al. (1994); Baynes and Thorpe (1999)). Microglia have scavenger receptors that are induced by injury (Bell et al. (1994); Grewel et al. (1997)). Little is known about the levels of AGEs during aging in brain, and the immunohistochemical findings do not give any absolute quantification (Li et al. (1994, 1995)). Although advanced diabetics have low levels of AGE in their peripheral tissues, local levels could be high enough to stimulate macrophages (Baynes and Thorpe (1999)). Studies are needed on effects of aging on the threshold response of microglia to oxidized proteins and lipids, and on the oxidative responses induced by activating scavenger receptors.

7. Influence of diet on microglial activation

In view of the potential role of oxidized proteins in glial activation, it is of great interest that many age changes in mRNA in the brain and other organs can be attenuated by caloric restriction (CR). Restricting the *ad libitum* food intake by 10–40%, and with sufficient micronutrients, gives robust increases of lifespan and slowing of many aging processes, as first demonstrated in laboratory rats 65 years ago (1935 report reviewed in McCay et al. (1989)). Whether begun early in life or at middle-age, CR increases mean and maximum lifespan, reduces and delays spontaneous abnormal growths and cancers, kidney degeneration, and many other age-related diseases (Weindruch and Walford (1988)).

The mechanisms involved in CR are poorly understood, despite intensive investigation, since so many theories of aging are fitted by the physiological changes of CR. For example, CR reduces not only O_2 consumption on a whole-animal basis, but also lowers thyroid hormone levels and body temperature, suggesting a lower metabolic rate (Weindruch and Walford (1988)). CR reduces blood glucose levels, increases insulin sensitivity, and preserves certain immunological functions of peripheral blood lymphocytes. By lowering blood glucose, CR also decreases the accumulation of AGE and other oxidized products, e.g. in skin collagen of C57BL/6 mice (Sell et al. (2000)). Owing to this abundance of effects, it is difficult to distinguish primary from secondary effects of CR on aging retardation.

In a current hypothesis that is gaining favor, CR acts through a global metabolic response that increases metabolic efficiency, lowers the production of toxic byproducts of metabolism, and enhances specific stress-adaptation responses (McCarter (1995); Sohal and Weindruch (1996); Frame et al. (1998); Masoro (1998)). Global stress adaptations, such as that mediated by the oxyR regulon, are well characterized in bacteria (Christman et al. (1985)) and are likely to be found in mammals. Other evidence for the role of stress responses in longevity comes from recent work in Drosophila and Caenorhabditis that show strong genetic determinant of aging rates. Specifically, aging in Drosophila is retarded by increasing the expression of antioxidant enzymes (Orr and Sohal (1994)), by genetic selection for long-lived strains (Hutchinson and Rose (1991)), or by the direct selection for resistance to oxidative stress (methuselah strain) (Lin et al. (1998)). Other links of metabolic control to aging in Caenorhabditis are the mutations in the insulin-related transcription factor DAF-16 that modify the lifespan (Ogg et al. (1997)). Moreover, DAF-2 mutations, another gene involved in metabolic control, increase resistance to thermal exposure and oxidative stress (Lithgow et al. (1995)). Returning to mammals, the beneficial effects of CR in a hippocampal model of oxidative injury is associated with the induction of specific glucose-regulated molecular chaperones, such as GRP78 (Lee J et al. (1999)). Taken together, these results suggest that both CR in mammals and specific genetic alterations in invertebrates may retard the aging process through shared mechanisms of improved metabolic efficiency and increased resistance to oxidative stress.

The oxidative-stress hypothesis of aging predicts that CR would lower steady-state levels of oxidative stress and retard age-associated increases in oxidative damage (Sohal and Weindruch (1996)). Experiments on rodents support these predictions. Long-term CR attenuated age-associated increases in rates of mitochondrial ROS production and oxidative damage (Lass et al. (1998)). The organs showing most oxidative damage and greatest effect of CR were brain, heart and skeletal muscle. To state the obvious, these organs are largely composed of long-lived, post-mitotic cells that may be the critical targets for the accrual of oxidative damage during aging. Mitochondria of aging mice (C57BL/6) have increased O_2 consumption and H_2O_2 production, and increased carbonyl content, all of which are attenuated by CR (Sohal et al. (1994)). Functional consequences of oxidative damage to proteins are implied by correlations of the severity of oxidative damage in various regions of mouse brain with age-related impairments of cognitive and motor functions (Forster et al. (1996)). CR retarded these losses and diminished the oxidative damage to proteins in pertinent brain regions (Dubey et al. (1996)). Most recently, the levels of oxidized adducts in skin predict the lifespan (early deaths) in *ad libitum*-fed and CR mice (Sell et al. (2000)).

Experiments with CR in aging rats (Brown Norway x Fischer 344) F1 males link CR, neuroinflammation and ROS. Using *in situ* hybridization to evaluate cell changes in mRNAs in rat hippocampus, CR attenuated increases with age in GFAP transcription (Yoshida et al. (1996)). These findings extended to other brain regions, and to additional markers of astrocytic and microglial activation including apolipoprotein E, apolipoprotein J (clusterin), complement 3 receptor (OX42), and OX6 (MHC class II antigen) (Morgan et al. (1999)). All markers were elevated in the corpus callosum during aging and were attenuated by CR, but other regions showed marked dissociation of the extent and direction of changes. Astrocytic activation, as measured with GFAP expression, increased with age in the corpus callosum, basal ganglia and hippocampus. Generally, CR attenuated the age-related increase in GFAP mRNA and immunoreactivity. CR also reduced the age-related increase in apolipoprotein J and E mRNA and heme oxygenase-1 immunoreactivity in the basal ganglia and corpus callosum. The age-associated microglial activation measured by OX6 and OX42 immunoreactivity was reduced by CR in most subregions. The localized subsets of glial age changes and effects of CR constitute a mosaic of aging consistent with the regional heterogeneity of changes.

These effects of CR in attenuating preselected sets of mRNAs were extended using microarray technology. Of the largest mRNA age changes (>1.7-fold), about 30% (34/114) were either completely or partly prevented by CR. The effect of CR on age-associated alterations in gene expression was highly selective, according to the type of gene. For example, CR influenced only 20% of the mRNA decreases of genes involved in neuronal growth and plasticity, whereas CR prevented 50% of the induced stress-response- and 65% of immune-response-related genes. CR had similar effects on mRNAs in cerebral cortex and cerebellum (Lee CK et al. (1999, 2000)). However, mRNAs involved in neurotransmitter systems were less modified by CR. This finding is in agreement with the limited effects of CR on the age-associated loss of striatal dopamine receptors (May et al. (1992)). In many brain aging changes, CR delays, but does not ultimately block, changes (Roth et al. (1984)). These effects of CR on immune- and stress-related mRNAs agree with indications that CR reduced autoimmunity (Weindruch and Walford (1988)) and oxidative damage (Sohal et al. (1994)). Since we only examined 5–10% of the mouse brain inventory of mRNAs, as many as 1000–2000 brain-expressed mRNAs may increase or decrease by twice or more during aging.

8. Towards a unifying hypothesis: aging, oxidative stress, and inflammation in macrophages

The neuroinflammatory processes observed in aging can be considered in the context of the similar, and more intense, changes in Alzheimer brain. In particular, both aging and Alzheimer's disease involve activation of astrocytes and microglia. Moreover, senile plaques include numerous inflammatory proteins, possibly produced by resident neurons and glia, in which many inflammatory-system mRNAs have been localized by *in situ* hybridization (reviewed in Akiyama et al. (2000); Finch and Longo (2000)). We hypothesize that amyloid-containing senile plaques are not a general cause of glial activation because similar glial activation occurs during aging in rodents that do not accumulate brain amyloid deposits unless given human trangenes for Alzheimer's disease.

Moreover, many tissues throughout the body show many manifestations of inflammatory processes during aging, for example the inflammatory processes in vascular disease (e.g. Wong et al. (2000); Yasojima et al. (1998); Ross, (1999)). Accumulations of oxidized proteins (AGEs) fit well because macrophages and endothelia have scavenger receptors that respond to AGEs (Section 6). These processes are progressive in most tissues from birth onwards, and may have profound importance to brain aging because AGEs activate microglia/macrophages through scavenger receptors. Thus, two domains of inflammatory processes may be considered: processes that are widespread in the body (AGE accumulation; macrophage activation) in distinction to organ-specific focal processes in the brain, or other particular tissues (astrocyte activation).

In nonneural tissues, the role of inflammation in cellular damage during "normal aging" is unclear, although local inflammatory mechanisms are well recognized in atherogenesis. Systemic IL-6 and other acute phase proteins are often elevated in aging rodents (e.g. Ye and Johnson (1999); Sierra et al. (1992)) and macrophages show increased phagocytic activities (e.g. Wustrow et al. (1982)). Several studies of community-dwelling elderly show increases of IL-6, and other cytokines that are predictors of subsequent decline in physical and mental functions (e.g. Ferrucci et al. (1999); Bruunsgaard et al. (1999)). These analyses used samples from EPESE (NIA-funded, Established Populations for the Epidemiological Study of the Elderly). The wide individual differences in inflammatory markers may represent degrees of underlying disease processes, e.g. atherosclerosis or rheumatoid arthritis.

Systemic inflammation can modify outcomes of aging through promoting the chronic increase of inflammatory proteins known to form amyloids, such CRP and SAA as CRP and SAA (Finch et al. (2000)). These deposits could, in turn, perpetuate inflammatory responses by activating phagocytes. The factors that govern individual accumulations of amyloid are unknown. The sporadic occurrence of amyloids during aging suggests a chaotic process with transients that escape overall homeostatic regulation. There continues to be uncertainty respecting the interactions of neuroinflammatory changes with the decline of instructional immunity, e.g. the smaller primary immune responses in aging rodents and aging humans (Haynes et al. (2000); Globerson and Effros, (2000)).

The oxidants generated during inflammation may also contribute to the damage of nonneural tissue by increasing the formation of extracellular amyloid deposits and of glycoxidized proteins. However, the role of nonneural amyloids in tissue damage has not been proved experimentally, and may prove to be as complex as the issue of amyloid neurotoxicity in Alzheimer's disease. The specific mechanisms by which aging processes are pro-amylodogenic are likely to vary widely between tissues and be sensitive to the genotype. Whatever the case, there is a general association of inflammatory processes of aging with amyloid formation in many tissues besides the brain. A fundamental chemical aspect of aging, oxidative damage, may be *the* shared substrate for inflammatory processes that, depending on the tissue and the genotype, promote specific pathogenic pathways.

In conclusion, during "normal aging," human and rodent brains consistently show general features of chronic inflammation. Similar, but much more specific, age-dependent inflammatory changes are also observed in nonneural tissues. In essence, we propose a convergence of the inflammatory and oxidative damage hypotheses of aging. While the role of inflammation in diseases ranging from AD to atherosclerosis is becoming clearer, our understanding of whether, and how, inflammation contributes to age-dependent changes is in an early stage.

These concepts provide a basis for the outlines of a global hypothesis of aging in which chronic, initially low-grade inflammatory processes progress during aging to become pro-amyloidogenic in different tissues. Individual outcomes of aging may depend on how early stage "smouldering gero-inflammatory processes" are kindled by the external environment, according to the genotype and species, as well as by physiological influences such as diet and hormones. The processes of aging in microglia thus may be a microcosm of more general changes in macrophages that participate in, and possibly drive, pathological aging throughout the body. To evaluate this hypothesis, we shall need to learn much more about whether aging alters the toxic and trophic activities of microglia in particular, and of tissue macrophages in general.

Acknowledgments

We are grateful for support for the NIH. (C.E.F, AG 13499 and AG 09793; R.W., AG 11915; T.A.P., CA 78723) and the John Douglas French Alzheimer's Foundation (C.E.F.).

Note A: Studies of age changes in gene expression will benefit from use of fewer rodent genotypes. As a precedent, one may consider the roaring success of microbial genetics, which derived from the use of a small number of bacteriophage strains, as enforced by Max Delbruck at the Cold Spring Harbor Laboratory (Fischer and Lipson, (1988)). Of course, immunogeneticists did not need such pressures to exploit the opportunities of inbred mice! The interested reader may notice that the C57BL/6J and related substrains have long been used for studies in the neurobiology of aging because of long lifespans and general good health (Finch et al. (1969); Chaconas and Finch, (1973); Finch, (1973); Severson and Finch, (1980); May et al. (1987)). This strain was also used to study the regulation of immunological aging by caloric restriction (Volk et al. (1994); Weindruch et al. (1979)), and for the microarray analysis of mRNA changes (Lee CK et al. (1999, 2000)). A few of the studies reviewed here also used C57BL/6 and related strains. Nonetheless, striking differences in neuron cell numbers and functions can arise in a few generations (see Finch and Kirkwood (2000), Table 2, footnote f, and Finch, (1990) p. 319).

Note B: The debate about neuron loss with "normal aging" continues, with technical advantages argued for both conventional (Landfield et al. (1996)) and stereologic histometry (West, (1999)). We suggest that hippocampal neuron loss during aging could be sporadic, possibly in association with chronically elevated plasma glucocorticoids that can cause loss of hippocampal pyramidal neurons during aging (e.g. Landfield et al. (1978; 1981); Meaney et al. (1988); Reagan and McEwen, (1997)). Thus, variations in the exposure to stress could be a sporadic cause of neuron death. These variations are not traceable to genotype, but can reflect uterine fetal position and stressful social interactions (Finch and Kirkwood, (2000); Meaney et al. (1988)). It may not suffice to identify individuals with elevated resting glucocorticoids, which show inverse correlations with astrocytic hypertrophy (Landfield et al. (1978)) and neuron loss (Meaney et al. (1988)). Special strategies are needed to characterize the stress history of individual animals and humans because of the formidable efforts required for detailed microscopic studies. Such individual history is rarely available in experimental studies on aging, but is not impossible to include in the case records of human aging (Morris et al. (1996); Harris et al. (1999)).

References

Akiyama H, Barger S, Barnum S, Bradt B, Bauer J, Cole GM, Cooper NR, Eikelenboom P, Emmerling M, Fiebich BL, Finch CE, Frautschy S, Griffin WS, Hampel H, Hull M, Landreth G, Lue L, Mrak R, Mackenzie IR, McGeer PL, O'Banion MK, Pachter J, Pasinetti G, Plata-Salaman C, Rogers J, Rydel R, Shen Y, Streit W, Strohmeyer R, Tooyoma I, Van Muiswinkel FL, Veerhuis R, Walker D, Webster S, Wegrzyniak B, Wenk G, Wyss-Coray T (2000). Inflammation and Alzheimer's disease. *Neurobiol Aging* 21:383–421.

Aldskogius H, Kozlova EN (1998). Central neuron-glial and glial-glial interactions following axon injury. *Prog Neurobiol* 55:1–26.
Anderson KJ, Scheff SW, DeKosky ST (1986). Reactive synaptogenesis in hippocampal area CA1 of aged and young adult rats. *J Comp Neurol* 252:374–384.
Andersson PB, Perry VH, Gordon S (1992). The acute inflammatory response to lipopolysaccharide in CNS parenchyma differs from that in other body tissues. *Neuroscience* 48:169–186.
Andrew W, Cardwell ES (1940). Neuronophagia in the human cerebral cortex in senility and pathological conditions. *Arch Pathol* 29:400–414.
Barker CF, Billingham RE (1977). Immunologically priveleged sites. *Adv Immunol* 25:1–54
Baynes JW, Thorpe SR (1999). Role of oxidative stress in diabetic complications: a new perspective on an old paradigm. *Diabetes* 48:1–9.
Bell MD, Lopez-Gonzalez R, Lawson L, Hughes D, Fraser I, Gordon S, Perry VH (1994). Upregulation of the macrophage scavenger receptor in response to different forms of injury in the CNS. *J Neurocytol* 23:605–613.
Brawer JR, Schipper H, Naftolin F (1980). Ovary-dependent degeneration in the hypothalamic arcuate nucleus. *Endocrinology* 107:274–279.
Bruce-Keller AJ, Geddes JW, Knapp PE, McFall RW, Keller JN, Holtsberg FW, Parthasarathy S, Steiner SM, Mattson MP (1999). Anti-death properties of TNF against metabolic poisoning: mitochondrial stabilization by MnSOD. *J Neuroimmunol* 93:53–71.
Bruunsgaard H, Andersen-Ranberg K, Jeune B, Pedersen AN, Skinhoj P, Pedersen BK (1999). A high plasma concentration of TNF-alpha is associated with dementia in centenarians. *J Gerontol A Biol Sci Med Sci* 54:M357–M64.
Butterfield DA, Howard BJ, Yatin S, Allen KL, Carney JM (1997). Free radical oxidation of brain proteins in accelerated senescence and its modulation by N-tert-butyl-alpha-phenylnitrone. *Proc Natl Acad Sci USA* 94:674–678.
Canady KS, Hyson RL, Rubel EW (1994). The astrocytic response to afferent activity blockade in chick nucleus magnocellularis is independent of synaptic activation, age, and neuronal survival. *J Neurosci* 14:5973–5985.
Chaconas G, Finch CE (1973). The effect of ageing on RNA-DNA ratios in brain regions of the C57BL-6J male mouse. *J Neurochem* 21:1469–1473.
Christman MF, Morgan RW, Jacobson FS, Ames BN (1985). Positive control of a regulon for defenses against oxidative stress and some heat-shock proteins in Salmonella typhimurium. *Cell* 41:753–762.
Colman PD, Kaplan BB, Osterburg HH, Finch CE (1980). Brain poly(A)RNA during aging: stability of yield and sequence complexity in two rat strains. *J Neurochem* 34:335–345.
Cutler RG (1975). Transcription of unique and reiterated DNA sequences in mouse liver and brain as a function of age. *Exp Gerontol* 10:37–60.
Day JR, Min BH, Laping NJ, Martin GIII, Finch CE (1992). New mRNA probes for hippocampal responses to entorhinal cortex lesions in the adult male rat: a preliminary report. *Exp Neurol* 117:97–99.
DiPatre PL, Gelman BB (1997). Microglial cell activation in aging and Alzheimer disease: partial linkage with neurofibrillary tangle burden in the hippocampus. *J Neuropathol Exp Neurol* 56:143–149.
Dubey A, Forster MJ, Lal H, Sohal RS (1996). Effect of age and caloric intake on protein

oxidation in different brain regions and on behavioral functions of the mouse. *Arch Biochem Biophys* 333:189–197.

Eglitis MA, Mezey E (1997). Hematopoietic cells differentiate into both microglia and macroglia in the brains of adult mice. *Proc Natl Acad Sci USA* 94:4080–4085.

Ferrucci L, Harris TB, Guralnik JM, Tracy RP, Corti MC, Cohen HJ, Penninx B, Pahor M, Wallace R, Havlik RJ (1999). Serum IL-6 level and the development of disability in older persons. *J Am Geriatr Soc* 47:639–646.

Fiala M, Zhang L, Gan X, Sherry B, Taub D, Graves MC, Hama S, Way D, Weinand M, Witte M, Lorton D, Kuo YM, Roher AE (1998). Amyloid-beta induces chemokine secretion and monocyte migration across a human blood—brain barrier model. *Mol Med* 4:480–489.

Finch CE (1990). *Longevity, senescence, and the genome.* University of Chicago Press. Second printing, 1994.

Finch CE (1973). Catecholamine metabolism in the brains of ageing male mice. *Brain Res* 52:261–276.

Finch CE, Felicio LS, Mobbs CV, Nelson JF (1984). Ovarian and steroidal influences on neuroendocrine aging processes in female rodents. *Endocrine Rev* 5:467–497.

Finch CE, Foster JR, Mirsky AE (1969). Ageing and the regulation of cell activities during exposure to cold. *J Gen Physiol* 54:690–712.

Finch CE, Kirkwood TBL (2000). *Chance, development, and aging.* Oxford University Press.

Finch CE, Longo V (2000). The Gero-inflammatory manifold. In: *Neuroinflammatory mechanisms in Alzheimer's disease: basic and clinical research,* Chapter 1 (Rogers J, ed.). Basel: Birkhäuser-Verlag.

Finch CE, Longo V, Miyao A, Morgan TE, Rozovsky I, Soong Y, Wei M, Xie Z, Zanjani H (2000). Amyloids, inflammatory mechanisms in Alzheimer disease, and aging. In: *Molecular mechanisms in neurodegenerative diseases,* Chapter 2 (Chesselet M-F, ed.), pp 87–110. Totowa NJ: Humana Press.

Finch CE, Sapolsky RM (1999). The evolution of Alzheimer disease, the reproductive schedule, and apoE isoforms. *Neurobiol Aging* 20:407–428.

Fischer EP, Lipson C (1988). *Thinking about science: Max Delbrück and the origins of molecular biology.* New York and London: W. W. Norton & Company.

Forster MJ, Dubey A, Dawson KM, Stutts WA, Lah H, Sohal RS (1996). Age-related losses of cognitive function and motor skills in mice are associated with oxidative protein damage in the brain. *Proc Natl Acad Sci USA* 93:4765–4769.

Frame LT, Hart RW, Leakey JE (1998). Caloric restriction as a mechanism mediating resistance to environmental disease. *Environ Health Perspect* 106 Suppl 1:313–324.

Gehrmann J, Mies G, Bonnekoh P, Banati R, Iijima T, Kreutzberg GW, Hossmann KA (1993). Microglial reaction in the rat cerebral cortex induced by cortical spreading depression. *Brain Pathol* 3:11–17.

Geinisman Y (1999). Age-related decline in memory function: is it associated with a loss of synapses? *Neurobiol Aging* 20:353–356.

Geinisman Y, Bondareff W, Dodge JT (1977). Partial deafferentation of neurons in the dentate gyrus of the senescent rat. *Brain Res* 134:541–545.

Geinisman Y, Detoledo-Morrell L, Morrell F, Heller RE (1995). Hippocampal markers of age-related memory dysfunction: behavioral, electrophysiological and morphological perspectives. *Prog Neurobiol* 45:223–252.

Geula C, Wu C-K, Saroff D, Lorenzo A, Yuan M, Yanker BA (1998). Aging renders the brain vulnerable to amyloid β-protein neurotoxicity. *Nat Med* 4:827–831.

Globerson A, Effros RB (2000). Ageing of lymphocytes and lymphocytes in the aged. *Immunol Today* 21:515–521.

Goldsmith SK, Wals P, Rozovsky I, Morgan TE, Finch CE (1997). Kainic acid and decorticating lesions stimulate the synthesis of C1q protein in adult rat brain. *J Neurochem* 68:2046–2052.

Gordon MN, Schreier WA, Ou X, Holcomb LA, Morgan DG (1997). Exaggerated astrocyte reactivity after nigrostriatal deafferentation in the aged rat. *J Comp Neurol* 388:106–119.

Goss, JR, Finch CE, Morgan DG (1990). GFAP RNA prevalence is increased in aging and in wasting mice. Brief Communication. *Exp Neurol* 108:266–268.

Graeber MB, Bise K, Mehraein P (1993). Synaptic stripping in the human facial nucleus. *Acta Neuropathol* (Berl) 86:179–181.

Grewal RP, Yoshida T, Finch CE, Morgan TE (1997). Scavenger receptor mRNAs in rat brain microglia are induced by kainic acid lesioning and by cytokines. *Neuroreport* 8:1077–1081.

Guttmann CR, Jolesz FA, Kikinis R, Killiany RJ, Moss MB, Sandor T, Albert MS (1998). White matter changes with normal aging. *Neurology* 50:972–978.

Harris TB, Ferrucci L, Tracy RP, Corti MC, Wacholder S, Ettinger WH Jr, Heimovitz H, Cohen HJ, Wallace R (1999). Associations of elevated interleukin-6 and C-reactive protein levels with mortality in the elderly. *Am J Med* 106:506–512.

Hauss-Wegrzyniak B, Vannucchi MG, Wenk GL (2000). Behavioral and ultrastructural changes induced by chronic neuroinflammation in young rats. *Brain Res* 859:157–166.

Hauss-Wegrzyniak B, Vraniak P, Wenk GL (1999). The effects of a novel NSAID on chronic neuroinflammation are age dependent. *Neurobiol Aging* 20:305–313.

Haynes BF, Markert ML, Sempowski GD, Patel DD, Hale LP (2000). The role of the thymus in immune reconstitution in aging, bone marrow transplantation, and HIV-1 infection. *Annu Rev Immunol* 18:529–560.

Hoff SF, Scheff SW, Cotman CW (1982). Lesion-induced synaptogenesis in the dentate gyrus of aged rats: II. Demonstration of an impaired degeneration clearing response. *J Comp Neurol* 1982 205:253–259.

Hutchinson EW, Rose MR (1991). Quantitative genetics of postponed aging in Drosphila melanogaster. I. Analysis of outbred populations. *Genetics* 127:719–727.

Johnson SA, Lampert-Etchells M, Rozovsky I, Pasinetti G, Finch C (1992). Complement mRNA in the mammalian brain: responses to Alzheimer's disease and experimental lesions. *Neurobiol Aging* 13:641–648.

Johnstone EM, Chaney MO, Norris FH, Pascual R, Little SP (1991). Conservation of the sequence of the Alzheimer's disease amyloid peptide in dog, polar bear, and five other mammals by cross-species polymerase chain reaction analysis. *Brain Res Mol Brain Res* 10:299–305.

Kato H, Walz W (2000). The initiation of the microglial response. *Brain Pathol* 10:137–143.

Kerr DS, Campbell LW, Applegate MD, Brodish A, Landfield PW (1991). Chronic stress-induced acceleration of electrophysiologic and morphometric biomarkers of hippocampal aging. *J Neurosci* 11:1316–1324.

Kohama SG, Anderson CP, Osterburg HH, May PC, Finch CE (1989). Oral administration of estradiol to young C57BL/6J mice induces age-like neuroendocrine dysfunctions in the regulation of estrous cycles. *Biol Repro* 41:227–232.

Kosel S, Egensperger R, Bise K, Arbogast S, Mehraein P, Graeber MB (1997). Long-lasting perivascular accumulation of major histocompatibility complex class II-positive lipophages in the spinal cord of stroke patients: possible relevance for the immune privilege of the brain. *Acta Neuropathol* 94:532–538.

Lamar CH, Hinsman EJ, Henrikson CK (1976). Alterations in the hippocampus of aged mice. *Acta Neuropathol* (Berl) 36:387–391.

Lampert-Etchells M, Pasinetti GM, Finch CE, Johnson SA (1993). Regional localization of cells containing C1qb and C4 mRNAs in the frontal cortex during Alzheimer disease. *Neurodegeneration* 2:111–121.

Landfield PW, Braun LD, Pitler TA, Lindsey JD, Lynch G (1981). Hippocampal aging in rats: a morphometric study of multiple variables in semithin sections. *Neurobiol Aging* 2:265–275.

Landfield PW, McEwan BS, Sapolsky RM, Meaney MJ (1996). Hippocampal cell death. *Science* 272:1249–1251.

Landfield PW, Rose G, Sandles L, Wohlstadter TC, Lynch G (1977). Patterns of astroglial hypertrophy and neuronal degeneration in the hippocampus of aged, memory-deficient rats. *J Gerontol* 32:3–12.

Landfield PW, Waymire JC, Lynch G (1978). Hippocampal aging and adrenocorticoids: quantitative correlations. Science 202:1098–1102.

Lass A, Sohal BH, Weindruch R, Forster MJ, Sohal RS (1998). Caloric restriction prevents age-associated accrual of oxidative damage to mouse skeletal muscle mitochondria. *Free Radic Biol Med* 25:1089–1097.

Lawson LJ, Perry VH, Gordon S (1993). Microglial responses to physiological change: osmotic stress elevates DNA synthesis of neurohypophyseal microglia. *Neuroscience* 56:929–938.

Lee CK, Klopp RG, Weindruch R, Prolla TA (1999). Gene expression profile of aging and its retardation by caloric restriction. Science 285:1390–1393.

Lee CK, Weindruch R, Prolla TA (2000). Gene-expression profile of the ageing brain in mice. *Nat Genet* 25:294–297.

Lee EY, Lee SY, Lee TS, Chi JG, Choi W, Suh YH (2000). Ultrastructural changes in microvessel with age in the hippocampus of senescence-accelerated mouse (SAM)-P/10. *Exp Aging Res* 26:3–14.

Lee J, Bruce-Keller AJ, Kruman Y, Chan SL, Mattson MP (1999). 2-Deoxy-D-glucose protects hippocampal neurons against excitotoxic and oxidative injury: evidence for the involvement of stress proteins. *J Neurosci Res* 57:48–61.

Li JJ, Surini M, Catsicas S, Kawashima E, Bouras C (1995). Age-dependent accumulation of advanced glycosylation end products in human neurons. *Neurobiol Aging* 16:69–76.

Li JJ, Voisin D, Quiquerez AL, Bouras C (1994). Differential expression of advanced glycosylation end-products in neurons of different species. *Brain Res* 641:285–288.

Lin YJ, Seroude L, Benzer S (1998). Extended life-span and stress resistance in the Drosophila mutant methuselah. *Science* 282:943–946.

Lindsey JF, Landfield PW, Lynch G (1979). Early onset and topographical distribution of hypertrophied astrocytes in hippocampus of aging rats: a quantitative study. *J Gerontol* 34:661–671.

Lithgow GJ, White TM, Melov S, Johnson TE (1995). Thermotolerance and extended lifespan conferred by single-gene mutations and induced by thermal stress. *Proc Natl Acad Sci USA* 92:7540–7544.

Liu Y, Jacobowitz DM, Barone F, McCarron R, Spatz M, Feuerstein G, Hallenbeck JM, Siren AL (1996). Quantitation of perivascular monocytes and macrophages around cerebral blood vessels of hypertensive and aged rats. *J Cereb Blood Flow Metab* 14:348–352.

Long JM, Kalehua AN, Muth NJ, Calhoun ME, Jucker M, Hengemihle JM, Ingram DK, Mouton PR (1999). Stereological analysis of astrocyte and microglia in aging mouse hippocampus. *Neurobiol Aging* 19:497–503.

Mackenzie IR, Munoz DG (1998). Nonsteroidal anti-inflammatory drug use and Alzheimer-type pathology in aging. *Neurology* 50:986–990.

Mander TH, Morris JF (1994). Perivascular microglia in the rat neural lobe engulf magnocellular secretory terminals during osmotic stimulation. *Neurosci Lett* 180:235–238.

Masoro EJ (1998). Influence of caloric intake on aging and on the response to stressors. *J Toxicol Environ Health B Crit Rev* 1:243–257.

Matyszak MK, Lawson LJ, Perry VH, Gordon S (1992). Stromal macrophages of the choroid plexus situated at an interface between the brain and peripheral immune system constitutively express major histocompatibility class II antigens. *J Neuroimmunol* 40:173–181.

May PC, Lampert-Etchells M, Johnson SA, Poirier J, Masters JN, Finch CE (1990). Dynamics of gene expression for a hippocampal glycoprotein elevated in Alzheimer's disease and in response to experimental lesions in rat. *Neuron* 5:831–839.

May PC, Severson JA, Osterburg HH, Finch CE (1987). Compartmentalization of calmodulin and tubulin in the male C57BL/6J mouse brain: heterogeneity of age changes in calmodulin compartments. *Neurobiol Aging* 8:131–137.

May PC, Telford N, Salo D, Anderson C, Kohama SG, Finch CE, Walford RL, Weindruch R (1992). Failure of diet restriction to retard age-related neurochemical changes in mice. *Neurobiol Aging* 13:787–791.

McCarter RJ (1995). Role of caloric restriction in the prolongation of life. *Clin Geriatr Med* 11:553–565.

McCay CM, Crowell MF, Maynard LA (1989). The effect of retarded growth upon the length of life span and upon the ultimate body size. 1935. *Nutrition* 5:155–171.

Meaney MJ, Aitken DH, van Berkel C, Bhatnagar S, Sapolsky RM (1988). Effect of neonatal handling on age-related impairments associated with the hippocampus. *Science* 239:766–768.

Miller MM, Gould BE, Nelson JF (1989). Aging and long-term ovariectomy alter the cytoarchitecture of the hypothalamic-preoptic area of the C57BL/6J mouse. *Neurobiol Aging* 10:683–690.

Miyata T, Inagi R, Iida Y, Sato M, Yamada N, Oda O, Maeda K, Seo H (1994). Involvement of beta 2-microglobulin modified with advanced glycation end products in the pathogenesis of hemodialysis-associated amyloidosis. Induction of human monocyte chemotaxis and macrophage secretion of tumor necrosis factor-alpha and interleukin-1. *J Clin Invest* 93:521–528.

Mooradian AD (1994). Potential mechanisms of the age-related changes in the blood-brain barrier. *Neurobiol Aging* 15:751–755.

Morgan DG, Gordon MN (1996). Aging and molecular biology. In: *The lifespan development of individuals. Behavioral, neurobiological, and psychosocial perspectives, a synthesis* (Magnusson D, ed.), pp. 469–487. Cambridge University Press.

Morgan TE, Nichols NR, Pasinetti GM, Finch CE (1993). TGF-β1 mRNA increases in macrophage/microglia cells of the hippocampus in response to deafferentation and kainic acid-induced neurodegeneration. *Exp Neurol* 120:291–301.

Morgan TE, Rozovsky I, Sarkar DK, Young-Chan CS, Nichols NR, Finch CE (2000). Transforming growth factor-b1 (TGF-b1) induces TGF-b1 and TGF-b1 receptor mRNAs and reduces complement C1qB mRNA in rat brain in microglia. *Neuroscience* 101:313–321.

Morgan TE, Xie Z, Goldsmith S, Yoshida T, Lanzrein A-S, Stone D, Rozovsky I, Perry G, Smith MA, Finch CE (1999). The mosaic of brain glial hyperactivity during normal aging and its attenuation by food restriction. *Neuroscience* 89:687–699.

Morris JC, Storandt M, McKeel DW Jr, Rubin EH, Price JL, Grant EA, Berg L (1996). Cerebral amyloid deposition and diffuse plaques in "normal" aging: evidence for presymptomatic and very mild Alzheimer's disease. *Neurology* 46:707–719.

Nichols NR, Day JR, Laping NJ, Johnson SA, Finch CE (1993). GFAP mRNA increases with age in rat and human brain. *Neurobiol Aging* 14:421–429.

Nichols NR, Masters JN, Finch CE (1994). Cloning of steroid-responsive mRNAs by differential hybridization. In: Neurobiology of steroids (De Kloet R, ed.). *Meth Neurosci* 22:296–313.

Ogg S, Paradis S, Gottleb S, Patterson GL, Lee L, Tissenbaum H, Ruvkun G (1997). The Fork head transcription factor DAF-16 tranduces insulin-like metabolic and longevity symbols in C. elegans. *Nature* 389:994–999.

Ogura K, Ogawa M, Yoshida M (1994). Effects of ageing on microglia in the normal rat brain: immunohistochemical observations. *Neuroreport* 5:1224–1226.

Ohe Y, Ishikawa K, Itoh Z, Tatemoto K (1996). Cultured leptomeningeal cells secrete cerebrospinal fluid proteins. *J Neurochem* 67:964–971.

Orr WC, Sohal RS (1994). Extension of life-span by overexpression of superoxide dismutase and catalase in Drosophila melanogaster. *Science* 263:1128–1130.

Pasinetti GM, Hassler M, Stone D, CE Finch (1999). Glial gene expression during aging in rat striatum and in long-term responses to 6-OHDA lesions. *Synapse* 31:278–284.

Pasinetti GM, Johnson SA, Rozovsky I, Lampert-Etchells M, Morgan DG, Gordon MN, Morgan TE, Willoughby DA, Finch CE (1992). Complement C1qB and C4 mRNA responses to lesioning in rat brain. *Exp Neurol* 118:117–125.

Pasinetti GM, Nichols NR, Tocco G, Morgan T, Laping N, Finch CE (1993). Transforming growth factor-β1 (TGF-β1) and fibronectin mRNA in rat brain: responses to injury and cell-type localization. *Neuroscience* 54:893–907

Perlmutter LS, Scott SA, Barron E, Chui HC (1992). Mhc class II-positive microglia in human brain: association with Alzheimer lesions. *J Neurosci Res* 33:549–558.

Perry VH, Matyszak MK, Fearn S (1993). Altered antigen expression of microglia in the aged rodent CNS. *Glia* 7:60–67.

Poirier J, Hess M, May PC, Finch CE (1991). Cloning of hippocampal poly(A)RNA sequences that increase after entorhinal cortex lesion in adult rat. *Mol Brain Res* 9:191–195.

Pow DV, Perry VH, Morris JF, Gordon S (1989). Microglia in the neurohypophysis associate with and endocytose terminal portions of neurosecretory neurons. *Neuroscience* 33:567–578.

Raivich G, Bohatschek M, Kloss CUA, Werner A, Jones LL, Kreutzber GW (1999). Neuroglial activation repertoire in the injured brain: graded response, molecular mechanisms and cues to physiological function. *Brain Res Rev* 30:77–105.

Raivich G, Jones LL, Kloss CUA, Werner A, Neumann H, Kreutzver GW (1998). Immune surveillance in the injured nervous system: T-lymphocytes invade the axotomized mouse facial motor nucleus and aggregate around sites of neuronal degeneration. *J Neurosci* 18:5804–5816.

Rasmussen T, Schliemann T, Sorensen JC, Zimmer J, West MJ (1996). Memory impaired aged rats: no loss of principal hippocampal and subicular neurons. *Neurobiol Aging* 17:143–147.

Reagan LP, McEwen BS (1997). Controversies surrounding glucocorticoid-mediated cell death in the hippocampus. *J Chem Neuroanat* 13:149–167.

Ross R (1999). Atherosclerosis-an inflammatory disease. *N Engl J Med* 340:115–126.

Roth GS, Ingram DK, Joseph JA (1984). Delayed loss of striatal dopamine receptors during aging of dietarily restricted rats. *Brain Res* 300:27–32.

Rozovsky I, Finch CE, Morgan TE (1998). Age-related activation of microglia and astrocytes: in vitro studies show persistence of phenotypes of aging, increased proliferation, and resistance to down-regulation. *Neurobiol Aging* 19:97–103.

Rozovsky I, Morgan TE, Willoughby DA, Dugich-Djordevich MN, Pasinetti GM, Johnson SA, Finch CE (1994). Selective expression of clusterin (SGP-2) and complement C1q and C4 during responses to neurotoxins *in vivo* and *in vitro*. *Neuroscience* 62:741–758.

Sell DR, Kleinman NR, Monnier VM (2000). Longitudinal determination of skin collagen glycation and glycoxidation rates predicts early death in C57BL/6NNIA mice. *FASEB J* 14:145–156.

Schipper HM (1996). Astrocytes, brain aging, and neurodegeneration. *Neurobiol Aging* 17:467–480.

Schipper H, Brawer JR, Nelson JF, Felicio LS, Finch CE (1981). Role of the gonads in the histologic aging of the hypothalamic arcuate nucleus. *Biol Reprod* 25:413–419.

Scott SA, Mandybur TI (1996). Astrocytic and microglial alterations in the aged mouse brain. *In: Pathobiology of the aging mouse,* Vol. 2 (Mohr U, Dungworth DL, Capen CC, Carlton WW, Sundberg JP, Ward JM, eds.), pp 39–52. Washington, DC: ISLI Press.

Severson JA, Finch CE (1980). Reduced dopaminergic binding during aging in the rodent striatum. *Brain Res* 192:147–162.

Shaw JA, Perry VH, Mellanby J (1994). MHc class II expression by microglia in tetanus toxin-induced experimental epilepsy in the rat. *Neuropathol Appl Neurobiol* 20:392–398.

Sheffield LG, Berman NE (1998). Microglial expression of Mhc class II increases in normal aging of nonhuman primates. *Neurobiol Aging* 19:47–55.

Sierra F, Coeytaux S, Juillerat M, Ruffieux C, Gauldie J, Guigoz Y (1992). Serum T-kininogen levels increase two to four months before death. *J Biol Chem* 267:10665–10669.

Sobin SS, Bernick S, Ballard KW (1992). Histochemical characterization of the aging microvasculature in the human and other mammalian and non-mammalian vertebrates by the periodic acid-Schiff reaction. *Mech Ageing Dev* 63:183–192.

Sohal RS, Agarwal S, Candas M, Forster MJ, Lal H (1994). Effect of age and caloric restriction on DNA oxidative damage in different tissues of C57BL/6 mice. *Mech Ageing Dev* 76:215–224.

Sohal RS, Weindruch R (1996). Oxidative stress, caloric restriction, and aging. *Science* 273:59–63.

Soong NW, Hinton DR, Cortopassi G, Arnheim N (1992). Mosaicism for a specific somatic mitochondrial DNA mutation in adult human brain. *Nat Genet* 2:318–323.

Streit WJ, Sparks DL (1997). Activation of microglia in the brains of humans with heart disease and hypercholesterolemic rabbits. *J Mol Med* 75:130–138.

Streit WJ, Walter SA, Pennell NA (1999). Reactive microgliosis. *Prog Neurobiol* 57:563–581

Sturrock RR (1988). An ultrastructural study of intraventricular macrophages in the brains of aged mice. *Anat Anz* 165:283–290.

Stuesse SL, Cruce WL, Lovell JA, McBurney DL, Crisp T (2000). Microglial proliferation in the spinal cord of aged rats with a sciatic nerve injury. *Neurosci Lett* 287:121–124.

Thomas WE (1999). Brain macrophages: on the role of pericytes and perivascular cells. *Brain Res Brain Res Rev* 31:42–57.

Thomas WE (1992). Brain macrophages: evaluation of microglia and their functions. *Brain Res Brain Res Rev* 17:61–74.

Topple A, Fifkova E, Baumgardner D, Cullen-Dockstader K (1991). Effect of age on blood vessels and neurovascular appositions in the CA1 region of the rat hippocampus. *Neurobiol Aging* 12:211–217.

Vaughan DW, Peters A (1974). Neuroglial cells in the cerebral cortex of rats from young adulthood to old age: an electron microscope study. *J Neurocytol* 3:405–429.

Vijayan VK, Cotman CW (1983). Lysosomal enzyme changes in young and aged control and entorhinal-lesioned rats. *Neurobiol Aging* 4:13–23.

Volk MJ, Pugh TD, Kim M, Frith CH, Daynes RA, Ershler WB, Weindruch R (1994). Dietary restriction from middle age attenuates age-associated lymphoma development and interleukin 6 dysregulation in C57BL/6 mice. *Cancer Res* 54:3054–3061.

Weindruch RH, Kristie JA, Cheney KE, Walford RL (1979). Influence of controlled dietary restriction on immunologic function and aging. *Fed Proc* 38:2007–2016.

Weindruch R, Walford RL (1988). *The retardation of aging and disease by diet restriction.* Springfield, IL: CC Thomas.

Wekerle H, Linnington C, Lassmann H, Meyermann R (1986). Cellular immune reactivity within the CNS. *Trends Neurosci* 9:271–277.

West MJ (1999). Stereological methods for estimating the total number of neurons and synapses: issues of precision and bias. *Trends Neurosci* 22(2):51–61.

Wong ML, Xie B, Beatini N, Phu P, Marathe S, Johns A, Gold PW, Hirsch E, Williams KJ, Licinio J, Tabas I (2000). Acute systemic inflammation up-regulates secretory sphingomyelinase *in vivo:* a possible link between inflammatory cytokines and atherogenesis. *Proc Natl Acad Sci USA* 97:8681–868.

Woods AG, Guthrie KM, Kurlawalla MA, Gall CM (1998). Deafferentation-induced increases in hippocampal insulin-like growth factor-1 messenger RNA expression are severely attenuated in middle aged and aged rats. *Neuroscience* 83:663–668.

Wustrow TP, Denny TN, Fernandes G, Good RA (1982). Changes in macrophages and their functions with aging in C57BL/6J, AKR/J, and SJL/J mice. *Cell Immunol* 69:227–234.

Yan SD, Schmidt AM, Anderson GM, Zhang J, Brett J, Zou YS, Pinsky D, Stern D (1994). Enhanced cellular oxidant stress by the interaction of advanced glycation end products with their receptors/binding proteins. *J Biol Chem* 269:9889–9897.

Yasojima K, Schwab C, McGeer EG, McGeer PL (1998). Human heart generates complement proteins that are upregulated and activated after myocardial infarction. *Circ Res* 83:860–869.

Ye SM, Johnson RW (1999). Increased interleukin-6 expression by microglia from brain of aged mice. *J Neuroimmunol* 93:139–148.

Yoshida T, Goldsmith S, Morgan TE, Stone D, Finch CE (1996). Transcription supports age-related increases of GFAP gene expression in the male rat brain. *Neurosci Lett* 215:107–110.

Xie Z, Morgan TE, Finch CE (2000). Neurotoxicity of activated microglia originated from young and old rats. *Soc Neurosci Abs* 26:726.4.

Zhu Y, Roth-Eichhorn S, Braun N, Culmsee C, Rami A, Krieglstein J (2000). The expression of transforming growth factor-beta1 (TGF-beta1) in hippocampal neurons: a temporary upregulated protein level after transient forebrain ischemia in the rat. *Brain Res* 866:286–298.

Index

acetylcholine, IL-2 and, 103
ACTH (adrenocorticotrophic hormone), as mediator of neural-immune-endocrine communication, 104
activated microglia, 2, 3, 141–143, 154, 190, 210
activation of microglia, 2, 5, 8–10, 13, 84–85, 170, 171
  aging and, 275, 289–291
  amyloid β and, 259–260
  diet and, 291–293
  focal ischemia, 135–137
  global ischemia, 140–141
  non-traumatic, 155, 158
  regulation, 198–202
adrenocorticotrophic hormone (ACTH), as mediator of neural-immune-endocrine communication, 104
AGE (glycated adducts), 291
aging
  Alzheimer's disease process, 286
  changes in microglia and, 278–285
  inflammatory processes and, 294–295
  messenger RNA and, 287–289, 290
  microglial activation and, 275, 289–291
  oxidative-stress hypothesis, 292–293
  response of microglia to injury, 285–287
Alzheimer's disease, 1
  aging process and, 286
  amyloid β, 10, 70, 245–247, 278
  cytokines and, 93–94
  microglia and, 1, 10–11, 278
  NSAIDs and, 261–262
ameboid microglia, 2, 15, 132
  brain development and, 188–189
  visual system development and, 15, 16, 23, 27, 28–29

ameboid microglial cells
  phagocytosis of cell debris by, 25
  radial migration, 20–21
  tangential migration, 18
  visual system and, 15, 16
ammonium, 71
amyloid β (Aβ), 10, 70
  Alzheimer's disease and, 10, 70, 245–247, 278
  carrier-associated amyloid β, 249
  clearance *in vitro,* 247–250
  clearance *in vivo,* 250–256, 260–264
  cytokine release and, 85
  cytokines and, 262–263
  insulin degrading enzyme (IDE) and, 248
  lysosome and, 249
  matrix metalloproteases (MMPs) and, 247–248
  microglia and, 249–250, 256–260
  neprilysin and, 248
  NSAIDs and, 261–262
  steroids and, 263–264
  vaccine, 260–261
amyloid β receptors, 256–260
amyloid precursor protein (APP), 246, 253
anti-inflammatory drugs, 261–262, 287
ApoE, 249
ApoJ, 249
APP (amyloid precursor protein), 246, 253
APP-transgenic mice, 253–256, 258, 260
astrocytes
  activation of, 290–291
  aging and, 283, 290–291
  microglial cytokines, 101–102, 158–159
astrogliosis, 101, 212–213
avian visual system, development of, 7, 15–30
axonal growth, microglia and, 26–27

307

axonal regeneration, 218–219
axotomized facial motor nucleus, 166–168
  GLT-1, 198
  IFNγ, 179
  IL-1, 178–179, 180
  IL-6, 177, 180
  inflammatory cytokines and, 175–180
  MCSF, 175, 176–177, 180, 181
  microglia and, 168–174, 277–278
  perivascular macrophages, 175
  TGFβ1, 178, 193–194
  TNFα, 179–180

basic fibroblast growth factor (bFGF), 191, 228
BAY K 8644, 50
BDNF (brain-derived neurotrophic factor), 100, 194
bFGF (basic fibroblast growth factor), 191, 228
biopolymers
  microglia combined with, 234
  phospholipid modified, 231–233
  polysaccharide scaffolds, 233, 237
  as substrate for axon repair, 229–238
birds, visual system development and, 7, 15–30
blood-brain barrier, aging and, 284
brain. *See also* central nervous system; neuropathology
  aging in, 275–296
  cerebral ischemia, 107–108, 125–144
  cytokines, effect on, 102–103, 109
  macrophages in, 213–215
  microglia in, 188, 275–286
brain-derived neurotrophic factor (BDNF), 100, 194
brain injury, TNFα and, 98

Ca^{2+}-activated K^+ channels, 38, 40, 43–45
Ca^{2+} binding proteins, calcium signaling, 72
Ca^{2+} buffers, calcium signaling, 71
Ca^{2+}-release-activated Ca^{2+} channels (CRAC channels), 48, 50
Ca^{2+} stores, calcium signaling, 71–72
calcitonin gene-related peptide (CGRP), 199–200
calcium ion channels, 48, 50, 67, 69, 72

calcium signaling
  amyloid β, 70
  Ca^{2+} binding proteins, 72
  Ca^{2+} buffers, 71
  Ca^{2+} extrusion, 60
  Ca^{2+} influx, 59
  Ca^{2+} release, 60
  Ca^{2+} sequestration, 60
  Ca^{2+} stores, 71–72
  chemokine receptors, 64–65
  cholinergic receptors, 66
  complement receptors, 63
  endothelin receptors, 64–65
  glutamate receptors, 65
  ion channels, 67, 69
  mechanisms of, 58–62
  P2 receptors, 62–63
  platelet-activating factor (PAF), 65–66, 68
  prion proteins, 70
  receptors, 62–67
  signal generation, 60–62
  thrombin receptors, 66, 68
caloric restriction, microglial activation and, 291–294
cAMP response element binding transcription factor (CREB), 201
carbachol, 68
carrier-associated amyloid β, 249
caspase-1, 107
CCR2, 171
CCR3, 99
CCR5, 99
CCR receptors, 99
CD8+ microglia/macrophages, cerebral ischemia and, 132–134
cell death, in visual system development, 23–26, 29
central nervous system (CNS)
  axonal regeneration, 218–219
  axonal repair, 228–238
  cytokines and, 80, 102–105
  immune surveillance, 81–82
  ion channels, 37, 39
  microglia as macrophages, 80–85
  microglial activation without tissue injury, 155
  pathology of and, cytokines, 107

central nervous system trauma, 84, 209, 210
  axonal repair, 228–238
  cerebral ischemia, 125–144, 155
  spinal cord injury, 152–161, 209–219
  TNFα and, 98
cerebral cortex, age-related changes in microglia, 278–282
cerebral ischemia
  cytokines and, 107–108
  focal ischemia, 125, 126–140, 144
  global ischemia, 140–141
  glutamate and, 197
  microglia and, 3, 126–144
  microglial activation, 135–137, 140–141
  microglial response, 137–140
  pathophysiology of, 125–126
  penumbra, 126
  transient, 155
CGA (chromogranin A), 68, 70–71
CGRP (calcitonin gene-related peptide), 199–200
charybdotoxin, 42
chemokine receptors, 63–64
chemokines, 86, 88, 90, 99–100
chloride channels, 50–52
cholesterol levels, microglial activation and, 9–10
cholinergic receptors, 66
chondroitin sulfate proteoglycan (CSPG), 216–217
chromogranin A (CGA), 68, 70–71
ciliary neurotrophic factor (CNTF), 212
c-Jun N-terminal kinase (JNK), 201
ClC-2 channel, 51
ClC-3 channel, 51
clusterin, 293
CMM (conditioned microglial medium), 191–194
CNM (conditioned neuronal medium), 200
CNS. *See* central nervous system
CNTF (ciliary neurotrophic factor), 212
complement receptors, 63
conditioned microglial medium (CMM), 191–194
conditioned neuronal medium (CNM), 200
cortical spreading depression (CSD), focal cerebral ischemia, 137

CRAC channels (Ca^{2+}-release-activated Ca^{2+} channels), 50–51
CREB (cAMP response element binding transcription factor), 201
CSD (cortical spreading depression), focal cerebral ischemia, 137
CSPG (chondroitin sulfate proteoglycan), 216–217
CX_3CR1, 99, 100, 171
CXCR receptors, 99
cycloxygenase-2, 197
cytokine receptor complexes, 87
cytokine receptors, 87–88, 89, 106
cytokine-receptor signaling, 89–90
cytokines, 4
  actions of, 102
  Alzheimer's disease and, 93–94
  amyloid β and, 262–263
  chemokines, 86, 88, 90, 99–100
  CNS, effect on, 80, 102–105
  CNS damage and, 107–108
  CNS pathology and, 107
  constitutive production, 91
  experimental limitations of studies, 82–84
  facial axotomy and, 175–180
  function in microglia, 79–80, 102
  induction, 85, 88–89
  interferons, 86, 90, 97–98
  interleukins, 86, 87, 89, 90
  as mediators of neural-immune-endocrine communication, 103
  in microglia, 91–110, 159
  as neuroregulatory factors, 102–103
  nomenclature of, 86–87
  receptors. *See* cytokine receptors
  signaling by, 89–90
  structure of, 87
  TNF, 92, 98–99

delayed rectifier K^+ channels, 37, 38, 39, 42–43
dexamethasone (DXM), amyloid β and, 263
diet, microglial activation and, 291–294
diltiazem, effect of, 50
dLGN (dorsal lateral geniculate nucleus), 28
dorsal lateral geniculate nucleus (dLGN), 28
DXM (dexamethasone), amyloid β and, 263

EAAT1, 198
EAAT2, 198
eicosanoids, 197
endocrine system, cytokines as mediators of neural-immune-endocrine communication, 104
endothelial cells, microglial cytokines and, 101–102
endothelin receptors, 64–65, 68
endotoxin, 71
eotaxin, 68
ERK (extracellular signal regulated kinase), 201
estradiol-17β, amyloid β and, 263
estrogen, amyloid β and, 263
extracellular matrix, microglia and, 216–217
extracellular signal regulated kinase (ERK), 201

facial axotomy, 166–168
   GLT-1, 198
   IFNγ, 179
   IL-1, 178–179, 180
   IL-6, 177, 180
   inflammatory cytokines, 175–180
   MCSF, 175, 176–177, 180, 181
   microglia and, 168–174, 277–278
   perivascular macrophages, 175
   TGFβ1, 178, 193–194
   TNFα, 179–180
fetal macrophages, 17
fetal tissue transplantation, 227
focal cerebral ischemia, 125, 126–140, 144
   microglial activation, 135–137
   microglial response, 137–140
   necrosis, 126–135
"four vessel" occlusion model, 126
fractalkine, 68, 85, 100
fractalkine receptor, 85
functional plasticity, 154

GLAST, 198
glial cell line-derived neurotrophic factor (GDNF), 212
gliosis, 209, 210–211
global cerebral ischemia, microglial activation, 140–141
GLT-1, 198

glucocorticoid, amyloid β and, 263
glutamate, and cerebral ischemia, 197
glutamate receptors, 66
glycated adducts (AGE), 291
G-protein-activated K^+ channels, 38, 40, 46
GROα, 99

Hayflick phenomenon, 284
H^+ channels. *See* proton channels
heart disease, microglial activation and, 9–10
HERG-like K^+ channels, 38, 46
HGF (hepatocyte growth factor), 193
hippocampus, age-related changes in microglia, 278–282
histamine, 68
hyper-ramified microglia, 154
hypothalamus, age-related changes in microglia, 282–283

IκB kinase (IKK), 90
ICAM1, 169
ICE (interleukin-1 converting enzyme), 107
IDE (insulin degrading enzyme), amyloid β and, 248
IFNγ, 97, 101, 179
IFNγ-inducing factor (IGIF), 95
IFN receptors, 89
IGIF (IFNγ-inducing factor), 95
IKK (IκB kinase), 90
IL-1R accessory protein-like molecule (IL1RAPL), 94–95
IL-1R-activating kinase (IRAK), 90
IL1RAPL (IL-1R accessory protein-like molecule), 95
immune system, cytokines as mediators of neural-immune-endocrine communication, 104
inclusion bodies, in microglia, 283–284
infection, cytokines and, 107–108
inflammatory processes, aging and, 294–295
injury, 209, 210
   aging and microglial response to, 285–287
   axonal repair, 228–238
   axotomized facial motor nucleus, 166–181, 193
   brain injury, 98

cerebral ischemia, 107–108, 125–144, 155
cytokines and CNS damage, 107–108
cytokines as mediators of neural-immune-endocrine communication, 104
in developing visual system, 28–29
microglia and, 4, 9
microglial activation and, 277
spinal cord injury, 152–161, 209–219
TNFα and, 98
insulin degrading enzyme (IDE), amyloid β and, 248
"insulyin," 248
interferon regulatory factor (IRF), 125
interferons, 86, 90, 97–98, 101, 179
interleukin-1 converting enzyme (ICE), 107
interleukins, 86, 87, 89, 90
    IL-1, 91, 93–95, 101, 178–179, 180, 212–213, 262
    IL-1β, 71, 144, 194, 236, 260
    IL-2, 96, 101, 103
    IL-3, 191
    IL-4, 96
    IL-6, 92, 93–94, 96–97, 101, 177, 180, 262
    IL-8, 99
    IL-10, 96
    IL-12, 92
    IL-13, 96
    IL-15, 96
    IL-18, 95
inward rectifier K^+ channels, 38, 41–42, 44
ion channels, 3, 36–53
    Ca^{2+}-activated K^+ channels, 38, 40, 43–45
    calcium channels, 48, 50, 67–69
    chloride channels, 50–52
    delayed rectifier K^+ channels, 37, 38, 39, 42–43
    G-protein-activated K^+ channels, 38, 40, 46
    HERG-like K^+ channels, 38, 46
    inward rectifier K^+ channels, 38, 41–42, 44
    potassium channels, 38, 39–46
    proton channels, 46–47
    single-channel current, 36, 37
    sodium channels, 48
    voltage-operated channels (VOCs), 67, 72
ION (isthmo-optic nucleus), 28–29
IP-10, 99
IRAK (IL-1R-activating kinase), 90
IRF (interferon regulatory factor), 125
ischemia. *See* cerebral ischemia
isthmo-optic nucleus (ION), 28–29

JAK-STAT principle, 89
Janus kinases (JAK), 89
JNK (c-Jun N-terminal kinase), 201

K^+ channels. *See* potassium channels
kaliotoxin, 42
Kir2.1 channel, 41
Kv1.3 channels, 42–43

laminin, 229
lipofuscin, 283
lipopolysaccharide (LPS), 68, 71, 88, 236
lipoprotein receptor related protein (LRP), 249
lipoprotein receptors, amyloid β uptake and, 258–259
lipoteichoic acid, 88
lipoxygenase, 197
LPS (lipopolysaccharide), 68, 71, 88, 236
LRP (lipoprotein receptor related protein), 249
L-type voltage-gated Ca^{2+} channels, 50
lysosomes, amyloid β and, 249

macrophage-colony stimulating factor (MCSF), 68, 175, 176–177, 180, 181
macrophage inhibitory factor (MIF), 218
macrophages, 80–85
    facial axotomy and, 175
    microglia-derived brain macrophages, 213–215
    spinal trauma, 158
    visual system development, 17–18, 26
margatoxin, 42
matrix metalloproteases (MMPs), amyloid β and, 247–248
MCP-1 (monocyte chemoattractant protein-1), 68, 131

MCP-2 (monocyte chemoattractant protein-2), 99
MCSF (macrophage-colony stimulating factor), 68, 175, 176–177, 180, 181
messenger RNA, aging and, 287–289, 290
2-Methacryloyloxyethyl phosphorylcholine (MPC), 232–238
MHC class I, 172, 174
MHC class II, 173
microglia
  activated, 2, 3, 141–143, 154, 190, 210
  activation. See activation of microglia
  age-related changes in, 278–285
  Alzheimer's disease and, 1, 10–11, 278
  amyloid β and, 249–250, 256–260
  amyloid β receptors, 256–260
  axonal growth and, 26–27
  in the brain, 188, 275–277
  calcium signaling in, 58–72
  cellular properties of, 83
  cerebral ischemia and, 3, 126–144
  conditioned microglial medium (CMM), 191–194
  cytokines in, 4, 79–110, 159
  dendritic lineage, 22
  experimental limitations of studies, 82–84
  extracellular matrix and, 216–217
  facial axotomy and, 168–174, 277–278
  functional plasticity, 154
  functions of, 2–8, 142–143, 188, 201
  inclusion bodies in, 283–284
  injury and, 4, 9
  interactions with neurons, 188–190
  ion channels, 36–53
  as macrophages, 80–85
  MPC-surface-modified implants and, 234–238
  neurons, interaction with, 188–190
  neurotoxic action of, 3, 142, 196–198, 201
  neurotrophic action of, 3, 5, 191–194, 201, 228–229
  neurotrophin release from, 194–196
  pharmacological intervention and, 108–109
  phenotypes and markers, 277
  precursors, 2, 17
  regulation of, 198–202
  spinal cord injury and, 153–166, 209–219
  synaptic remodeling, 6–7
  tissue development, 7
  tissue repair and, 3, 5, 7–8
  transplantation, 8, 217–218, 234
  visual system development and, 7, 15–30
microglial precursors, 2, 17
microgliosis, regeneration and, 217
MIF (macrophage inhibitory factor), 218
MIP-1, 63
MIP-1α, 99
MIP-1β, 99
MIP-2, 99
MMP-9, 247–248
MMPs (matrix metalloproteases), amyloid β and, 247–248
monocyte chemoattractant protein-1 (MCP-1), 68, 131
monocyte chemoattractant protein-2 (MCP-2), 99
monocytes, as microglia precursors, 2
MPC (2-methacryloyloxyethyl phosphorylcholine), 232–238
mRNA (messenger RNA), aging and, 287–289, 290
Müller glial cells, 25
multiple sclerosis, 101
MyD88, 90

Na^+. See sodium channels
necrosis, focal cerebral ischemia, 126–135
neprilysin, 248
nerve growth factor (NGF), 26, 100, 143, 189, 194, 228
nervous system, cytokines and, 79, 90
neurodegenerative processes, 210
  cytokines and, 107–108
  microglial activation and, 277–278
neuroendocrine activities, cytokines as mediators of neural-immune-endocrine communication, 104
neuronal death
  phagocytosis of cell debris, 24–25
  in visual system development, 23–24, 26, 29
neurons, interaction with microglia, 188–190

neuropathology
  cytokines and, 107–108
  microglia and, 108–109
neurothrophin-3 (NT-3), 100, 143, 194, 228
neurothrophin-4 (NT-4), 100
neurotoxic action, microglia, 3, 142, 196–198, 201
neurotransmitters, cytokines and, 103
neurotrophic action, microglia, 3, 5, 191–194, 201, 228–229
neurotrophins, release from microglia, 194–196
neutral endopeptidase, 248
NGF (nerve growth factor), 26, 100, 143, 189, 194, 228
nifedipine, effect of, 50
nitric oxide, 263
nonsteroidal anti-inflammatory drugs (NSAIDs), amyloid β and, 261–262
non-traumatic activation, 144, 158
norepinephrine, 68
noxiustoxin, 42
NSAIDs, amyloid β and, 261–262
NT-3 (neurothrophin-3), 100, 143, 194, 228
NT-4 (neurothrophin-4), 100

oligodendrocytes
  microglial cytokines, 101–102
  visual system development and, 27
OPN (osteopontin), 144
optic pathways. See visual system development
optic tectum, 18
osteopontin (OPN), 144
oxidative-stress hypothesis, aging, 292–293

P2 receptors, 62–63
PAF (platelet-activating factor), 65–66, 68
paravascular microglia, 21–23
parenchymal microglia, 81
patch-clamp technique, 36, 48
phagocytosis
  of amyloid, 250–251
  systemic tissue macrophages, 256
  visual system development, 24–25
phorbol esters inward rectifier channels and, 41–42

phospholipids, synthetic, 231–232
plasminogen, 192
platelet-activating factor (PAF), 65–66, 68
PLGA (poly-(DL-lactide-co-glycolide)), 230
poly-(2-hydroxyethyl methacrylate) (HEMA), 230
poly-(DL-lactide-co-glycolide) (PLGA), 230
polylactides, 228, 230
polymers, as substrate for axon repair, 229–238
poly-(organo)phosphazenes, 230
polysaccharide scaffolds, 233, 237
polyurethanes, 230
potassium channels (K^+ channels), 38, 39–46
  Ca^{2+}-activated K^+ channels, 38, 40, 43–45
  delayed rectifier K^+ channels, 37, 38, 39, 42–43
  G-protein-activated K^+ channels, 38, 40, 46
  HERG-like K^+ channels, 38, 46
  inward rectifier K^+ channels, 38, 41–42, 44
prion proteins, calcium signaling, 70
prostaglandin E_2, 43
proteases, 85
proteoglycans, 88
proton channels, 46–47

radial migration, ameboid microglial cells, 20–21
RAGE (receptor for advanced glycation end product), 249, 260
ramified microglia, 15, 18, 154, 168, 189–190, 209
RANTES, 68, 99
reactive microglia, 190, 210, 212
reactive oxygen species (ROS), 196, 263, 293
receptor for advanced glycation end product (RAGE), 249, 260
receptors
  for amyloid β, 256–260
  chemokine receptors, 64–65
  cholinergic receptors, 66
  complement receptors, 63
  endothelin receptors, 64–65, 68
  glutamate receptors, 65

receptors (*continued*)
  P2 receptors, 62–63
  platelet-activating factor (PAF), 65–66, 68
  thrombin receptors, 66
regeneration, axonal, 218–219
regulation
  of amyloid β clearance *in vivo*, 260–264
  of microglia, 198–202
resting microglia, 15, 154, 210
resting ramified microglia, 276
retina, developing, 16, 17, 18–19, 28, 189
ROMK1 channel, 41
ROS (reactive oxygen species), 196, 263, 293

SCI. *See* spinal cord injury
scorpion toxins, 42
SDF-1α, 68
segmented poly(etherurethane) (SPU), 233
self-repair, spinal cord injury, 215–216
signaling
  by cytokines, 89–90
  cytokines as mediators of neural-immune-endocrine communication, 103
  TNF-receptor signaling, 90
signal transducer and activator of transcription (STAT) proteins, 89
single-channel conductance, 37
single-channel current, 36, 37
SOCs (store-operated channels), 69
sodium channels, 48
spinal cord, demyelination, 218–219
spinal cord injury (SCI), 152–153, 209–219
  axonal repair, 228–238
  microglia and, 153–161
  regeneration and, 217
  secondary injury, 153
  self-repair, 215–216
  transplantation of microglial cells into spinal cord, 217–218, 234
SPU (segmented poly(etherurethane)), 233
STAT proteins, 89
stellate microglia, 132
steroids, amyloid β and, 263–264
store-operated channels (SOCs), 69

stroke
  cytokines and, 107–108
  microglial activation and, 277
synaptic remodeling, microglia, 6–7

tangential migration, ameboid microglial cells, 18
teichoic acid, 88
TGF-β1, astrogliosis and, 212–213
TGFβ (transforming growth factor-beta), 96, 178
thrombin, 68
thrombin receptors, 66
thrombospondin (TSP), 216, 228–229
TNFα (tumor necrosis factor-α), 92, 98–99, 101, 144, 179–180, 194, 197, 201
TNF-R, 99, 180
TNF receptor-associated factor 6 (TRAF6), 90
TNF-receptor signaling, 90
tollip, 90
TRAF6 (TNF receptor-associated factor 6), 90
transforming growth factor-beta 1 (TGFβ1), 178
transplantation
  fetal neural tissue, 227
  of microglial cells into injured spinal cord, 217–218, 234
  microglia/macrophage grafts, 8
trauma. *See* injury
TSP (thrombospondin), 216, 228–229
tumor necrosis factor-α (TNFα), 92, 98–99, 101, 144, 179–180, 194, 197, 201
"two vessel" occlusion model, 126

vaccine, amyloid β, 260–261
vascularization, visual system development, 27–28
verapamil, effect of, 50
vertebrates, avian visual system development and, 7, 15–30
visual system development, 15–16
  axonal growth, 26–27
  cell death in, 23–24, 26, 29
  early macrophages, 17–18
  induction of cell death, 26

microglial progenitors, 18–24
oligodendrocytes and, 27
phagocytosis of cell debris, 24–25
response to injury, 28–29
spread of microglial cells, 16–17
vascularization, 27–28
VLA-4, 131

VOCs (voltage-operated channels), 67, 72
voltage-dependent Cl^- channels, 51–52
voltage-independent Cl^- channels, 51
voltage-operated channels (VOCs), 67, 72

white matter, age-related changes in microglia, 278–282